Praise for *A Man and His Ship*

"A terrific book! By entertaining, informing, and ultimately inspiring, *A Man and His Ship* transforms its readers into passengers traveling across an ocean and through time. A skilled verbal navigator, Steven Ujifusa has charted an efficient and yet immensely satisfying course through a sea of facts, images, and stories."

— David Macaulay, best-selling author of *Cathedral, Castle,* and *The Way Things Work*

"Steven Ujifusa has done something remarkable in his book, *A Man and His Ship*: he has brought back an era of American dominance in shipbuilding through the life of one of its giants: William Francis Gibbs. In some ways, Gibbs was the Steve Jobs of his era—a perfectionist with few people skills who nevertheless was single-handedly able to change his industry by the power of his vision and overwhelming professional competence. We need more public historians like Ujifusa working in business history. Using the highest research standards, he has written a great book that tells great story."

— G. Richard Shell, Thomas Gerrity Professor, The Wharton School of Business and author of *Bargaining for Advantage*

"Few of man's creations possess even half the romance of the passenger ships that once steamed across the world's oceans, especially the North Atlantic. That is why Steven Ujifusa's *A Man and His Ship* is such a compelling work."

— John Steele Gordon, *The Wall Street Journal* (best nonfiction of 2012)

"Much of Ujifusa's book is a portrait in determination, as Gibbs's plans for his big ship are continually tossed about in political, economic and personal squalls. A less single-minded man may have given up at numerous times."

— Stephen Heyman, *The New York Times Style Magazine*

"In his debut, Ujifusa harks back to a time when men were men, and transatlantic ships were serious business. . . . Written with passion and thoroughness, this is a love letter to a bygone time and the ships that once ruled the seas."

— *Publishers Weekly* starred review

"Ujifusa describes the construction of the ship in engrossing detail and provides informative digressions on the golden age of ocean travel, when liners carried millionaires, celebrities, and desperate refugees."

— *Booklist*

"The sea inspires obsessions in determined men, from Captain Ahab to Admiral Rickover. Steven Ujifusa introduces us to another—the naval architect William Francis Gibbs. His ingenious design of mass-producible Liberty ships helped win World War II, but Gibbs's obsession was to build the world's fastest, safest, and most elegant Atlantic liner. He ultimately succeeded, but in a decade his masterpiece was obsolete and unprofitable. Ujifusa narrates this tragedy well, in all its technical, political, and human dimensions."

— Admiral Dennis C. Blair, U.S. Navy (Ret.),
Former Director of National Intelligence

"A fascinating historical account. . . . A snapshot of the American Dream culminating with this country's mid-century greatness."

— *The Wall Street Journal*

"Few of man's creations possess even half the romance of the passenger ships that once steamed across the world's oceans, especially the North Atlantic. That is why Steven Ujifusa's *A Man and His Ship* is such a compelling work."

— *The Wall Street Journal*

"A marvelous narrative of America's premier naval architect."

—Barrett Tillman, author of *Enterprise*

"A *Man and His Ship*, a hugely entertaining re-creation of the age of the ocean liner, will leave older readers nostalgic, younger readers envious, and all of them engrossed in the drama of William Francis Gibbs as he fights to build the greatest ship of them all, the S.S. *United States*. The Cunard Line once boasted that 'getting there is half the fun.' Now Steven Ujifusa has given us the other half."

— A. J. Langguth, author of *Driven West*

"A delightful account of the era of grand ocean liners and the brilliant, single-minded designer who yearned to build the greatest ocean liner of all."

—*Kirkus*

"A fitting memorial to our greatest naval architect."

—*National Review*

AMERICA'S GREATEST NAVAL ARCHITECT AND

HIS QUEST TO BUILD THE SS *UNITED STATES*

STEVEN UJIFUSA

A Man and His Ship

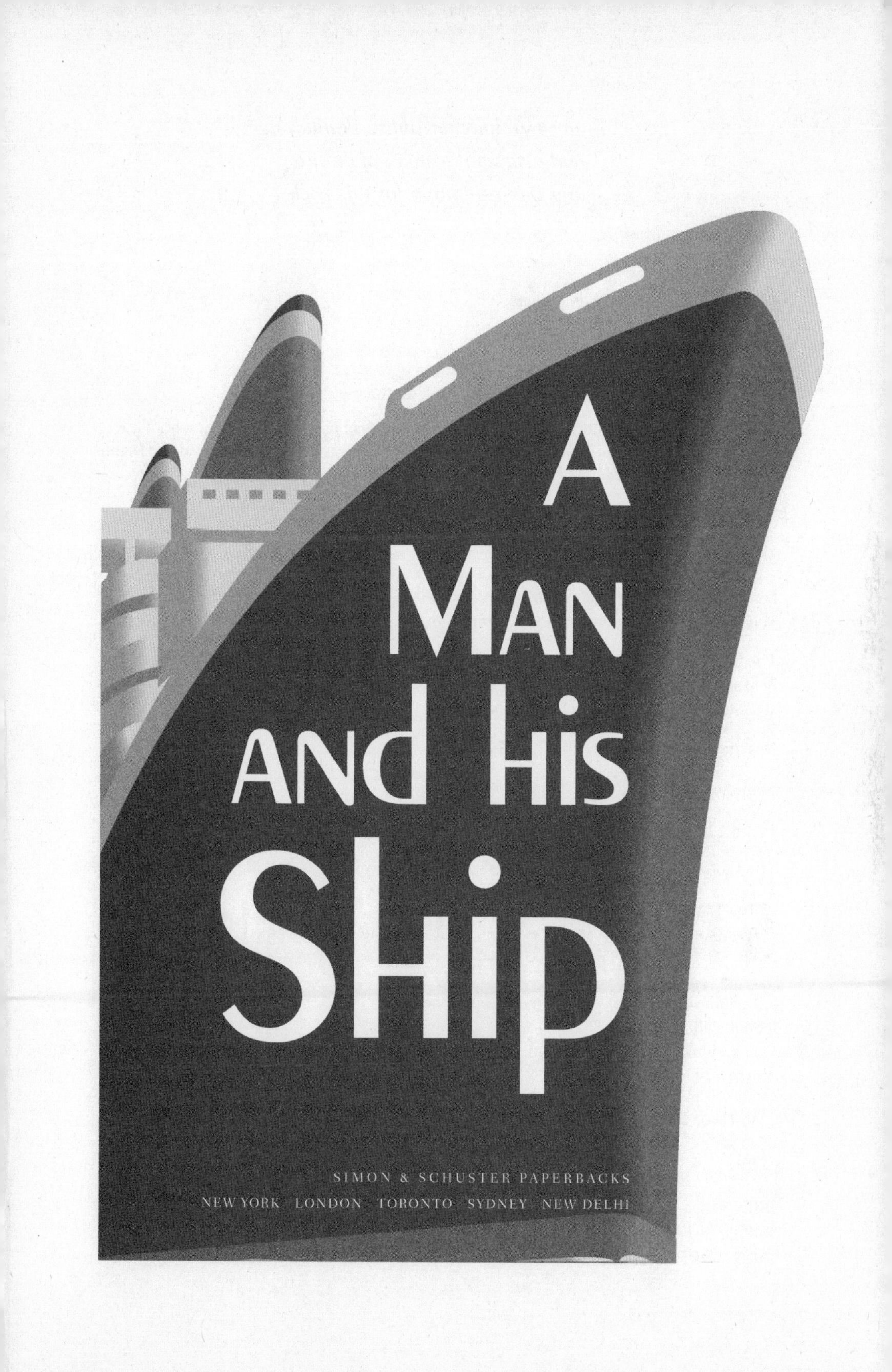

SIMON & SCHUSTER PAPERBACKS
NEW YORK LONDON TORONTO SYDNEY NEW DELHI

To my grandmother Judith Follmann,
world traveler, woman of culture,
and the inspiration for this book

Simon & Schuster Paperbacks
1230 Avenue of the Americas
New York, NY 10020

First Simon & Schuster Paperbacks edition June 2013

Designed by Akasha Archer

Manufactured in the United States of America

10 9 8 7 6 5 4 3 2 1

The Library of Congress has cataloged the hardcover edition as follows:

Ujifusa, Steven.
A man and his ship : America's greatest naval architect and his quest to build the S.S. United States / Steven Ujifusa.
p. cm.
Includes bibliographical references and index.
(ebook : alk. paper)
1. Gibbs, William F. (William Francis), 1886–1967. 2. Naval architects—United States—Biography. 3. Gibbs & Cox—History—20th century. 4. United States (Steamship) 5. Ocean liners—United States—History—20th century. I. Title.
VM140.G52U55 2012
623.8092—dc23
[B] 2011049883

ISBN 978-1-4516-4507-1
ISBN 978-1-4516-4509-5 (pbk)
ISBN 978-1-4516-4508-8 (ebook)

CONTENTS

PROLOGUE

The Way It Was

The transatlantic ocean liner possessed a mystique now lost to the world. For the first half of the twentieth century, ships named *Mauretania, Bremen, Normandie,* and *Queen Mary* were known and loved by tens of millions of people on both sides of the Atlantic. When a big liner arrived in New York City for the first time, thousands lined the Hudson to watch a man-made object—one that seemed to have life and soul—move serenely upriver. Their eyes were following something simply massive—she could be up to five city blocks long and twelve stories high, her deep-throated whistles bellowing in response to a cheering crowd. Sculpted hull, gleaming paint, and raked-back smokestacks conveyed beauty, power, and speed.

In the New York newspapers, the shipping news doubled as society news, as readers learned if Greta Garbo, Cary Grant, Margaret Truman, Vincent and Brooke Astor, or the Duke and Duchess of Windsor were aboard one of the ocean liners arriving or leaving that day. When a great ship left for Europe, it was an occasion awash in champagne and laughter. On board, first-class passengers enjoyed public rooms and private quarters that were decorative showplaces for the world's most talented designers, men and women who created some of the most stunning interiors ever built on land or sea. En route, high standards of service for the ship's most privileged passengers meant money for its owners and prestige for the nation whose flag she flew. Ships connected businessmen to transatlantic partners, diplomats to their posts, jazz artists to European audiences, students to adventures, immigrants to American jobs, and refugees to freedom. During two devastating world wars, liners converted to troopships carried millions of GIs to the front, and then brought them home again in triumph.

To the public, the ocean liner—once the only way to get across the Atlantic—was the epitome of glamorous travel. She also represented the pinnacle of technology—the most complex and powerful machine on earth. Deep inside her hull were engines capable of propelling a thousand-foot-long mass of steel through the giant waves of the North Atlantic at nearly 40 miles per hour. The liner that crossed the Atlantic the fastest captured a prize called the Blue Riband. A winner became the most famous ship in the world—until a faster rival bested her.

From the 1860s to the 1950s, all of the liners that captured the Blue Riband flew European flags, as a passive America seemed to accept the superiority of foreign engineering, manufacturing, and managerial prowess. One American did not, and this is the story of his quest to build the fastest, most beautiful, and safest ocean liner ever—the ship that was to become one of the greatest engineering triumphs in American history.

BOOK I

The Man and the Vision

SIZE, LUXURY, AND SPEED

The first time he saw an ocean liner, little Willy Gibbs knew what he wanted to do with his life.

On a rainy November 13, 1894, twenty-five thousand people waited outside the gates of Philadelphia's Cramp Shipyard on the banks of the Delaware River. They were there to see a marvel of the age: the steamship *St. Louis,* one of the largest ocean liners in the world and America's brand-new entry into the transatlantic passenger trade. When the gates opened, people surged toward the ship. She was 550 feet long and decorated from stem to stern with flags of the world, with the American Stars and Stripes flying high above the bow.*

The owner of the new ship, Philadelphia businessman Clement Griscom, was on his way to the shipyard with the christening party, headed by the bulky U.S. president, Grover Cleveland and the elegant, much younger first lady, Frances Cleveland. A chuffing Pennsylvania Railroad locomotive pulled the presidential train right up to the Cramp Shipyard gates. Stepping out, the first lady took Griscom's arm, and the

---

* *St. Louis* was almost two modern football fields long, and at 11,500 gross tons, the third-largest ship in the world, after Cunard's 12,000-gross-ton *Campania* and *Lucania* of 1893.

group of dignitaries walked to the launching platform, joined along the way by shipyard owner Charles H. Cramp.

Among those watching the scene was the forty-eight-year-old William Warren Gibbs, a crafty, aggressive financier who was said to sit on more boards of directors than any other man in America. On this blustery fall day, he had brought his two young sons—eight-year-old William Francis and six-year-old Frederic—to watch the launch of the great liner.

Self-made William Warren Gibbs was one of Philadelphia's most daring entrepreneurs. His physical appearance matched his temperament: he was lean, with fierce, defiant eyes, and a dark, pointed beard. A farm boy from the small town of Hope, New Jersey, he had arrived in the city thirty years before with little more than a skill for persuasion, but went on to become a multimillionaire laying gas lines and selling electric batteries. The United Gas Improvement and Electric Storage Battery companies had also enriched many of the city's leading citizens. When he brought his sons to see the launch of *St. Louis,* he was rumored to be worth $15 million, a stupendous pile of money in 1894.[1] His sometime partner in the gas business was a well-connected member of an old Philadelphia family: *St. Louis*'s owner Clement Griscom, president of the International Navigation Company, a shipping firm he founded with the help of the mighty Pennsylvania Railroad.

William Warren Gibbs might have looked at ships with an eye for profit. But for his eight-year-old son William Francis, seeing a great ship was pure poetry. During summer days at the family's summer home on the New Jersey shore, the boy scanned the horizon for funnels, masts, and black smudges of coal smoke, and then sketched what he saw. He knew that as he looked north, ocean liners, growing bigger and faster every year, were sailing in and out of the great port city of New York. Little Willy yearned for a closer look at one of these ocean greyhounds.

And now, at his father's side, he had his chance—*St. Louis* was a liner of vigorous beauty, her graceful hull draped with red, white, and blue bunting.

The shipyard was also a marvel to behold. William Cramp & Sons had been building cargo ships, passenger liners, and warships for over sixty years. The proud standard bearer for Philadelphia's industrial might, Cramps employed more than five thousand workers, many of them immigrants from Ireland and Italy.[2] In the yard was a towering crane, perched atop a floating barge, that could pick up a seventy-ton boiler and deftly swing it into the hull of a ship over three hundred feet away. Muscular riveters put hulls together by hammering red-hot rivets into steel and iron plates. Roaring orange fires glowed from forges where men shaped mammoth pistons, propellers, and funnels with the precision of watchmakers.

As the music from the band faded into silence, the little boy and the crowd around him awaited the launching of the great ship. Mounting the platform, Cramp handed the first lady a bottle of champagne. The hydraulic rams then hit the ship a bit too early, and the hull started to creep down the ways. Startled, the first lady called out, "I christen thee *St. Louis*!" and smacked the bottle across the prow before it slid out of reach.[3] Picking up speed, the ship roared down the tallow-greased slipway toward the Delaware River, kicking up billows of acrid smoke and, upon hitting the water, sending waves smashing against the banks. Once fully in the river, heavy chains slowed her to a stop. Tugboat crews secured their lines, and she was towed to the fitting-out basin. Her sister ship, *St. Paul,* remained on an adjacent slipway, to be christened in April of the following year.

At a luncheon after the launch, frock-coated dignitaries toasted the glory of the new American flagship and the presumed rebirth of the nation's preeminence on the North Atlantic. The American merchant marine—before the Civil War a vast fleet of clippers, whaling ships, and sailing packets—had been in steady decline for decades. An American steamship had not held the transatlantic speed record for half a century. The culprits were lack of government support, a shortage of private capital, and cheaper, subsidized foreign competition.

But to President Cleveland and Clement Griscom, *St. Louis* represented the dawn of a new era of American maritime might. "We may well be proud because we have launched the largest and most powerful steamship in the Western hemisphere," the president declared, "built on American plans, by American mechanics, and of American materials." The two ships would "furnish the revival and development of American commerce and the renewed appearance of the American flag in foreign ports."[4]

To little William Francis Gibbs, the launching of the new ship on that drizzly November day marked the start of a lifelong love affair. He would grow up to build a ship much bigger, faster, and grander than the magnificent *St. Louis.*

"That was my first view of a great ship and from that day forward I dedicated my life to ships," William Francis Gibbs later recalled. "I have never regretted it."[5]

The size, beauty, and luxury of the nineteenth-century ocean liner captivated the public, but even more alluring was speed. "Speed is the only thing which they talk, think, or dream of anywhere between Sandy Hook and Roche's Point," the *New York Times* said about American passengers in 1883. "Whenever their vessel distances some other steamship which is bound in the same direction, they are thrown into ecstasies."[6] Shipbuilders were just as obsessed with speed. "Each successive lowering of the record," boasted Philadelphia's Cramp Shipyard, "marks a triumph for the designer and builder, a fame world-wide, and substantial benefits to mankind."[7]

This speed record was known as the "Blue Riband of the Atlantic," a mythic-sounding prize developed in the middle of the nineteenth century and awarded to the fastest steamship sailing between the old and new worlds. Because actual miles traveled varied from voyage to voyage, the unspoken rule was to award the Blue Riband not to the ship making the quickest trip, but to the one achieving the highest average speed in nautical miles per hour, or knots (1.15 land miles per hour).* Prevail-

---

* One knot is the equivalent of 1.15 land miles per hour. Thirty-three knots, or 33 nautical miles per hour, translates to about 37 miles per hour. But land measures don't give a full picture of the

ing winds and currents made the westbound crossing, from Europe to America, more difficult than the one eastbound. So the Blue Riband was divided into two prizes: one for the westbound record and another for the eastbound; the former, more arduous crossing, carrying more prestige. There was no set course, but the generally accepted rule was that the clock started when the ship left its last port of call and achieved full cruising speed—usually off the southwest tip of England—and ended at the entrance to New York harbor, either at Sandy Hook or Ambrose Lightship, when she had to slow down. What started as an advertising ploy quickly grew into an international contest into which steamship companies, engineers, and governments poured talent and money.

It was the advent of steam that allowed oceangoing passenger ships to keep regular schedules, and the first commercially successful steamship was an America creation. In 1807 Robert Fulton's steamboat *Clermont,* plying the Hudson River, cut the travel time from New York to Albany from three days down to thirty-two hours, and could make regularly scheduled departures regardless of wind and currents. Twelve years later, an enterprising group of businessmen from Savannah, Georgia, outfitted a small sailing ship with a crude steam engine and paddle wheels and sent her across the Atlantic. *Savannah*'s epic voyage, even if only made partially under steam, was a landmark in maritime history, but American businessmen decided that steamships were best used on inland and coastal routes. The ocean remained the domain of the sailing ships, most notably the clippers, which journeyed around Cape Horn to the gold fields of California and the tea and spice hubs of East Asia.

It was a different story for the British, whose fortunes were tied to the overseas wealth generated by its far-flung colonies. For the

---

speed involved when moving through water. The fastest nineteenth-century sail-powered clipper ships could average 16 knots; the great aircraft carriers of World War II averaged about 32 knots. The Cunard liner *Queen Mary* held the Blue Riband for the fastest Atlantic crossing from 1938 to 1952, averaging just over 30 knots.

British government, supporting this new transportation technology—steamships that could carry passengers, mail, and cargo on a regular schedule—was a matter of imperial necessity. Not only that, but there were fortunes to be made carrying immigrants in steerage to the United States and Canada. In 1839, Samuel Cunard, an enterprising colonial who moved to England from Nova Scotia, finagled a British government subsidy of £60,000 a year to start a transatlantic steamer line that would carry the mails.[8] Cunard's first ship, *Britannia*—a two-hundred-foot-long, wooden paddle-wheel steamer with a top speed of 8.5 knots (about 10 land miles per hour)—made its first voyage between Liverpool, Halifax, and Boston in July 1840. Service to New York began in 1857. Cunard's ships cut a typical Atlantic crossing time down from two months under sail to a mere two weeks under steam. For Samuel Cunard, safety and reliability trumped luxury. Cabins were cramped, furnishings plain, and cuisine bland at best. As his partner Charles MacIver once made clear to an unhappy passenger, "Going to sea is a hardship."[9]

The transition from sail to steam made crossings faster, but not necessarily more pleasant; the North Atlantic was still arguably the most treacherous body of water on the planet. Except for a brief sunny summer interlude, passengers boarding a liner in Liverpool bound for New York expected gray skies, heaving seas, and blustery winds during most of the voyage. During the depths of winter, spitting rain, howling gales, and monster waves would punish the steamship, send furniture and clothing flying through the air, and make everyone on board seasick. And then when the ship reached the Grand Banks, off the Canadian coast, thick fog would often roll in, making visibility close to zero. Most passengers were more than happy to stumble ashore after enduring two weeks of cramped quarters and nausea aboard a Cunard steamer.

After a decade of Cunard supremacy on the North Atlantic, one American tried to outdo the British in speed and luxury. During the 1850s, a New Englander named Edward Knight Collins was on the receiving end of a mail subsidy from Washington for steamship service between New York and Liverpool: a princely $385,000 a year. Three of his luxurious ships took the new Blue Riband, making 13 knots and beating the British ships by an average of seven hours.[10] But after two of his money-losing ships sank and drowned hundreds of passengers, Congress killed the line's subsidy. Urging the move was Collin's unsubsidized rival in the transatlantic steamship business, a brash New Yorker named Cornelius Vanderbilt, known by the public as "the Commodore." The Collins Line collapsed without the subsidy, but Vanderbilt's transatlantic line also failed—he sold his ships and purchased the New York Central Railroad. While Washington gave away millions of acres of land out West to the railroads, the American merchant marine got little support.

After the Civil War, European companies dominated the transatlantic route. Supported by state mail contracts and construction subsidies, Cunard and its competitors made spectacular profits from carrying immigrants in cramped squalor and wealthy travelers in opulence. In 1871, Thomas Ismay, a cantankerous but shrewd Liverpool businessman, founded the White Star Line in a partnership with the Irish shipbuilder Harland & Wolff. Ismay's company showcased British white-glove service to wealthy Americans, pleasing even the finicky historian Henry Adams, who marveled at how the transatlantic liner represented human progress.[11] Another was a haughty, Hartford-born banker named John Pierpont Morgan. The young man was so impressed with the White Star Line that he dreamed of one day buying it.

Across the North Sea, two German companies, Norddeutscher Lloyd, based in Bremen, and the Hamburg-Amerikanische Packetfahrt-Actien-Gesellschaft, known as HAPAG, captured the lion's share of the immigrant trade. HAPAG in particular profited immensely from transporting "huddled masses," many of them Jewish pogrom refugees, from Eastern Europe to America. Packing immigrants into steerage bunks at twenty-five dollars a head translated to spectacular profits. The wünderkind of the German shipping business was a brilliant, diminutive Jew named Albert Ballin. Appointed HAPAG's managing director in 1899, Ballin was a self-made man who strove to make his ships perfect. When ships like his *Kaiserin Auguste Victoria* or *Amerika* were in port, he prowled all over them, making note of the slightest deficiencies in service or appearances. He even hired the Ritz-Carlton company to operate specialty first class restaurants on board.[12]

The heated competition between Great Britain, Germany, and

eventually France during the late nineteenth century spurred great technological advances, as lines plowed their vast profits into building bigger and faster ships. Progress was astounding. Liners grew from 3,000 gross tons to 10,000 gross tons, and lengths doubled from 300 feet to nearly 600.*

Ship construction moved from wood to iron, auxiliary sails were dropped, and paddle wheels were abandoned for the screw propeller. By the 1880s, first-class passengers could dine and read in public rooms lit by electricity, and enjoy hot and cold running water in their cabins. Engine technology also advanced rapidly. By the 1870s, British engineers had perfected the so-called compound engine. Here steam would pass through a series of three or four cylinders before being ejected into the condenser. So-called triple and quadruple expansion engines allowed for more steam pressure, which meant more speed.[13] As a result, travel time between Liverpool and New York was cut from two weeks to just over six days, and top speeds approached 20 knots, double the speed of the first paddle-wheel liners. Ships were now boasting two screw propellers rather than just one.

Following the Civil War, a lone American steamship company was left to brave Atlantic waters. Philadelphia's Clement Griscom had attracted the backing of the Pennsylvania Railroad and John D. Rockefeller's Standard Oil Company, which saw a transatlantic service as a means to pick up where railroads ended, at the water's edge. In 1871, Griscom's American Steamship Company commissioned Cramp Shipyard to build four liners to carry passengers, oil, and bulk cargo between Philadelphia and Liverpool. Griscom, a bewhiskered, florid-faced Quaker, belonged to all the right Philadelphia clubs and had married a member of the famous Biddle family, but he also socialized with nouveau riche entrepreneurs like Peter Widener and drank a pint of champagne during the workday. Bored with the "proper Philadelphian" professions of medicine and law, Griscom intended to make a big splash on the world stage with

---

* For ships, a gross ton has nothing to do with weight. It is a cubic volume measure of a ship's enclosed space (some areas are exempted from the calculation). In general terms, 1 gross ton (GT) equals 100 cubic feet of volume. In practical terms, the Pilgrim ship *Mayflower* was about 180 GT; the Civil War ship *Monitor* was 2,391 GT; *Titanic* was 46,329 GT.

his ships. To do it, he needed vast quantities of capital, as well as political support from Washington.

At first, Griscom faced near failure. His American Steamship Company lost so much money that the Pennsylvania Railroad refused to provide the additional cash Griscom needed to stay in business. An appeal to the federal government also failed. To keep going, the Philadelphia shipper negotiated a $100,000-a-year mail subsidy from the Belgian government for a new venture: the International Navigation Company, also known as the Red Star Line.[14] Still somehow in business, Griscom then purchased the moribund British Inman Line in 1886 and ordered two new ships, *City of Paris* and *City of New York,* from a Scottish yard. At 10,000 gross tons each and with service speeds of over 20 knots, they were the largest and fastest ships of the day. Both captured the Blue Riband with ease, but they did so as "British ships," thanks to American navigation laws that protected American industry. In a maneuver meant to protect American shipbuilders from cheaper foreign competition, Congress forbade foreign-built ships from flying the Stars and Stripes. The law backfired, however, as American ship owners either sold out their shipping interests (like Vanderbilt) or operated foreign-built fleets under foreign flags (like Griscom). When it came to building and operating ships, America remained at a major competitive disadvantage—European governments subsidized their passenger fleets, while the United States did not.

Griscom was undeterred and decided to use his now-famous ships to leverage a mail subsidy out of Washington. He told Congress that if his liners were granted American registry, he would kill the Inman Line, make the two vessels part of the fleet of a renewed American Line, and then build two new ships in an American yard. Congress agreed and passed a bill in 1892 that gave Griscom a mail subsidy of $12,400 per crossing.[15] The following year, the crews of the renamed *New York* and *Paris* raised the American flag on their sternposts, and at the William Cramp & Sons Ship and Engine Building Company in Philadelphia, workers laid the keel plates for two new ships—*St. Louis* and *St. Paul.*

The American public hoped that their new transatlantic liners would take the Blue Riband from the current holder, the Cunard liner

Lucania, whose best time was 5 days, 7 hours, and 23 minutes—just 5 hours and 46 minutes faster than Griscom's 1892 winner, *Paris.*[16] But neither *St. Louis* nor *St. Paul* could match *Lucania*'s pace. The British ship kept the prize until 1897, when a great four-funneled beast from imperial Germany, *Kaiser Wilhelm der Grosse,* snatched it away. The Norddeutscher Lloyd ship had managed an average speed of 22.5 knots, nearly three knots faster than the American ships and more than a half a knot faster than *Lucania.* There was consternation in the British public as the Germans humbled the nation's best engineers.

But the German triumph did not deter Clement Griscom. Teaming up with financier J. P. Morgan, the Philadelphian hoped to buy every single transatlantic shipping line, European and American, and merge them into a gigantic shipping trust based in New York. The two men made a perfect team. Like Griscom, Morgan was from an old-line family that had been wealthy for generations. His huge physique reflected a gargantuan appetite for food, rare books, art, and mistresses. A hideous outbreak of rhinophyma left his nose bloated and purple, a condition that made him avoid photographers. But Morgan possessed a genius for deal making. The financial mastermind of the American industrial trusts, Morgan believed that consolidation was the future of American business. To him, investing in steamships was a good deal more interesting than steel, sugar, and oil—they awakened a lust in him equal to his passion for art and women. The transatlantic liner was the era's ultimate status symbol, and Morgan vowed to own as many of them as he could. And to get his hands on them, he needed Griscom's shipping savvy.

To break the European grip on the transatlantic trade, Griscom and Morgan would use the House of Morgan's financial muscle to force all parties, American and foreign, into the trust. It would be called the International Mercantile Marine, a company that under another name would grow to become the largest and greatest American shipping firm, one that would be closely identified with the career of the young boy so awed by the christening of *St. Louis.*

ESCAPING THE RICH BOYS

After the launch of the great *St. Louis* in November 1894, the bookish, sickly William Francis Gibbs began thinking and reading of nothing but ships. Back at his father's mansion on North Broad Street, his bedroom was cluttered with technical publications that he quietly read into the night. Most he had borrowed from his father's study.

A reclusive, high-strung child, little Willy Gibbs was tongue-tied and unnerved by the parade of business and social callers to his parents' home. Once when an engineer who worked for his father came by, he found the Gibbs boy engrossed in a periodical.

"What are you reading, Francis?" the engineer asked.

The boy silently handed over an issue of *Cassier's,* a sophisticated engineering trade journal.

"Good Lord!" the visitor exclaimed.

William Francis snatched the issue back and resumed his reading.[1]

Born on August 24, 1886, Willy grew up in the "let-the-poor-child-alone system of education," a journalist wrote later, probably based on the child's own adult recollections. His mother seemed to spoil him. "A sneeze was always good for a week at home. A slight cough was good for a month. A pronounced bark enabled him to stay away from school for an entire year."[2] He also forged a close bond with his younger brother,

Frederic, born when William Francis Gibbs was still a twenty-month-old toddler. Both were fascinated not just with ships, but also firefighting equipment. They found a fellow enthusiast in the family's coachman, a former Philadelphia fireman. If he heard about a fire somewhere in the city, he would rouse the boys from their sleep, hitch up the horses, and gallop with them into the dark corners of the industrial city. "Lumberyard fires are almost always terrific!" Gibbs later recalled.[3]

His parents, on the other hand, were not interested in showing Little Willy the rougher side of Philadelphia. They made sure their six children—William Francis had four older sisters in addition to his younger brother—grew up in Gilded Age opulence. Father William Warren Gibbs was also an avid reader of engineering journals, but he scanned them looking for ideas to back. He was not an inventor, but a promoter, a wheeler-dealer who had a knack for associating with other men of wealth and ambition. With his cronies, the entrepreneurs Peter Widener and William Elkins, Gibbs shrewdly chose some "respectable" men to sit on the board of the United Gas Improvement Company, which by the 1890s had enriched many of Philadelphia's oldest families and earned an infamous reputation for political corruption.[4] He also possessed a smooth tongue and legendary powers of persuasion. The former provost of the University of Pennsylvania, the patrician Dr. William Pepper Jr., once tried to solicit a donation to his Ivy League institution from William Warren Gibbs. A fifty-thousand-dollar contribution would help Gibbs's social standing, the distinguished physician hinted. But the middle school dropout deftly turned the tables on Pepper, convincing him to purchase stock in a new venture: Marsden Cellulose, which sold ground-up coconut shells to insulate battleships.[5] The *New York Times* reported that the smelly material was "unhealthful and breeds vermin."[6]

Mother Frances used her husband's vast wealth to throw lavish parties. At one of her fêtes, the ballroom of a Philadelphia hotel became a "magnificent apartment of the period of Louis XIV . . . enhanced by profuse decorations of palms, ferns, and American beauty roses."[7] To cap off a season at their summer home in Spring Lake on the New Jersey shore, the Gibbs family hosted a "bal poudre"—the guests wore powdered wigs and hair.[8] When the Gibbs's eldest daughter, Augusta May, married sportsman William Hamilton Tevis Huhn in 1899, a reporter marveled at the wedding gifts on display at the Gibbs home, noting that "one large room could not hold the great amount of cut glass, silver, in plate and in service, rare china, ornaments and bric-a-brac."[9]

But their efforts at social acceptance were in vain: Philadelphia snobbery held that the Gibbs family mansion on rich and opulent North Broad Street was on the wrong side of Market Street, the demarcation line separating "fashionable society" from the nouveaux riche.

For the Philadelphia elite, family pedigree was all that mattered, and lack of it sometimes drove turn-of-the-century strivers out of the city altogether. One poor but supremely talented Harvard Law School graduate, raised "north of Market" by his widowed mother, was warned by an Old Philadelphia lawyer: "They'll never take you seriously in this town—in New York your grades will count for something."[10] Young John J. McCloy took the advice, got a job with New York corporate lawyer Paul Cravath, and rose to become "the Chairman of the Establishment" under the aegis of the Rockefeller family.

William Warren and Frances Gibbs, however, were determined to stay in Philadelphia and launch their children into proper society. With that in mind, in 1899 the Gibbs family acquired an address above all social reproach: a mansion fronting Rittenhouse Square, the most prestigious neighborhood in Philadelphia. Behind the doors of its high-stooped town houses, a visitor could hear the sounds of privilege: clocks ticking on marble mantelpieces, the clink of crystal, and the quick, soft steps of Irish maids.

The three-story, yellow brick mansion at 1733 Walnut Street, commanding the northeast corner of the square, was sold to William Warren Gibbs by banker Anthony Joseph Drexel Jr.[11] A prominent Rittenhouse Square address announced that after eighteen years in Philadelphia, the Gibbs family had finally arrived. That same year, they were listed in the Philadelphia *Social Register* for the first time.

The socially awkward young William Francis Gibbs took to the big yellow house immediately; its vast size offered him privacy and seclusion. For twelve hours a day he would read, doodle, and tinker. Since William Francis avoided social contact with schoolmates, his closest companion continued to be his younger brother, Frederic, who, like his father, had a talent for mathematics and finance. At the family's summer home on the New Jersey shore, he sat for hours on the veranda watching ships sail in and out of New York harbor. He and Frederic also traveled to Philadelphia's leafy suburbs to play tennis on the grass courts of the Germantown and Merion cricket clubs. When they were older, they would play against a talented boy named Bill Tilden.

"We played against Tilden, and sometimes we won!" William Francis would recall about his matches against the future Wimbledon champion.[12]

By the time they were teenagers, the tennis trophies cluttered the upstairs bedrooms of the Rittenhouse Square mansion. Frustrated, their mother Frances scolded her sons, insisting it was "a little vulgar to display them all."[13] No matter. A good lawn tennis player—someone who played by the rules without losing his cool—was synonymous with being a true gentleman, a perception that must have pleased the Gibbs parents.

The boys also got the chance to travel abroad. When he was twelve, his parents packed little Willy off to Europe with an older cousin. But what stood out in his mind were not the cathedrals and museums, but a library in Switzerland packed full of engineering publications. On a later trip home aboard the White Star Line's new liner *Celtic,* then the biggest ship in the world, William Francis and Frederic Gibbs constructed a house out of blocks in their first-class stateroom. The slow but plush British liner was so steady that the block house remained standing for the entire eight-day trip.[14] The young Gibbs was fully aware that smaller, more powerful ships than *Celtic* were making headlines by crossing the Atlantic in less than six days, and that newer European ships had far eclipsed the American Line's *St. Louis* in size, speed, and luxury.

Because of his poor health, Gibbs was not sent to one of the Episcopal boarding schools that served the children of the northeastern elite. Instead he was enrolled at the De Lancey School, an easy walk from the family home. De Lancey advertised itself as providing the rigors of a Groton education while allowing the children to live at home.[15] And for reclusive Willy, this was a good thing. In addition, it appears that De Lancey was a more intellectual place than its New England counterparts, which had been modeled after the muscular Eton and its playing fields.

Illness continued to dog Gibbs. He did not graduate from De Lancey until he was nearly twenty. In his class of seniors, nearly half went to Harvard, most of the rest to the University of Pennsylvania.[16] He applied to Harvard most likely at his father's insistence; it was, of course, a place better known for cultivating gentlemen than naval architects.

By today's standards, Gibbs's college application was abysmally unimpressive. His high school transcript was peppered with C's and D's. He flunked Latin, French, and oddly enough, advanced algebra. Nevertheless, a member of the selection committee of Harvard's Lawrence Scientific School stamped "Admitted" on his application card, on condition that he get a passing grade in a foreign language after enrollment.[17] That the boy came from a very rich family must have been decisive. But the Lawrence Scientific School, later Harvard's Graduate School of Engineering, might also have known something of his powerful, self-driven interest in engineering and ships.

In September 1906, the twenty-year-old William Francis Gibbs, now a gangly six feet, two inches tall, boarded a train for Boston. His father did not want his son to study engineering, because William Warren Gibbs believed engineers were impractical and inarticulate—qualities that would not recommend a man to the people who mattered. He wanted a different life for his eldest son: an elite university education, a respectable legal profession, and the social status the law would guarantee. In short, the life that entrepreneurial, risk-taking William Warren Gibbs had never had.

Strangely, William Francis's younger brother, Frederic, did not immediately continue on to college at eighteen. One possibility is that like William Francis, he was a late bloomer as a student. Or maybe their seemingly high-flying parents had their own reasons to conserve cash by the time William Francis left for Cambridge. In the end, Frederic never got a college education.

Gibbs was passionate about ocean liners, but did not seem to possess the technical aptitude, financial savvy, and force of will that was needed to build giant machines for hard-nosed shipping men, visionaries like Samuel Cunard, Albert Ballin, Clement Griscom, and J. P. Morgan.

His parents also must have worried that their reclusive son would have a tough time at Harvard, with its demanding academics and conformist social scene.

Freshman William Francis Gibbs entered Harvard when the famed William James was still chairman of its philosophy department. The great pragmatist crowed about the college, "Our undisciplinables are our proudest product!"[18]

But one new undergraduate was miserable in Cambridge.

It was the waning tenure of university president Charles Eliot, and Harvard had yet to complete its transition from a finishing school for rich boys from Boston, New York, and Philadelphia into a world-class teaching and research university. Many undergraduates coasted through classes, and spent hours drinking in the private "Gold Coast" dormitories and the elite social clubs lining Mount Auburn Street.

The young Philadelphian with an immensely wealthy father was immediately accepted as a respectable "Gold Coaster." Gibbs took up residence in Claverly Hall, a Georgian brick pile that boasted wood-paneled walls, a sweeping grand staircase, and a small electric elevator. But unlike most students in Claverly, Gibbs showed no interest in the hijinks and hilarity typical of the turn-of-the-century "Harvard Man." William Francis continued to spend his free hours much as he had in his family's Walnut Street mansion. He read technical journals, pored over blueprints of British battleships, and drew. The budding designer approached the plans of these ships with "great deference." But that could not stop him from carefully rearranging their engine spaces, or adding more watertight bulkheads, imagining what might be done to increase speed or keep a ship afloat if it were struck by enemy shells or torpedoes. "What's the next step?" he would ask himself as he examined each engineering masterpiece.[19]

The boy's growing understanding of design came at a time when naval ideas were in ferment. The rout of Russia's navy in the Russo-Japanese War, which had occurred just a year before Gibbs left for college, showed what could happen to an outmoded and unprepared fleet. Britain was steaming ahead with new technologies and strategies, but many American naval officers—certainly, some of the naval thinkers in the journals Gibbs was reading—felt the United States was not keeping up. They pointed to the Navy's failure to implement British naval advances to modernize its own fleet, and the serious lack of coordination between progressive Navy engineers and hidebound line officers. Studying the latest articles and ship blueprints in his dorm room, the intensely patriotic Gibbs began to see a role for himself in rebuilding the American Navy.

To many of his classmates, the preoccupied student from Philadelphia was a strange one. He was painfully shy, and later recalled that some of the loud-mouthed, arrogant scions of privilege "filled him with alarm."[20] Terrified of being bullied, he kept the door to his room locked to protect his blueprints and ship photographs from mockery and practical jokes.[21]

Photos of Gibbs taken during his time at Harvard show a young man in a long dressing gown with a sash and striped lapels, looking profoundly lonely. In one picture he stands against his dresser, his right arm draped over a pile of books. In another he sits in a chair next to his room's fireplace, his hands clasped over his lap. Instead of college pennants, the walls are covered with photographs and prints: warships with smoking funnels and fine automobiles. A basket with rolled-up blueprints sits by his desk at the window.[22]

In fact, Gibbs was not alone in rejecting conventional collegiate life. Classmate John Reed, who would play a role in the Russian Revolution of 1917, remembered students who "criticized the faculty for not educating them, attacked the sacred institution of intercollegiate athletics, sneered at the undergraduate clubs so holy that no one dared mention their names."[23] Other members of the famous Harvard Class of 1910 were poet T. S. Eliot and future political pundit Walter Lippmann.

Yet for most boys, Harvard was a place to solidify ties begun in a privileged childhood and to extend them into an even more privileged adulthood. This created a powerful ethos of social exclusion and conformity. Future Harvard professor Samuel Eliot Morison, Class of 1908, portrayed the college's social sorting system as a cruel machine, noting that, "once having 'made' a club, you could reassert your individuality; often by that time you had none."[24]

As a prep school–educated Protestant, Gibbs had the right background, but the final clubs—exclusive societies that set the college's social tone—showed no interest in a quiet oddball who shut himself up in his room. For his part, Gibbs relished his difference. Still, there were times he feared that he would be labeled "the eeriest of all the eccentrics in three centuries of Harvard life if it were known that he was busy improving British battleships."[25]

In the Class of 1910 album, Gibbs listed no extracurricular activities. But he did serve a socially obligatory stint of service as an usher at Saturday football games. There, with a red usher's ribbon pinned to his coat, he would shepherd wildly enthusiastic fans to their seats in the recently constructed Harvard Stadium, where they watched matches so fierce that the collisions sometimes killed players on the grassy field below.[26] But the only sport Gibbs played himself was tennis—a noncontact, solitary game of individual skill and determination.

The future naval designer had problems in his engineering classes, for reasons that Gibbs never later discussed. He had not been a successful high school student, if success were measured by grades. Years of intermittent schooling, when he was sick as a child, may have trumped the compulsion to do everything by the book. And for all his solitary studying, he did not impress his engineering professors. At the end of his first year, he received C minuses in "Descriptive Geometry" and "Mechanism—Study of Gearing and Gearing Mechanisms." His best grade was a B in "Steam Machinery," a subject he knew well from reading trade journals.[27] Underlying it all may have been a basic inadequacy: Gibbs had trouble with simple math. An observer noticed the following later in life: "Gibbs is afraid of arithmetic. An eight-year-old child can beat him at adding, subtracting, and dividing. He won't trust figures until a machine has gone over them three times."[28]

Gibbs would later insist he never took a formal course in naval architecture at Harvard or anywhere else: "I studied engineering at night out of books." For emphasis, he added, "That's the way to really learn things—by *yourself*."[29] Highly intelligent, with a powerful visual sense and an astonishing memory for details, he did eventually master the mathematics needed to pursue work as an engineer, something that required singular focus and will.

But the freshman boy was not yet an achiever. At the end of the 1906–1907 academic year, Harvard University recorder George Cram wrote Gibbs's father a stern letter, warning him that if William Francis did not get his engineering grades up and fulfill his language requirement, "he will have to register again as a first-year student."[30]

On October 30, 1907, in his second year, Gibbs withdrew from Harvard College. "On account of sickness," the college report noted, but that was almost certainly not the case. He and his brother, Frederic, had other plans for November 1907, with Gibbs thinking he could regroup academically later, after a little adventure.

William Francis also must have been worried about his family back in Philadelphia. His parents had swaddled him in protective luxury all his life, carefully ensuring that their son would only associate with people of their own class. But within a few years, the poor little rich boy would have to confront a very unpleasant reality: his father's success was not as solid as it appeared.

By the time George Cram's letter arrived on William Warren Gibbs's desk, the great Philadelphia promoter had big worries of his own. Throughout early 1907, the American stock market had been volatile and falling. Soon enough, the trouble on Wall Street impacted the banks and trust companies, creating a crisis that would come to be known as the Panic of 1907. It was not a good time for companies to be indebted, and the timing seems to have been especially bad for Gibbs's father.

Although by all appearances financially secure, William Warren Gibbs had skated on thin ice any number of times during his career. As early as 1891, he had found himself $3 million in debt after he failed in a bid to take control of the Philadelphia & Reading Railroad. But Gibbs eventually repaid the money, and went on to make more.[31] Then rumors began circulating that one of his enterprises was not all it seemed, and the *Philadelphia Inquirer* hinted that William Warren Gibbs might have crossed the line between aggressive tactics and fraud. "It is quietly likely," the paper reported, "that some of the shareholders of record of the Alkali Company [will] unite . . . and make a test case."[32]

William Warren had organized the American Alkali Company in 1899 to manufacture bleach and soda powder for use in paper mills. He purchased British patents and built a factory in Sault Ste. Marie, Canada. Two years after putting together the company and promoting its stock, he had retired from active management. The accusations began almost immediately. In the April 1902 shareholder lawsuit, plaintiffs alleged William Warren Gibbs had siphoned off an illegal $349,597 in cash, 15,900 shares of preferred stock, and 151,800 shares of common stock in a secret transaction with a British company.[33] The "fraudulent scheme" was "concocted by Gibbs," the *New York Times* wrote in March 1902, adding that "he misappropriated to himself and his appointees excessive salaries." Most damning was the allegation that the company's Canadian plant made nothing at all, and that the patents were worthless.[34]

The American Alkali Company ended up in receivership with $300 in assets.[35] By April 1905, William Warren Gibbs was forced to settle with the preferred shareholders and the receiver; his assets in the company, totaling $50,000, were put up for sale.[36] Philadelphia society must have whispered that William Francis Gibbs's father was a crook.

It is inconceivable that the son knew nothing of the scandal. How much he knew of his father's declining personal finances is less clear.

Life at elegant 1733 Walnut Street seems to have gone on as usual, at least for a while, but sometime around the time Gibbs left for Harvard, his parents mortgaged their grand mansion and asked friends for a $60,000 second mortgage to cover family expenses.[37] His father's slide was on.

As William Warren Gibbs's fortunes dimmed, his onetime gas company partner Clement Griscom's brightened, thanks to his close ties with J. P. Morgan. Their shipping trust, the International Mercantile Marine (IMM), was working to put American leadership—if not necessarily American ships—back in the transatlantic game. Their American board of directors was an all-star team of the nation's smartest, most aggressive businessmen. Among them was Peter Widener, by now the richest man in Philadelphia through his investments in streetcars, electric batteries, oil refining, and gas lines. Much of Widener's wealth had come from his partnerships with William Warren Gibbs. But neither Widener nor Griscom asked the former wizard of United Gas Improvement and the Electric Storage Battery Company, who was now tainted by the American Alkali scandal, to serve on the board of their new maritime trust.

IMM was put forward to American travelers and policy makers as a way to bring order to a chaotic shipping system routinely shaken by rate wars. Accordingly, Griscom argued that an international shipping monopoly would be a public benefit. "The object of combination is to try to give better transatlantic service at a decreased cost," he said, and promised daily, on-time departures from New York to Europe.[38] Yet rate stability (or as some asserted, driving competitors out of business) was not the only thing Morgan and Griscom wanted. Under the 1817 U.S. Navigation Act, only ships built in the United States could fly the American flag (and receive American subsidies). But it was still cheaper to build new ships abroad, and British-flagged vessels were much less expensive to staff and operate than American liners. Backed by Morgan's capital, the huge combination could build larger and faster ships at a pace that independent lines could not match.

Morgan and Griscom nearly got things right. Within a few years, many transatlantic passenger lines sold out to the trust. In addition to Morgan and Griscom's original American Line, IMM grew to include Belgium's Red Star Line and four British lines: Leyland; Shaw Savill & Albion; Dominion; and the famed White Star Line.

The French and German lines were next on Morgan and Griscom's list. The French Line was capable of resisting Morgan because it was a quasi-governmental agency. The two big German lines did not have such protection, although Kaiser Wilhelm II strongly supported keeping German ships German. So HAPAG's cagey Ballin struck a deal to keep his company independent and maintain control over his highly profitable immigrant business. The German lines would be part of a profit-sharing arrangement with IMM, while also agreeing that neither party would acquire stock in the other's enterprises. HAPAG and its rival Norddeutscher Lloyd would pay IMM a percentage of their dividends, and IMM would pay the two companies a fixed annual sum. The two sides agreed, in short, to take in each other's laundry.[39]

IMM's $32 million buyout of the White Star Line caused an uproar with the British public. They saw Morgan as robbing Britannia of its prized commercial fleet, considered one of the Empire's crown jewels. Not only that, but Morgan appointed White Star's chairman, J. Bruce Ismay, the new president of IMM.

Britain's one surviving big transatlantic line, the illustrious Cunard Steamship Company, refused to sell out, even though the company was strapped for capital. Chairman Lord Inverclyde "resolutely opposed any such absorption of the Cunard Line, which, more than any other, had become a national institution."[40] To keep IMM's hooks out of Cunard, Inverclyde turned to British prime minister Balfour and the Admiralty for help. The company and the government agreed that "no foreigner shall be qualified to hold office as a director of the company or to be employed as one of the principal officers of the company; and no shares of the company shall be held by, or in trust by, or be in any way under the control of any foreigner or foreign corporation, or any corporation under foreign control."[41] Now Britain's greatest transatlantic shipping company was Morgan-proof.

More help would be needed for Cunard to compete against Morgan's clout, and the British nation supplied it. Cunard received a low-interest government loan of £2.6 million to construct two large, fast passenger liners in British yards.[42] Each ship would be 790 feet long and of 30,000 gross tons, making them a third larger than any of their German competitors. In addition, the Admiralty would pay Cunard a handsome operating subsidy of £150,000 per year. Cunard, for its part, agreed that should war ever break out, the two superliners could be taken over by the Admiralty and converted into armed merchant cruisers. This would make the ships not only the biggest, fastest liners ever, but warships in disguise. There was little protest from British taxpayers; national honor was at stake.[43]

Cunard got to work building what would be the two greatest ships in the world. Their keels were laid in 1904 and they were launched in 1906. Both were christened with imperial-sounding names: *Lusitania,* after the Roman province of Portugal, and *Mauretania,* after the Roman province of North Africa.

A year later, William Francis Gibbs, free of college and accompanied by his brother, Frederic, was going to sail on both of them. Somehow, in spite of their family's financial problems, they got enough money out of their father for round-trip tickets to Europe.

MAURETANIA

On November 7, 1907, the Swan, Hunter & Wigham of Newcastle, England, shipyard team took the new Cunard flagship *Mauretania* out on her trial runs. Captain John Pritchard made a series of twelve-hour runs, up and down a 304-mile course along the Scottish coast. The new ship and crew fought fierce Atlantic winds, rising at one point to a force 7 gale, or wind speeds approaching 40 miles an hour. On the bridge, *Mauretania*'s chief designer, Andrew Laing, hovered closely around Captain Pritchard. After two runs with the wind and two against, Laing was jubilant. Over the course of nearly forty-eight hours, the four-funneled *Mauretania* had averaged 26.06 knots over 1,216 nautical miles.[1] This was nearly three knots faster than her German rivals.

The Scotsman Laing, the most accomplished marine engineer in Great Britain, had designed engines for record-breaking Cunarders for the last twenty-five years. He wanted this ship to take back the Blue Riband. His feeling was shared by *Mauretania*'s entire crew, from the officers on the bridge to the sweat-drenched stokers heaving coal into the ship's furnaces below. It was not, now, a matter of beating the Germans. One month earlier, on only her second run, *Mauretania*'s sister ship *Lusitania* (designed by Leonard Peskett and built by the rival John Brown yard in Scotland) had taken the Blue Riband—twice: eastbound, besting Norddeutscher Lloyd's *Kaiser Wilhelm II;* and westbound, beating HAPAG's *Deutschland.*

The crew of *Mauretania* felt that they had the better ship. "We'll lick the *Lucy,*" cried one of the stokers as he clanked his shovel against the boiler grate on the delivery run from Newcastle to Southampton, "even if we bust the *Mary* to do it!"[2]

Nine days after the trial runs, on November 16, 1907, William Francis and Frederic Gibbs leaned over *Mauretania*'s railings. As a hard rain fell, they looked up and down the Liverpool landing stage and saw it crowded with thousands of waving, cheering well-wishers.

They had been part of such a scene before, when they left New York on her sister *Lusitania* less than one month before. Although they had traveled on big ships as children, these Cunard liners set a new standard for size and luxury. The Gibbs brothers relished the newness and thrilled at the speed of *Lusitania.* They ran all over the ship, probably comparing her to the blueprints of the British warships William Francis had stashed in his dorm room. While other passengers remained in the warmth of the ship's paneled Georgian salons, he spent hours with the wind whipping in his face, leaning over the railing and watching *Lusitania*'s knifelike prow cut through the gray Atlantic at nearly 30 miles an hour.

Now, on the westbound maiden voyage of the bigger, more powerful *Mauretania,* they hoped to see another record fall. This crossing would be one to remember.

Below the Gibbs boys, *Mauretania*'s hundreds of electric lights cast a bright glow on the crowds and flickered off the gray harbor waters. Wisps of black coal smoke curled lazily from the ship's four red-and-black stacks, each of which towered several stories above her brilliantly illuminated uppermost decks. On board, first-class passengers unpacked in their staterooms and then explored the public rooms, all paneled in gilt-trimmed walnut, sycamore, and mahogany carved by three hundred artisans brought in from Palestine.[3] She looked like a British country house put to sea. Gone were the days of the nineteenth-century wooden paddle-wheel steamer, with its stiff bunks, chamber pots, and pervasive stink of bilgewater. On board *Mauretania,* men dressed for dinner in white tie and tails, women retrieved their jewels from the purser's office, and the first-class kitchen rivaled the one at the London Ritz. Sheets were changed each day, fresh flowers bloomed in the foyers, and an onboard printing press churned out menus and newspapers for two thousand passengers in three classes. Second-class passengers enjoyed spacious public rooms comparable to first-class spaces on older vessels. Those traveling steerage had private staterooms rather than open dormitories, as well as their own deck space.

The modern ocean liner, a swift and sleek city afloat, had arrived.

As *Mauretania* headed out to sea, William Francis and Frederic Gibbs inspected the new Cunard superliner. They marveled at the two-deck-high first-class dining saloon, the newfangled electric elevators, and the lounge with its frosted glass dome. Strains of Strauss's "Blue Danube" and Léhar's "Merry Widow" accompanied passengers at dinner. There was at least one Philadelphian on board whom the Gibbs brothers may have chatted with at dinner: Anthony Drexel Jr., the Philadelphia banker who had sold 1733 Walnut Street to their father; Drexel booked one of the liner's Regal Suites, which went for $1,500 per person one-way (almost $35,000 today).[4] First class aboard *Mauretania* was a luxurious, civilized sphere to which the Gibbs brothers had been accustomed since birth.

As splendid as the new ship was, what interested the Gibbs boys most was what was pulsing steadily beneath their feet. *Lusitania* and *Mauretania* were not only the two largest liners in the world, but the first Blue Riband contenders to use a new kind of engine: the steam turbine. And watching a great ship's engines, William Francis would later say, gave him the same thrill as listening to the music of Bach.[5]

Germany's four turn-of-the-century Blue Riband holders—*Kaiser Wilhelm der Grosse, Deutschland, Kronprinz Wilhelm,* and *Kaiser Wilhelm II*—all used massive quadruple reciprocating expansion engines, whose four steam chambers produced maximum power for an otherwise traditional piston engine. By 1900, reciprocating engines had hit a brick wall in terms of horsepower—they were getting too big and heavy, and it seemed that 23 knots was the maximum speed a big ship could reach. Even worse, as thumping pistons turned twin propeller shafts, the German ships shook, quite literally from stem to stern. Glasses crashed off tables and bunks trembled all night long. Many sleep-deprived travelers called the hard-riding HAPAG record breaker *Deutschland* the "Cocktail Shaker."[6]

The new British ships, on the other hand, took advantage of an experimental power plant, invented in 1884 by Charles Parsons. The steam turbine operated on the same principle as a windmill. High-pressure steam from the boilers would be blasted against a rotor lined with thousands of tiny blades. The spinning rotor would be connected to a propeller shaft. The result was a faster, more efficient engine that took up much less space.

Parsons demonstrated just how good his engine was when he crashed the 1897 Naval Review at Spithead. The event, meant to celebrate Queen Victoria's Diamond Jubilee, turned into a vehicle for Charles Parsons to promote his hundred-foot-long *Turbinia.* A brash Parsons zipped his nimble craft between battleships and cruisers at an incredible 34 knots (39 land miles per hour), evading a pursuing navy pilot boat. Even the aged Queen Victoria was astonished by the performance of the strange craft. Although many in the Royal Navy were outraged, others took notice and began to investigate installing turbines on new British warships.[7]

The steam turbine proved to be one of the greatest breakthroughs in engineering history, allowing ships to achieve unprecedented speeds. In 1906, the debut of the British battleship HMS *Dreadnought,* the first large warship to abandon the traditional reciprocating engine, intensified the ongoing naval arms race between Great Britain and Germany. With turbines connected to quadruple screws, *Dreadnought* could cruise at a steady 21 knots, making her the fastest battleship in the world. Yet even before launch of the *Dreadnought,* British engineers had already adopted turbine technology for use in passenger ships. In 1905, Cunard built two large sister ships: one was powered by reciprocating engines, while the other by Parsons turbines. The turbine vessel proved to be more than a knot faster, as well as quieter and more economical to operate.[8]

Cunard was finally sold on turbines. They were installed in both *Lusitania* and *Mauretania,* and there were more innovations as well. For the first time on a commercial liner, the ships would have not two, but four propellers. The arrangement meant engine torque would be more evenly distributed, lessening the risk of breakdown. Inspired to match the powerful physical appearance of the German ships, the designers gave each ship four raked funnels, wide and long enough to fit a locomotive inside. And while the engines of the Norddeutscher Lloyd flagship *Kaiser Wilhelm II* were rated at 45,000 horsepower, those of the Cunard ships were each rated at over 70,000.

The contract to build *Lusitania* was awarded to John Brown & Company of Clydebank, Scotland. *Mauretania*'s went to Swan, Hunter & Wigham in Newcastle-upon-Tyne, England, with engine construction subcontracted to Andrew Laing at Wallsend. The two-contract arrangement stoked Andrew Laing's competitive fires. From the start, it appeared that the Swan, Hunter–built ship would be the mechanically superior of the two. To make his engines lighter than *Lusitania's,* Laing chose a new steel alloy called Whitworth fluid-pressed steel for many of the turbine elements. Laing also eliminated extra parts to ensure that the assemblies would have "maximum strength and rigidity with a minimum weight." Each ten-foot-diameter rotor was strung with three million tiny blades. The completed rotor was then placed inside a cast-iron casing.[9] Because of clearances of only a tenth of an inch, the slightest defect along the rotor's forty-five-foot length would cause the blades to rub against the turbine casing and strip them, completely destroying the engine. But when run for the first time, less than a week before her November 7 trials, the turbines spun in perfect balance.[10]

Everyone aboard *Mauretania,* including the Gibbs brothers, thought that their ship would take the Blue Riband on her maiden voyage in November 1907. But as soon as she left the Irish coast behind, a terrific autumn gale hit the liner. Great waves lifted the bow sixty feet into the air only to send it smacking back down, unleashing gigantic bursts of spray. Despite her smoother turbine power plant, engineers at the trial runs noticed that vibration from *Mauretania*'s four thrashing propellers sent shudders throughout the entire ship, and so they had added stiffeners throughout the stern to solve the problem. The storm on this maiden voyage made any vibration problems irrelevant to her passengers—*Mauretania*'s relatively slim underwater hull made her a bad roller.

Nearly sixty years earlier, Charles Dickens had endured a winter gale aboard Cunard's first transatlantic liner, the wooden paddle wheeler *Britannia*. "Before it is possible to make any arrangement at all compatible with this novel state of things, the ship rights," the wretchedly seasick novelist wrote. "Before one could say 'Thank Heaven!' she wrongs again. . . . All is grand, and all appalling and horrible in the last degree. . . . Only a dream can call it up again in all its fury, rage, and passion."[11] Despite decades of progress in ship design, Dickens's description of life aboard a ship during a four-day gale could easily have been written of *Mauretania*'s maiden voyage. A rogue wave broke a row of first-class cabin windows, drenching clothes and bed linens. In the first-class lounge, a large walnut display case ripped away from the bulkhead and crashed to the floor, hurling glass shards everywhere.[12] For the seven hundred largely Irish immigrants in steerage, the experience was absolutely terrifying. Tightly packed into the forward part of the ship, they were subjected to the worst of the ship's motion as *Mauretania* heaved and plunged in the storm. Mothers held screaming babies tightly in their arms, as spilled porridge and vomit sloshed across tables and onto the floor. But it was even worse for the "black gang" down in the gloom of the boiler rooms, where dozens of sooty stokers struggled to feed the furnaces. When Captain Pritchard saw steam pressure gauges dropping, an officer would squawk into the ship's telephone, "More steam, more steam!" Chief Engineer John Currie strode up and down the boiler room alleys, yelling to his men as they heaved shovels of coal into the glowing fires. "Steady boys!" he bellowed. "Now keep her going and all work together."[13]

The second night out, with the storm going strong, passengers woke to a tremendous thud. Fearing the ship had hit something, they clambered up on deck to see that waves had torn loose the ten-ton spare anchor and sent it sliding across the forecastle deck. Pritchard slowed *Mauretania* and descended from the bridge in the driving rain and howling wind to face the problem. After an hour and a half of herculean effort, the crew resecured the anchor before it could gouge a hole in the deck. The soaked Pritchard returned to the bridge and *Mauretania* resumed her westward course.[14]

"Show me the captain of a transatlantic liner, and I'll show you a religious man," William Francis Gibbs would say later. "Out there in dirty weather, you feel like nothing, and you have to believe."[15] He might have been one of the awestruck passengers who watched *Mauretania's* crew struggle with the anchor that night; it is hard to imagine that the ship-obsessed twenty-one-year-old kept to his stateroom for much of the voyage at all. He would have known in advance all that had been written about *Mauretania;* this was his chance to investigate its workings firsthand. Exploring other designers' ships, even if he had to be covert, was a habit he was to keep all his life.

On the fourth day, the storm subsided and the sun shone on the gently rolling waves of the Atlantic. Pritchard ordered a full head of steam, and *Mauretania* picked up speed, smoke trailing from her four funnels. That day, the ship made 624 miles at an average speed of 25.83 knots, a world record for a single day's run.

On the fifth day, a dense fog rolled in. For safety, Pritchard slowed the liner to a crawl, and all hopes of breaking *Lusitania*'s record vanished. When she arrived in New York on the wet, dreary afternoon of November 22, she had completed the run in 5 days, 5 hours, and 10 minutes, which was four hours more than *Lusitania*'s maiden run. The average crossing speed was just 21.22 knots, only slightly faster than the old *St. Louis,* built a decade before.[16]

To express their gratitude, the passengers collected nine hundred dollars for the stokers and engineers. *Mauretania* had not won the Blue Riband, but she had proved her potential as a seagoing thoroughbred. After the ship tied up at Pier 54 on the Hudson River, the head of Swan,

Hunter shipyard boasted, "Not a bolt or screw about the engines was wrong. The vessel will continue to clip a little bit off the records each time she sails for some time to come."[17]

His prediction proved true. In 1909, after two years of swapping the record back and forth with her sister, *Mauretania* decisively captured the Blue Riband with a record eastbound crossing time of 4 days, 10 hours, and 51 minutes at an average speed of 26.06 knots. *Mauretania* would become the most famous and successful ship of its time, beloved by hundreds of thousands of travelers. For devotees of great ships, she was a model of elegance in form and function. "The *Mauretania* always fascinated me," said Franklin Delano Roosevelt, then a young New York lawyer, "with her yacht-like lines—her four enormous black topped funnels—her appearance of power and good breeding."[18]

It was an appreciation of marine design that the young Gibbs shared with the future president—a lover of ships, who would later work with his fellow Harvard man to rebuild the American Navy.

On November 22, 1907, as *Mauretania* eased into her New York pier for the first time, a financial storm was brewing only a few blocks away. Wall Street was about to change forever William Francis Gibbs's sense of place in the world. As the weary passengers disembarked at Pier 54, stevedores carefully unloaded $12.9 million of gold bullion from *Mauretania*'s strong room. Bound for the vaults of the Treasury Department in Washington, the gold was evidence of an international financial panic.[19]

The Panic of 1907 had begun earlier that year with industrial failures and a sharply falling stock market. By autumn, the prominent Knickerbocker Trust Company had failed, and one bank after another followed. The shock began to ripple beyond the financial markets. Angry swarms of depositors lined up outside of banks hoping to retrieve their savings before tellers ran out of cash. For William Francis Gibbs's father, a big risk taker, the panic was a fatal blow.

Exactly what happened to all William Warren Gibbs's supposed millions following the American Alkali scandal and the panic is unclear.

Appearances were kept up, though the family was in debt simply from living expenses. In September 1908 Gibbs reenrolled at Harvard and moved back into Claverly Hall. He switched his concentration from engineering to economics, and excelled. Indeed, he was doing so well that he was excused from some of his final exams.[20] At the end of his third year of studies, the economics department recommended that William Francis Gibbs receive a bachelor of science degree in economics, magna cum laude—on condition that he completed his pesky language requirement. It seemed that he had finally found himself academically.

But the spring of 1910 spelled the end of his time at Harvard. On April 8, one of the giant iron gates at the entry of the Gibbs mansion tore away from rusted hinges, fell toward the street, and crushed an eight-year-old newsboy to death. A photograph appeared in the papers the next day showing an elaborately scrolled gate resting on its side against one of the stone entrance portals. A young boy, probably a fellow "newsie," stands next to the gate, a sour expression on his face.[21] Shortly after the accident, the Gibbs family moved out of their Walnut Street mansion, decamping to a small rented house in the Main Line suburb of Haverford. The tragedy may have provided an excuse, but rumors floated through Philadelphia's drawing rooms that William Warren Gibbs had not left his prized home voluntarily.

In June 1910, William Francis Gibbs's original college class graduated. A Harvard dean reminded Gibbs that he still hadn't fulfilled his German requirement. He never did, and he later claimed he left Harvard because he flunked Latin. Most likely he dropped out because his father could no longer afford the cost of tuition and his fancy room in Claverly Hall.

The Gibbs family's losses were confirmed on December 2, 1910, when the Philadelphia's sheriff office seized 1733 Walnut Street and transferred it to the holders of the first mortgage of nearly $180,000—over $4 million today. The couple also defaulted on the second mortgage of $60,000 put up by friends.[22] The vaunted $15 million that the Gibbs family had once been said to possess had vanished. None of their rich friends seems to have come to their aid.

Sometime earlier that year, William Warren had pulled his oldest son aside for a talk. A humiliated father wanted to make sure that his son would become a respectable lawyer who would earn the steady, good salary needed to support his family. An unhappy son made a deal with his father. He would finish his undergraduate degree at Columbia and then attend its law school. To pay for both, he would work at the same time. He then agreed to practice law for a year and send money home. After a year, his father told him sternly, he could do what he wanted.[23]

During the years Gibbs spent at Columbia, 1910 to 1913, the former Gibbs mansion at 1733 Walnut Street sat vacant. Its windows, through which passersby could once see light streaming from crystal chandeliers, now showed nothing inside but a dark, skull-like emptiness. The abandoned house was finally sold to a developer, who wanted to build a luxury apartment building on the site.[24] The wreckers hacked away at the brick walls, tore out paneling, and broke up a fountain in the reception room, which according to a newspaper report had once "delighted all beholders with the beautiful effects produced by the water flowing over a bewildering number of incandescent bulbs."[25]

Twenty-six-year-old William Francis Gibbs did not appear to have looked back. He graduated from law school and went to work in a small New York firm that specialized in real estate law. Each month, he sent money home to support his father and the rest of the family: mother Frances, brother Frederic—who because of the family's financial ruin would never attend college—and two unmarried sisters, Bertha and Genevieve. But even as he served as the dutiful son, he was looking for a way to get out of being a lawyer before his promised year was up.

"My father was an entrepreneur," Gibbs confided to a friend many years later. "You know what that means? He wanted me to get into a solid profession—the law. He thought engineering was pretty unstable—said most of the engineers he knew were pretty impractical people." His friend noticed that Gibbs's voice hardened with belligerence as he remembered his father. But the son also conceded that he would never have amounted to anything if his father hadn't gone bankrupt and forced him to work for a living.[26] Gibbs also realized that in order to make something of himself, he would have to be a fighter. "Learn to withstand body blows because it's the man who's standing at the end of the fight that wins," he said.[27]

In time, William Francis Gibbs would demonstrate a business savvy beyond that of his promoter father. But despite his intense drive, he strove to remain scrupulously honest, and he turned a blind eye toward social acceptance for its own sake. Wealth and appearances, he learned, could be fleeting.

While the Gibbs family fortunes sank, the greatest disaster in the history of the North Atlantic blighted the reputation of America's most powerful financier, as error and arrogance claimed the finest ship in the world.

J. P. MORGAN'S *TITANIC*

The nation quickly rebounded from the financial panic, but was captivated by more gripping headlines only a few years later. A tragedy at sea showed that even the greatest ships, built by the most ambitious men, were not as invincible as the public assumed. The *Titanic* disaster would always haunt William Francis Gibbs, as well as the man who would one day become one of his dream ship's biggest financial backers.

As darkness fell over New York on the evening of April 15, 1912, a tall, solemn-faced twenty-one-year-old Harvard sophomore named Vincent Astor pushed his way through the crowd outside the offices of the White Star Line, the shipping company owned by the International Mercantile Marine and operated by Clement Griscom and J. P. Morgan. Astor had heard rumors in Cambridge that made him drop everything and take the first train to New York.

The airwaves of the North Atlantic were abuzz: the largest, most luxurious ship in the world had struck an iceberg the previous night. But White Star Line management assured the public that their brand-new flagship, although crippled, was still afloat. Yet as more wireless messages streamed ashore, it appeared that the unthinkable had occurred.

Vincent Astor was the son of Colonel John Jacob Astor IV, reputed to be worth $80 million. He was also the grandson of Caroline

Schermerhorn Astor, the imperious, diamond-encrusted hostess at the center of New York's "Four Hundred," a group of families she deemed acceptable in society. Back even further was the first John Jacob Astor, a barely literate German immigrant who had become the nation's first millionaire by making shrewd investments in Manhattan real estate. By 1912, in addition to owning the magnificent Waldorf-Astoria hotel, the Astors were also the biggest slumlords in New York City, owners of tenement apartment houses derisively known as "Astor Flats."[1]

Vincent Astor loved automobiles and airplanes more than fancy living, and found his grandmother's Fifth Avenue chateau a miserable place.[2] His father was an aloof, eccentric philanderer who, according to one report, "would set upon him with a shoe or strap."[3] His mother, Ava, verbally abused young Vincent. After spending most of his adolescence at boarding school and tinkering with motorcycles, Vincent had hoped to go to the U.S. Naval Academy. But his father sent him to Harvard instead. There, in his freshman year, Vincent had to endure a scandal that would have unseated most other families from good society: his forty-seven-year-old father divorced his mother and quickly married a nineteen-year-old named Madeleine Force. The Four Hundred was shocked, and the public could not get enough of it. Readers of the tabloids learned that John Jacob Astor IV and his wife, Madeleine, would honeymoon in Europe. They would return on the magnificent new flagship of the White Star Line: the Royal Mail Steamer *Titanic*.

Once inside the White Star building, Vincent Astor went straight to the office of IMM vice president Philip Franklin. He came out in tears. Vincent's father was presumed dead, while his pregnant young bride had survived. Gathering himself together, Vincent went to his father's office and donated ten thousand dollars to help destitute survivors.[4]

Titanic was born out of the blood feud between IMM and its German and British competitors. In 1907, the year Cunard's *Mauretania* and *Lusitania* made their maiden voyages, IMM founders J. P. Morgan and Clement Griscom teamed up with White Star president J. Bruce Ismay and Harland & Wolff shipyard president Lord Pirrie to push back. The three dreamed up a trio of three superliners that would be far bigger, more beautiful, and luxurious than anything afloat.

IMM needed the new ships. The huge holding company had been founded on the idea that combining international shipping lines would both stabilize rates and drive independent companies not part of the trust out of business. Quite the opposite happened. After IMM was formed, Cunard and HAPAG declared war on Morgan. In 1904, Cunard announced a "sweeping reduction in the price of eastbound first and second class tickets."[5] Morgan's White Star met Cunard's price reductions almost immediately. Then Griscom slashed fares for the American Line to a mere fifty-five dollars for a first-class berth. Germany's HAPAG, apparently cooperating with Morgan, followed suit by slicing their rates. The timing for all companies could not have been worse. The Panic of 1907 disrupted the flow of immigrants to America, which peaked that year at just over 1.7 million passengers.

Cunard's construction of the *Lusitania-Mauretania* duo using British government money had been an affront to Morgan's ego, and the rate war was the last straw. As the mastermind of America's biggest trusts, he was accustomed to getting his own way. Not only that, but it appeared that Ballin was in talks with the Kaiser about building a new trio of big ships for HAPAG. Cunard could hold a silly speed record if it wanted, but Morgan was determined to carry more passengers. But because he could not squeeze a subsidy out of Congress as his partner Griscom had a decade earlier, he decided to finance the construction of bigger ships out of his own deep pockets. The keel of IMM's White Star Line steamer *Olympic* was laid in 1908 in Belfast, Ireland. The construction of *Titanic,* her slightly larger and more refined second sister, began three months later. Their cruising speed would be 21 knots—too slow to capture the Blue Riband from the Cunarders, but they would be much cheaper to operate. White Star advertised them as the most modern and magnificent liners afloat, as well as the safest.

When *Titanic* set sail from Southampton, England, on her maiden voyage on April 10, 1912, the liner was carrying a total of 2,228 passengers and crew, or about two-thirds capacity. But she carried only twenty lifeboats: sixteen wooden craft and four canvas-sided collapsible rafts—equipment approved by the British Board of Trade shortly before departure. Four days later, she struck an iceberg and six of her sixteen watertight compartments were open to the sea. With so much water pulling the ship down by the bow, veteran captain Edward J. Smith and chief designer Thomas Andrews knew that *Titanic* was doomed. They also knew the lifeboats had seats for only 1,200 people.

News of the *Titanic* disaster struck IMM leadership like a hammer. By midnight of April 15, twenty-four hours after *Titanic* sank, wireless messages relayed from the small Cunard liner *Carpathia*, which had picked up all of the survivors, confirmed the worst. IMM vice president Franklin finally admitted the death toll was well over one thousand. "I thought her unsinkable," he said, crying. "I based my opinion on the best expert advice. I do not understand it."[6]

The rescue ship *Carpathia* did not arrive in New York until the evening of April 18. Officials, reporters, and families were there to meet her when she docked at New York's Pier 54 on that rainy night. Senator William Alden Smith of Michigan, head of the U.S. government disaster inquiry, kept a close eye on everybody who walked off the plain, sturdy vessel. First came *Carpathia*'s own passengers, well dressed but looking shaken. Then followed a steady stream of pale, bedraggled people, most of them women. Some were lucky to meet relatives, whom they tearfully embraced. Others were alone and destitute, not long ago in steerage. These were the 705 *Titanic* survivors.

J. Bruce Ismay, chairman of the White Star Line and IMM's president, was known to be a survivor, as he had climbed into the last lifeboat lowered from *Titanic* before she went down. When Ismay was not among those who got off the ship, Senator Smith boarded to find him. He asked *Carpathia*'s captain, Arthur H. Rostron, for Ismay and was led to the captain's stateroom. Knocking on the door, Smith found himself staring at Philip Franklin's haggard face. The IMM vice president had hurried from company headquarters and boarded the rescue ship as soon as she had docked. His mission: protect the White Star chairman from the prying press.

Faced not by a reporter but by a United States senator, Franklin still insisted that Mr. Ismay was "way too ill" for anyone to see him.

"I'm sorry," Smith barked, "but I will have to see that myself." He pushed past Franklin, and found the pale J. Bruce Ismay lying in the captain's berth. The slightly built Englishman appeared to be drugged. Smith gruffly introduced himself and announced that Ismay was to appear before the official American inquiry into the *Titanic* disaster the following morning at the Waldorf-Astoria hotel. Ismay quietly begged to be allowed to go back to England, but Smith said no. All surviving crew members were also served subpoenas to prevent them from sailing back to England, where they would be immune from any American legal action.[7]

Ismay spoke first at the Senate inquiry the next day. When asked if the number of lifeboats was standard British practice, Ismay responded, "I could not tell you that, sir. That is covered by the Board of Trade regulations. She may have exceeded the Board of Trade regulations, for all I know."[8]

As the proceedings continued, Senator Smith and his board learned that neither *Titanic*'s captain nor the officers gave any clear warning to the passengers that the ship was sinking until very late. To make matters worse, many of the officers had been afraid to load the lifeboats to their certified capacity of sixty-five fully grown men, fearing they would buckle and send their occupants into the water. Worse still, many of the passengers refused to leave. It seemed safer aboard the big, warm vessel than to get into a rowboat and be lowered sixty feet into the dark ocean. Many of the boats left half full. It was not until the ship's bow was awash that many of her passengers began fighting for a precious seat. Just then, a group of terrified steerage passengers emerged from below. The last lifeboat, bearing Ismay, left the ship at 2:05 A.M., a scant fifteen minutes before *Titanic* sank. The ship's captain, her chief designer, and more than 1,500 men, women, and children perished.

Fifty miles from the foundering *Titanic, Carpathia* had received the White Star liner's distress signals shortly after midnight. Her captain turned the ship around and raced to the scene, dodging icebergs all the way. *Carpathia*'s crew began picking up the 705 survivors as the first rays of a pink sun tinted the gray North Atlantic on the freezing morning of April 15. As the survivors rowed close to the rescue ship, those on board *Carpathia* could not help notice that many of the bobbing lifeboats were only partially filled. Five hundred more lives could have been saved.

Then there was the matter of the ship's speed at the time of the accident. Contrary to speculation in the press, there was no way *Titanic* could have captured the Atlantic speed record from *Mauretania.* But rumors still ran rampant that Ismay had put pressure on *Titanic*'s captain to maintain her top speed of 22.5 knots through a known ice field so that she could beat her older sister ship *Olympic*'s maiden voyage crossing time of just over five days. Above all, Smith wanted to find out if *Titanic* had been going too fast.

As the hearings went on, Philip Franklin realized that the more his British boss said, the worse he appeared. Already William Randolph Hearst's *New York American* had branded the IMM president as "J. Brute Ismay."[9]

While on the witness stand, Philip Franklin swore under oath that he first learned about the sinking at 6:16 P.M. on April 15, less than two hours after a young radio operator atop Wanamaker's department store named David Sarnoff picked up a faint message from *Titanic*'s sister ship, *Olympic:* "Please allay rumors that the *Virginian* has any of *Titanic*'s survivors. I believe they are all aboard the *Carpathia.*"[10] *Carpathia*'s captain had refused to answer incoming radio messages for much of her return voyage to New York, deciding instead to send out lists of the saved.

After hearing Franklin's testimony, Senator Smith concluded that IMM vice president Philip Franklin did not withhold news about *Titanic*'s sinking.

On May 25, as the hearings were winding down, Captain Herbert James Haddock of *Titanic's* sister ship, *Olympic,* phoned Franklin at White Star Line headquarters. Smith had showed up at the pier completely unannounced. He wanted to inspect *Olympic.*

Franklin said that he would get to Pier 59 as soon as he could. In the meantime, he ordered *Olympic*'s captain and crew to do whatever the senator asked.

Smith was not alone that morning. With him were Rear Admiral Richard M. Watt—chief constructor (head of design) for the United States Navy—and a stenographer. After taking sworn testimony from *Olympic*'s officers and crew, the senator walked the decks. He eyed the forty-three lifeboats and rafts set hard by each other along the boat deck. Before *Titanic* sank, *Olympic* carried just sixteen emergency craft. Smith pointed at one of the lifeboats and asked Captain Haddock if he could have it swung out. It was, and all watched it swaying in the spring breeze sixty feet above the Hudson. Haddock hoped that the senator was satisfied. He wasn't. Smith then asked the captain to load the boat with sixty-five men from *Olympic*'s crew and lower it into the Hudson.

Haddock froze for a moment, but then remembered Franklin's order: do whatever the senator asks. One by one, members of *Olympic*'s crew stepped across the gap and took a seat in the dangling lifeboat. Senator Smith took out his pocket watch. When the lifeboat splashed into the Hudson River, the senator noted that it had taken eighteen minutes to swing out, load, and lower the boat. He seemed satisfied.

As Smith and Watt continued to inspect the White Star ship, an out-of-breath Philip Franklin jogged up the gangway leading to *Olympic*. He rushed up the grand staircase, passing the grand clock with the three allegorical figures of "Honor" and "Glory" crowning "Time." Once on the boat deck, he could see light streaming in through a glass dome and onto the fine-grained panels, white tiled floors, and gilded balustrades—almost exactly as it once did on *Titanic*. Ahead he saw a knot of men standing around Senator Smith, who continued his probe.

The group descended into the depths of the vessel, first through luxurious public rooms and foyers, down deep-carpeted corridors, and then into the stark service areas. Finally, the men walked down a steel staircase into one of *Olympic*'s six boiler rooms, thirty feet below the waterline. As they entered the towering but dimly lit space, all gasped for air as acrid clouds of coal dust tore at their throats. The ship's boilers were still aglow and the temperature hovered around 100 degrees. Through the murk, the senator saw dozens of sweaty, soot-smeared men lined up and at work. They pushed their shovels into piles of coal, tossed their loads into the blazing furnaces, and then clanged their shovels on the grates after every scoop thrown in. Even though the ship was in port, the boilers still provided electrical power for the ship's mechanical systems.

"I found the head firemen of the *Titanic*," Senator Smith recalled, "and there in the grease and the heat, by a dim light and surrounded by his companions, he swore that he was the first man to see the water come through the sides of the stricken ship. He said that the tear extended through the side of the forward fire room . . . that the water came from a point about 20 feet below the sea level, and rushed like a mighty torrent into the ship."[11]

Frederick Barrett, one of the few stokers who survived, had been put right back to work on *Titanic*'s sister ship. Barrett told Smith that 24 of the ship's 29 boilers were fired at the time of collision, and that the ship was barreling along at "best speed she had ever shown."[12]

Smith finally had firsthand confirmation that *Titanic* was plowing ahead at full speed into the ice field.

Three days later, on May 28, 1912, Senator Smith presented his findings to a joint session of Congress. Despite the terrible loss of life, the American inquiry could not find IMM guilty of negligence, because the company had broken no existing laws. But this did not stop Senator Smith from blasting the arrogance of *Titanic*'s designers and owners.

Smith then called for the rebuilding of the American merchant marine and urged that "Americans must reenlist in this service; they must become the soldiers of the sea. . . . Their rights must be respected and their work carefully performed."[13]

After he finished, the senator put forward a bill that would create a new Maritime Commission for the purpose of drafting new legislation regulating safety at sea for all ships using American ports. Congress quickly passed it.[14]

Smith's bill also created the International Convention for the Safety of Life at Sea, charged with writing a treaty that would compel seafaring nations to set construction and safety standards for new passenger liners. President Taft, who lost a close advisor in the disaster, quickly signed the Smith bill into law.

When the International Convention for the Safety of Life at Sea (SOLAS) met in London in November 1913, Britain, the United States, and other seafaring nations agreed to require all vessels to carry enough lifeboats for all passengers and crew. The SOLAS treaty also required that masters reduce speed and change course in the event of ice warnings, and that all liners carrying more than fifty passengers have wireless sets manned around the clock. Finally, the treaty called for the creation of the International Ice Patrol. Funded by all participating nations, patrol ships would alert approaching vessels about icebergs that had drifted into Atlantic shipping lanes.[15]

The British public was incensed at an American effort to speak to what they believed, incorrectly, was the sinking of a British ship—she was in fact wholly owned by J. P. Morgan's American trust, and only flagged as British to avoid higher American operating costs. The *Saturday Review* wrote that Senator Smith of Michigan was "the man that we describe in England as an ass."[16]

The disaster devastated those who had invested in the International Mercantile Marine. Seven months after *Titanic* went down, Clement Griscom, the profane Quaker tycoon who launched the United Gas Improvement Company with William Warren Gibbs and who convinced J. P. Morgan to organize IMM, collapsed of a stroke and died at age seventy-two. Peter Widener suffered the greatest loss: his son George and grandson Harry (three years ahead of William Francis Gibbs at Harvard) were killed in the disaster.

J. P. Morgan, who had canceled his reservation aboard the *Titanic* at the last minute, died a morose and vilified man at age seventy-six in the spring of 1913. The last year of his life had seen not only the *Titanic* disaster, but the antitrust Pujo hearings in Washington, as a House committee investigated how the old man's bank did business. His son

Jack hovered around his ailing father, who seethed with rage all during the inquiry.

In the end, because Morgan ran the company that owned the ship, the public decided he owned the disaster. "The ocean was too big for the old man," a journalist concluded.[17] But the great financier's son, J. P. Morgan Jr., known as Jack, had been running the day-to-day affairs of the House of Morgan long before his father's death. It was Jack Morgan who forced the disgraced Ismay to resign as chairman of IMM. But the great shipping trust could not get past the disaster. In 1915 it defaulted on its bonds and collapsed into bankruptcy, largely from the financial repercussions of the *Titanic* disaster.

The New York Chancery Court appointed IMM vice president Philip Franklin as the receiver in charge of reorganizing the company. Working long hours and capitalizing on climbing shipping rates, he did his best to try to rejuvenate the firm. Franklin did such a good job at bringing IMM back from the brink of collapse that the directors, including Jack Morgan, appointed him president of the company.

A week after the *Titanic* sank, on April 22, 1912, the crew of the *Mackay-Bennett,* a cable repair ship Philip Franklin chartered to recover bodies from the disaster site, found a soot-covered corpse floating amid the wreckage. Most of the three hundred bodies found were bloated and ghoulish. Still, this one stood out. The undertaker described the corpse on a card before it was embalmed and placed in a coffin:

NO. 124—MALE—ESTIMATED AGE 50—LIGHT HAIR & MOUSTACHE.
CLOTHING—*Blue serge suit; blue handkerchief with "A.V."; belt with gold buckle; brown boots with red rubber soles; brown flannel shirt; "J.J.A." on back of collar.*
EFFECTS—*Gold watch; cuff links, gold with diamond; diamond ring with three stones; £225 in English notes; $2440 in notes; £5 in gold; 7s. in silver; 5 ten franc pieces; gold pencil; pocketbook.*
FIRST CLASS. NAME—J.J. ASTOR[18]

Vincent Astor met the *Mackay-Bennett* when it docked in Halifax, Nova Scotia, and brought his father home for burial in the churchyard of New York's Trinity Church. The gold pocket watch found on his father's body he kept. Restoring it to working order, Vincent would wear it on his own vest.

At twenty-one and now in possession of a $70 million inheritance, Vincent Astor then dropped out of Harvard to manage the family fortune. One of his first decisions was to sell off all of the family's tenement properties.[19]

In due course, Astor would become the lead investor in a ship designed by William Francis Gibbs that would become the anti-*Titanic* of maritime history—not just the fastest and most beautiful ship, but also the safest. For Gibbs, a ship's safety at sea would become a complete and lifelong obsession, one that would trump the quest for size, beauty, or luxury.

PIPE DREAMERS AT WORK

As the public followed the *Titanic* disaster, William Francis Gibbs was keeping his promise to his father—attending Columbia Law School and entering law practice in New York. He hated both.

During class, he spent most of his time trying to solve engineering equations in his notebook. What frustrated Gibbs was how his love of ships continued to run hard against his own mathematical limitations. The fear of math, which had dogged him in high school and led to his poor undergraduate engineering grades, continued to bother him. His complete lack of interest in the law, however, spurred him to keep plugging away at the math culled from his collection of engineering journals. Moreover, the *Titanic* disaster had captured the public's imagination. Ocean liner safety was now front-page news, which prompted Gibbs to look for a way out of being a lawyer.

But during his three years at Columbia, he began to display a sense of humor, and even play to the crowd. When a law professor once asked him what he thought about a classmate's explanation of a case, he put his pencil down, rose to his feet, and mimicked the phrase the professor used to embarrass students who were not prepared: "The former speaker's comments are interesting but immaterial and completely irrelevant."

The entire class burst into guffaws. Even the professor cracked a smile.

Gibbs then sat down and went back to his engineering.[1]

After graduating from Columbia Law School in 1913, he took a job at a real estate law firm in New York. He hated practicing law even more than studying it, but he dutifully sent money home to his cash-strapped family. The weekends provided his only release from this life of drudgery. Every Friday, he boarded a train for Philadelphia, and headed to his parents' modest house on the Main Line. There, with his brother, Frederic, he set to work designing his dream ship: a one-thousand-foot American superliner, intended to be the fastest and best ever built, intended to surpass the ill-fated *Titanic* in every respect. Analyzing the design flaws of the era's most modern liners, Gibbs sketched out his ship's hull and power plant, his long fingers flying across the blueprints. Not far away, Frederic sat at a typewriter and banged out pages of financial analysis. They did their work in a cramped attic study. In the summer, under the hot roof, they could hear the Main Line trains pulling out of Haverford station, and the crack of the bats from the Merion Cricket Club across the street.

Life at the Haverford house was hardly serene. In July 1911, their father collapsed and was rushed to the hospital. The *Philadelphia Inquirer* reported that he underwent "a serious operation for internal troubles," and that his family was "not permitted to see or talk to him because the physicians deemed that it would be taxing the patient's strength."[2] William Warren Gibbs recovered, but two years later the family had another scare when the family's rented Haverford house caught fire. Motorists leapt out of their cars to help the Gibbs brothers drag out what they could. The household was eventually put back together.[3] The Gibbs brothers managed to salvage their plans and get back to work.

Like the garage inventors of the computer age, the brothers did not let surroundings distract them. Within a year or so, they were ready to present preliminary drawings for a duo of ocean liners. Judged by the standards Gibbs would set in the years to come, the designs were awkward and derivative, a hodgepodge of visual features from predecessors, and a four-stacked silhouette echoing British liners. The prototype would be 1,001 feet long, 119 feet longer than *Titanic.* But what really excited the twenty-nine-year-old, first-time designer were the engines. Even by current standards, they were monsters. Not only would the ships be big; they would be very fast—much faster than *Mauretania.*

Financing the project appeared impossible, especially since they had no formal training in ship design. Nevertheless, the Gibbs brothers felt this was their chance to change the direction of their lives and their country's merchant marine. Passenger ships flying the American flag, such as the nearly twenty-year-old *St. Louis,* were obsolete and unable to compete for passengers. To realize their improbable dream, the brothers needed to sell their project to people with the money and know-how to finance its construction and operation.

There was only one American company that could do it: the International Mercantile Marine, then struggling to get out of bankruptcy. Somehow they had to get a meeting with the man in charge: John Pierpont Morgan Jr., son of the man who had financed the ill-fated *Titanic.* Clement Griscom, the company's other cofounder and their father's onetime business partner, was dead—the Gibbs brothers had to get to Morgan on their own.

As Gibbs's superliner was just beginning to take form on his drafting table, Europe was on the precipice of a major conflict, one that would unleash the full force of mechanized warfare. Britain and Germany knew that their big, fast passenger liners could be converted into powerful military assets: troop transports and hospital ships. Across the ocean, American shipping interests like IMM feared a world war would disrupt trade and trigger the seizure of their European-flagged vessels.

Despite the war clouds, Germany's mighty HAPAG, ignoring the possibility that the North Atlantic would become a war zone, dazzled the world with the first of three superliners, ships bigger and more luxurious than any the British had built. Even while Albert Ballin knew that Europe was moving toward war, all three of his ships were given provocative, even militaristic, names: *Imperator* (Latin for "Emperor"); *Vaterland* (Fatherland) and *Bismarck* (named for the nineteenth-century German "blood and iron" chancellor). The first ship, *Imperator,* was at 52,000 tons easily the largest ship in the world. Commissioned just a year after *Titanic*'s sinking, she had enough lifeboats for all of her nearly five thousand passengers and a massive searchlight to help spot icebergs. But critics were quick to point out that all the luxurious marble and heavy wood in her upper decks made her a terrible roller.

By building *Imperator* and her sisters, Ballin reneged on the profit-sharing agreement he had made with J. P. Morgan, an arrangement that had long irked German nationalists. Kaiser Wilhelm II called it a "scheme by the American plutocracy to prostrate Germany, if not Europe itself."[4]

When the second sister, *Vaterland,* arrived in New York for the first time in May 1914, a group of cadets from the New York Maritime Academy at Fort Schuyler toured the vessel. As they passed by Commodore Hans Ruser, *Vaterland*'s captain, Ruser turned to one of his officers and snickered in German, "These boys, of course, will never have a ship like this."[5]

One of the cadets, Harry Manning, was German-born, and overheard Ruser. Manning was furious. He would always remember the sting of the commodore's arrogance. Years later he would have the satisfaction of helping William Francis Gibbs get the last word.

Commodore Ruser might have regretted arrogance of any sort as he prepared *Vaterland* for her return trip to Hamburg. As the ship's band serenaded hundreds of well-wishers on the New Jersey pier, a mechanical failure caused *Vaterland*'s astern turbines to engage at full speed. The giant ship—with over two thousand passengers aboard—shot backward across the Hudson. Ruser's crew somehow brought it to a halt before her stern rammed into the New York side of the river.[6] When *Vaterland* returned to New York in late July, company headquarters ordered she remain there. War was imminent. Better the German flagship remain in a neutral American port than risk a mad dash across the Atlantic to Germany. Or so they thought.

When war did break out in early August 1914, the British Admiralty commandeered most liners for wartime trooping and hospital ship duty. By the end of the year, Cunard's greyhound *Lusitania* was the only big British liner left in commercial service, and was run at low speed to save coal. The outlook for passenger business, especially for big liners, was bleak.[7] Even so, Jack Morgan's international shipping trust, IMM, continued to thrive under Philip Franklin, who loved the cutthroat, high-stakes game of shipping. He saw that with careful management, IMM might be in position to survive the economic crisis of war. The company's diverse holdings could help it weather the temporary loss of its commandeered British ships. In fact, the boom in wartime cargo shipments to Britain became a nicely profitable business. In 1916, the shipments netted a profit of $26 million, a fourfold increase from the year before.[8]

Still, Franklin knew that IMM would have to avoid the mistake Morgan himself once made—basing future financial health on atypically high yearly revenues. For IMM's longer-term outlook, Franklin decided the company needed to shift its focus away from Europe and become an "American," not an "international," mercantile marine company.

Cunard, IMM's biggest British competitor, was having problems of its own. Most of their vessels were requisitioned for war service, but their two biggest ships proved to be problematic as warships because of their heavy fuel consumption. As a result, *Lusitania* remained in regular service and *Mauretania* became a hospital ship, and later a troopship. Both were designed to British Admiralty specifications, requiring bulkheads that ran lengthwise, parallel to the keel, and athwart ship, or side to side. At least one official at the British Board of Trade had argued that such compartmentalization made ships less stable when breached, and made lowering all the lifeboats impossible.[9]

He turned out to be correct. On May 7, 1915, Captain Walter Schwieger of German submarine U-20 sighted *Lusitania* off the west coast of Ireland, loafing along at 18 knots to make Liverpool on the tide. Schwieger launched a single torpedo that slammed into *Lusitania*'s hull, sending up a plume of steam and flames. Shortly after the strike, a much larger explosion blew her bottom out. The ship's power and steering then failed, trapping dozens in jammed elevator cages. In twenty minutes, *Lusitania* smashed onto the sea floor, a twisted wreck. Only 6 of her 48 lifeboats got away. Of the 2,000 passengers and crew on board, nearly 1,200 died. One hundred and twenty-eight were Americans, including millionaire Alfred Vanderbilt.

What caused the second explosion became a matter of intense controversy. The British press charged (incorrectly it proved) that it was a second torpedo. The Germans said that munitions—possibly purchased by the House of Morgan acting as a British agent—exploded in her cargo holds. Whatever the case, *Lusitania*'s longitudinal compartmentalization—bulkheads that ran parallel to the keel—hastened her end and those of the people who drowned.

The sinking of *Lusitania* galvanized American public opinion against Germany and spurred legislation providing government support for American shipbuilding. The Shipping Act of 1916, signed into law by President Woodrow Wilson, authorized the building of commercial ships to transport American goods abroad in times of war. The bill also created a five-man governing body, known as the United States Shipping Board, which would operate the government-owned ships as well as regulate freight and passenger rates.[10]

The desire for an American-flagged superliner was made all the more urgent by the loss of *Titanic*'s youngest sister ship on November 21, 1916. *Britannic* had been built as the last word in safety, with extra compartmentalization and an extra-high double hull. But while steaming through the Aegean Sea to pick up wounded soldiers from the Gallipoli campaign, *Britannic*—converted into a British hospital ship before carrying a single paying passenger—ran into a mine. She capsized and sank in less than an hour, killing 30 of her 1,100 crew and medical staff.

Britannic's loss was another blow to IMM's balance sheet. The company was counting on passenger revenue from the big liner to help pay off its debt once hostilities ended. Never again would ships financed by American capital be at the mercy of the British Admiralty for use in European wars.

* * *

As the Great War raged in Europe and the high seas, the Gibbs brothers continued to slave away in their stifling Haverford attic, which became cluttered with magazines, books, and discarded drawings. As their initial design neared completion, Frederic Gibbs sat hunched over a typewriter banging out financial projections. Mild, shy Frederic, now completely unable to attend college, decided to fully support his older brother's dream. The best way to do so, he reasoned, was to give his natural mathematical and business skills to the project, skills that William Francis sorely lacked.

Gibbs would later say of Frederic: "Everything I have done he has provided the sinews of war for."[11]

Frederic reasoned that the way to attract wary potential investors was to make a deal with a railroad. He found there was plenty of room to build a completely integrated sea-to-rails facility on the eastern tip of Long Island, at Montauk. Passengers, mail, and freight could be quickly transferred to the Pennsylvania Railroad's trains, which would then speed to New York or other destinations. Such a terminal on Long Island would cut nearly twelve hours off any transatlantic voyage.

As Frederic laid out the financial math, his brother did more engineering calculations. To move a ship as big as his at 30 knots, the engines would have to churn out 180,000 horsepower—more than double *Mauretania*'s maximum output. Not only had no engineer built such a powerful steam plant, but the astronomical fuel cost for each voyage would make the liner a coal-guzzling, money-losing monster.

There was only one system, electric drive, that might work, and there was only one person in the country who might listen to the possibility of its use on a ship. That was the chief engineer of General Electric, William LeRoy Emmet, promoter of the GE-Curtis turbo generator system used in the newest American electric power plants. The more efficient Curtis turbine was a direct challenge to the British Parsons's bigger, slower turbine, which had recently displaced reciprocating piston engines on ships. It had never been used on an ocean liner, but Annapolis-educated Emmet might be open to trying.

Emmet wanted something big: to break Parsons's stranglehold on ship engines and sell his improved version to the U.S. Navy. The principle he was touting with missionary fervor was "turbo-electric" power. Rather than using steam to turn a ship's rotor directly, the Curtis turbine would use steam to turn a rotor, spinning at 2,000 revolutions per minute, to power a massive electric generator. This would in turn power a huge electric motor that would turn a ship's propeller shaft. The motor could move at a rate slow enough for the propellers to effectively grip the water, avoiding a vibration-causing phenomenon known as cavitation—the ship's screws turning so fast that they were generating bubbles in the water around them, creating a non-uniform medium in which to turn.

Emmet argued that electric propulsion was best suited for "vessels requiring very large power and high rates of speed reduction"—meaning either a battleship or a transatlantic ocean liner.[12] But even after demonstrating the efficiency of turbo-electric propulsion in two experimental naval vessels, Emmet had no luck convincing anybody in Washington to adopt the system. President Wilson's secretary of the Navy, Josephus Daniels, told Emmet to stop bothering him. So did Assistant Secretary of the Navy Franklin Delano Roosevelt.

As an alternative to turbo-electric power, Navy engineers tried geared turbines, which worked somewhat like an automotive gearbox. The power from the rapidly spinning rotor would then be transfered to a set of gears, reducing revolutions and preventing cavitation. The Navy design had strong bureaucratic allies. But someone among the higher-ups was looking ahead, and finally gave Emmet's new technology a chance on the battleship USS *Pennsylvania*. Her sister ship USS *Arizona* would use geared turbines. Not until ship trials held in 1916 would Emmet know that his turbine would prove to be a big part of the Navy's future.

Gibbs had a college friend who could provide an introduction to Emmet.[13]

Emmet was ready to talk ships when the Gibbs brothers asked for a meeting. The prospect of electrically powered battleships excited Emmet, but a transatlantic superliner set his imagination on fire. He invited the two to GE's Schenectady, New York, headquarters and sat down with them in his office at the plant sometime in 1915. The brothers bore a name that would definitely have been known to Emmet. Their father, William Warren Gibbs, had founded a major company, Exide, that sold batteries to Emmet at GE. William Warren had long ago severed his Exide connection, and it is not known if he provided a letter of introduction for his sons—or if it would have helped if he had. Sitting across from the bespectacled, mustachioed Emmet, William Francis Gibbs was on his own.

If Gibbs still had any of his old shyness, he couldn't let it show now. This was the chance he really needed, and the earnest young dreamer laid out his case. By his calculations, he told Emmet, a liner using GE's electric drive could produce 20 percent more speed than the famed 26-knot *Mauretania*.[14]

Emmet was stunned. After years of dealing with the Navy bureaucracy, here was a young man who not only understood marine design, but had a vision for something really grand. The nervy Harvard dropout and failed lawyer appeared to have the real makings of a naval architect. And unlike many other engineers, he was unafraid to defy convention. The presentation had sold Emmet on more than the ship: it had sold him on Gibbs as well.[15]

William Francis and Frederic Gibbs walked out of General Electric headquarters two very happy young men. Emmet told them that General Electric would help design the 180,000-horsepower electric turbines for their proposed superliners. The commitment from GE would also help get the brothers a meeting with the man best qualified to vet their hull designs: Rear Admiral David W. Taylor, chief of the Navy's Bureau of Construction and Repair.

Taylor was well fitted to judge naval design. The son of a hardscrabble farmer in Louisa County, Virginia, he was so brilliant that he entered Randolph-Macon College at age thirteen. Five years later, he received an appointment to the U.S. Naval Academy, and graduated at the top of his class. After his sea duty, the Navy Department sent him as part of a select group to study marine engineering at the Royal Naval

College in Greenwich, England, where he graduated with the highest marks ever earned by any student, foreign or British, up to that time.[16]

Back in the United States, the rising naval officer devised a system of calculating ship stability and buoyancy so lucid and accurate that it became standard practice throughout the Navy. He also pioneered the U.S. Navy's work with experimental models in hull design—looking for that "single, correctly-designed hull" that would have optimal ratios for beam-to-draft and speed-to-length. During his experiments, Taylor came up with the idea of the bulbous bow, a protrusion that stuck out from the vessel's stem below the waterline. Rather than cutting through the water, the bulbous bow would push the water away from the hull, reducing resistance. Everything else being equal, a properly designed bulbous bow allowed a ship's engines to be 5 percent more efficient.[17] Taylor's magnum opus, published in 1910, was *The Speed and Power of Ships*. William Francis almost certainly had a marked-up copy of Taylor's book in his attic study. Now, at his first meeting with the famed engineer, he handed over his drawings for evaluation by a master.

Admiral Taylor immediately recognized that the plans were the work of an inspired amateur, but he was taken by the encyclopedic knowledge the Gibbs brothers had about the shipping business, and impressed further when he learned that William Emmet of General Electric had offered to design the liner's power plant. And Taylor personally admired the young men's daring in an area close to his heart—taking on the giant European liners. The admiral was well-known to be unhappy about American commercial shipbuilding, which lagged so far behind what was happening in Germany and Great Britain.

And so after tweaking the drawings, Taylor said he was willing to build a ¹⁄₂₄th scale model of the Gibbs vessel for testing in the U.S. Navy Experimental Model Basin at the Washington Navy Yard, with the admiral's engineering staff providing full technical support for the engineering of the liner's hull. Built in 1898 under Taylor's direct supervision, the towing tank was 470 feet long and topped by a truss-and-glass ceiling. A motorized beam, set on parallel tracks, pulled a miniature hull through waves created at the far end of the tank. Engineers on catwalks would then evaluate the model's performance in a variety of simulated sea conditions. Taylor also set up a key meeting with Secretary of the Navy Daniels, who had rebuffed many of Emmet's earlier entreaties.[18]

Admiral Taylor, a man of full face and warm eyes, soon became William Francis Gibbs's informal mentor and surrogate father, providing him the support and guidance that his own father never did. For his part, the aspiring naval designer admired the man who he said "had the rare advantage of a brilliant mind and a natural talent for expressing himself in concise scientific language. He was never satisfied until he had reached perfection in exposition and he avoided always the pitfall of stating opinions that were not completely buttressed by the facts."[19]

After his meetings with Emmet and Taylor, Gibbs finally felt free to resign from his law firm to work on the superliner project full-time with Frederic. He had fulfilled his agreement with his father for a year of unhappily practicing law. Packing his belongings, he left New York and joined Frederic in Philadelphia.

A month later William Francis walked into the huge Navy Experimental Model Basin in Washington Naval Yard. Floating serenely in the basin was a 41.7-foot-long pine model of the Gibbs design called "Proposed American Passenger Steamship." It would carry the Navy project number S-171.[20]

The model showed some changes made by Admiral Taylor to Gibbs's drawings—the ship looked lower, sleeker, more modern. The superstructure now had three decks high rather than four. A knob-like cruiser stern replaced the overhanging counter-stern, and the bow projected forward slightly. But the basic hull design remained what Gibbs drew—rather than the round, full lines of other large passenger liners, the ship's bottom was sharply cut out at both the bow and stern, reducing the ship's underwater volume. Taylor did not add a bulbous bow, probably because he did not feel quite ready to use it on such a large craft.

The tests began. The model was towed the length of the tank in a variety of simulated conditions: flat calm, moderate seas, and gale conditions. Despite its fine lines, there was none of the nauseating corkscrewing that afflicted lean-hulled record breakers such as *Mauretania.* Nor did the model roll drunkenly like top-heavy German ships. And most important, when towed at a simulated maximum output of 180,000 horsepower, the model moved effortlessly through the water at the equivalent of 33 knots, an astonishing six knots faster than *Mauretania's* top sustained speed.[21]

But model tests were just one step on a long road. The U.S. Navy was not in the business of building commercial ships. The brothers needed financing from a private source. With Emmet's and Taylor's design modifications in hand, and Frederic's financing plan for a railroad tie-in, the Gibbs brothers decided to go see Ralph Peters, the president of the Long Island Rail Road, a subsidiary of the Pennsylvania Railroad. Because the brothers thought a letter from two unknowns would end up in the trash, they decided to show up at Peters's office in person, and *make* him listen to them.

"Mr. Peters is very busy today," his secretary told the two strangers, not looking up from her typewriter.

"Tell him some men want to talk about ships from Montauk to England," Gibbs insisted.

The secretary told the Gibbs brothers to wait outside.

For the last ten years, various idea men had hectored Peters about building a new terminal in Montauk. For most of that time, Peters was preoccupied with the completion of the Pennsylvania Station and rail tunnels under the East River. But by 1916, with trains moving in the tunnels, Peters needed a new project. He knew all along that all of the freight and passenger traffic coming into New York could be monopolized if the piers could be directly tied into the Pennsylvania Railroad and the LIRR.

After his secretary told him about his uninvited visitors, Peters strode out of his office, his rosy face showing a big smile.

As Peters thumbed through a black leather-bound volume containing the proposal, he said he loved the idea of a Montauk terminal. But when he saw the superliner design, his eyes bulged. It was huge. Not only that, but its proposed turbo-electric power plant had the stamp of approval from William Emmet and its hull by none other than Admiral David W. Taylor.

The railroad man stared intently at the two young men and asked what engineering qualifications they had.

They said none.

Peters picked up the phone and called the office of J.P. Morgan and Company, saying he wanted to speak to Jack Morgan.[22]

Approaching sixty in 1916, "Jack"—or J. P. Morgan Jr.—had led a charmed life. Following his graduation from Harvard in 1889, he floated effortlessly to the top of the House of Morgan. Despite not being especially close to his father, he had inherited $69 million when the older Morgan died in 1913. He also received the legacy of three generations of New England and New York banking prominence, and close ties with international heads of state and business.

When Peters's call came, Jack Morgan was in no mood to meet with anybody. He had just survived an assassination attempt. On July 3, 1915, a former Cornell University professor named Frank Holt drove through the gates of J. P. Morgan Jr.'s estate at Glen Cove, Long Island. Wielding two revolvers, Holt assaulted Jack Morgan and the British ambassador Sir Cecil Spring-Rice. A bullet struck Jack in the hip, and another tore through his thigh. The butler ran up from behind Holt and smashed a lump of coal against the side of his head, after which the police arrived and arrested the would-be assassin.

Newspapers readers learned that Holt had set off a bomb at the U.S. Capitol the day before he showed up on Long Island. The former professor, a German sympathizer, was outraged over the House of Morgan acting as Great Britain's American financial agent.[23] Much of the war supplies traveled across the Atlantic on vessels owned by Morgan's shipping trust: the International Mercantile Marine.

Jack Morgan made a complete recovery. "The experience was a very disagreeable one," Jack wrote to his friend Owen Wister. "I was singularly fortunate."[24]

Jack Morgan's place of business was at 23 Wall Street. The House of Morgan lay hidden in a forest of skycrapers at the southern tip of Manhattan, an austere gray limestone building only five stories high. It looked more like a temple than a bank. Yet its power inspired near-religious awe on Wall Street, for it underwrote securities for the biggest trusts in the world, including the International Mercantile Marine.

It was here that William Francis and Frederic Gibbs had their audience with a recovered Jack Morgan and Philip Franklin of IMM.

"My brother and I proceeded immediately to lay out our key designs and blueprints," Frederic recalled later. "As we did, my brother explained each special feature."

The drawings called for the two largest and fastest ships ever constructed. Each would cost about $30 million, more than three times the amount J. P. Morgan Sr. had paid to build *Titanic* several years earlier.

The young designer then launched into the second part of his presentation: the Montauk sea-land terminal would cost about $15 million, and would have easy access to ground transportation, as well as plenty of room to expand.

Suddenly Jack Morgan got up from his seat and walked out of his office. Franklin scurried after his boss, and a door slammed behind them.

The Gibbs brothers sat in Morgan's office, as their confusion turned into anxiety. "That wait seemed like eternity," Frederic recalled. "It was about 20 minutes, but each minute seemed like an hour. We stood and looked at each other. I rolled up some of my biggest blueprints. My brother looked at his watch. Neither of us said a word. For the rest of my life I never have endured a wait such as that one."[25]

The Gibbs brothers' stomachs were turning when Morgan and Franklin walked back into the office. Jack sat down at his desk and stared at the Gibbs brothers.

As William Francis Gibbs started to say something, Morgan raised his hand and intoned, "Very well, I will back you. How much money do you need to work up final plans?"[26]

William Francis Gibbs, aged twenty-nine, and Fredric Gibbs, aged twenty-seven, had convinced two of the savviest businessmen in the country to finance a hugely expensive superliner project. And they were to start to work immediately.

The Gibbs brothers moved out of their family's Haverford home and into a New York apartment at 31 East Forty-Ninth Street, just off Fifth Avenue. A few months later, they were at the Washington Navy Yard again, staring at a modified scale model of their ship, floating in the Navy's test tank. This time the Navy model builders had added four propeller bossings (winglike structures enclosing the shafts) to simulate the drag created by the vessel's quadruple screws. The test proved once again that at 180,000 horsepower, a 30-knot ship was possible.

With Morgan's bankroll behind him, Gibbs recruited an engineering and design team to work on S-171 at Franklin's IMM. As work continued into early 1917, Gibbs could look out of his office at 11 Broadway and see several large German passenger ships tied up across the Hudson River. As seagulls wheeled around their masts, marooned German sailors paced the decks, wondering if they would ever make it home. These HAPAG and Norddeutscher Lloyd ships had been stuck at their Hoboken, New Jersey, piers since war broke out almost three years earlier. The German liner that caught William Francis Gibbs's eye towered over the rest. She was the largest ship in the world, *Vaterland,* flagship of the mighty HAPAG and the pride of Albert Ballin and imperial Germany.

PRIZES OF WAR

As the Gibbs brothers won the support of the House of Morgan, America moved closer and closer to war. On January 31, 1917, a starving, blockaded Germany resumed unrestricted submarine warfare, which meant that all ships carrying supplies to Great Britain could be sunk without warning. On a single day in March, submarines sank three American-flagged vessels carrying supplies to Britain, and the public wanted revenge. In Hoboken, the pro-German charity balls aboard the interned HAPAG flagship *Vaterland* once attended by the likes of anti-British newspaper publisher William Randolph Hearst, ceased. Her palatial public rooms fell silent.

On April 6, 1917, President Wilson asked Congress for a declaration of war on Germany, after which two hundred American soldiers stormed aboard *Vaterland* to seize what was considered a prize of war. Sixty policemen guarded the pier entrance. Other military units seized several other idle large German liners docked in Hoboken and Boston, imprisoning their now-enemy crews and hoisting the Stars and Stripes on their fantails.[1]

"You will never run her!" shouted *Vaterland*'s chief engineer, Otto Wolf, as he was hauled off to Ellis Island to be detained along with the rest of the German crew. When asked if he had sabotaged *Vaterland*'s engines, Wolf was reported to have laughed. "Ruin those engines? I didn't have to. They were ruined before she ever started on her first trip out of Hamburg. I will take my hat off to the Yankee engineer that can ever make that rubbish do decent work."[2]

The Germans indeed had done some hasty sabotage. Some shipboard machinery had been sliced with hacksaws. Telegraphs had been smashed, and blueprints destroyed. However, an expert from the Brooklyn Navy Yard also determined the engines were poorly designed and that, "the major part of the damage appears to have been due to faulty operation." The public was not told, but damage from the accidental backing into the Hudson three years earlier had been so bad that on her last crossing, *Vaterland* was limping along on three propellers.[3]

Despite her condition, the American government still needed the ship to take troops to the front. During the next few months, construction crews repaired *Vaterland*'s engines, installed rows of standee bunks, smashed partitions to create open dormitories, and carted away truckloads of furniture. Workers then looted anything of value left—table linens, silverware, paintings, faucets, marble sinks, brass bedsteads, and a bronze bust of the hated Kaiser Wilhelm II. Portraits of German royalty were slashed with bayonets.[4] The outraged German public felt that the Americans had destroyed a German national treasure.

When asked for a new name for America's biggest war prize, President Wilson replied, "Why, that's easy, *Leviathan*. . . . It's in the Bible, monster of the deep."[5] The former German flagship would eventually transport nearly 120,000 American soldiers to the Western Front, sometimes carrying as many as 14,000 men per voyage. Crew, doughboys, officers, and dockworkers affectionately called her "the Big Train."

Overseeing troop and cargo transport to Europe was a new government body, the Shipping Control Committee. To head it, President Wilson turned to an experienced elder from the industry: Philip Franklin, who promptly took leave from IMM and moved to the old HAPAG offices in New York, which had been seized by the government. There he spent hours bent over maps and charts, carefully allocating troops and cargo for ships bound for the European front.

Franklin took more than a professional interest in using passenger liners to transport American troops. In the spring of 1917, his twenty-one-year-old son John, then nearing the end of his junior year at Harvard, decided to leave school and serve his country. He was a popular varsity rower but was in dire academic straits. John Franklin saw a way out when Harvard handed out certificates to any student who left to enlist; he got one and did not look back. "It was the only kind of diploma I ever received," he said later.[6]

Philip Franklin hoped that the Army would give his son some direction in life.

John Franklin took basic training at Camp Plattsburgh, New York, with other young men from prominent New York families. He was then assigned to a unit in Hattiesburg, Mississippi, "an awful place," where the Army put Sergeant Franklin in charge of a pack of mules. "When the order was given to clean out the right or left hind foot," he wrote, "there was a murmur of profanity up and down the line."[7]

Eager to escape mule duty, Franklin volunteered for the 301st Battalion of the Army Tank Corps. Unlike Camp Plattsburgh, this was no silk-stocking outfit. While driving these newfangled contraptions up and down the muddy fields of Camp Meade, Maryland, Franklin grew to know the tough tank crews—men who were "big and powerful, ex-regular army sergeants, soldiers of fortune, taxi drivers, bums—all chosen for some particular attribute, all enthusiastic, and all imbued with the commendable but rather stupid idea of getting a crack at the Germans before the war's end." Some perhaps had a criminal past. In short, they were men much like those who worked on his father's ships. But Sergeant Franklin loved his unit. He was also impressed by the commanding officer of the 301st: a young captain named Dwight D. Eisenhower.

When the battalion was ready to deploy, Eisenhower become very upset when he learned that there was no ship available to carry his men to Europe. "I'm going to New York and see if I could get this outfit moved overseas," he announced to his men.

Sergeant Franklin approached his CO. "Sir," he said, "if you're going up to New York to get this outfit moved, you'd better take me with you."

"Just why should I take you with me?" Captain Eisenhower asked.

"Sir, my old man has a lot to do with moving troops."

"Is he in the Army?"

"No sir. He has too much sense to get mixed up with the Army."

"What's his job?"

"He is chairman of the Shipping Control Committee," Franklin answered.

"What the hell is that?"

"I don't know, sir," Sergeant Franklin replied. "But I understand he has a lot to do with moving troops."

Eisenhower and his subordinate boarded a train to Manhattan.

Philip Franklin, surprised to see his son and even more surprised to see his son's commanding officer, invited them into his office. "Sit down, boys," he said. "What can I do for you?"

The twenty-seven-year-old Eisenhower explained that the 301st Battalion was "the most valuable outfit in the American Army." He intended to lead it straight to Berlin, he continued. "It was essential to the outcome of the war that the outfit be shipped immediately."

Franklin then called his wife. "Laura, Jack and a Captain Eisenhower are here. They'll be coming to dinner with us tonight. And by the looks of them, you'd better have a good meal ready."

Eisenhower protested, saying that he had to return to Camp Meade by sundown.

"Captain," Philip Franklin replied firmly, "I gather from your conversation that you came here to get this outfit shipped overseas. I will not be able to give you any information until dinnertime. I'll see you boys at six o'clock."

The two soldiers soon found themselves in the parlor of the Franklin residence on East Sixty-First Street, sipping tea and munching on cinnamon toast with Mrs. Franklin. The plain-spoken Captain Eisenhower, who had grown up in a wood frame house in Abilene, Kansas, must have been taken aback when he was waited on by a uniformed butler.

Philip Franklin was home at six, as promised. "Well, Captain," he announced, "the *Olympic* got in this morning. . . . I've arranged for this outfit that you think so highly of to be assigned to her for transporation overseas."

On March 28, 1918, the men of the 301st Heavy Tank Battalion joined some six thousand other troops on *Olympic* bound for Europe. As a child, Sergeant Franklin had frequently traveled to Europe with his father on White Star ships, but the sight of this majestic, four-stacked liner dressed for war made a deep impression on the young soldier. To Franklin, "the old gal," *Titanic*'s sister, was "very dear to my heart."[8]

The war meant that public rooms were packed with standee bunks and that the kitchens served army meals. A stripped first-class stateroom accommodated Sergeant Franklin for the seven-day voyage. The rest of his regiment bunked in the pool, near the most fought-over real estate on board: toilets and showers.

Captain Dwight Eisenhower was not aboard. Shortly after getting back to Camp Meade, Eisenhower discovered that they were going to keep him there; his value as a man who trained other men was just too great. John Franklin remembered Eisenhower having tears in his eyes. "I presume a West Pointer who did not get overseas in the war, considered his career ruined," Franklin recalled.[9]

Six days out of New York, as *Olympic* approached the U-boat infested waters surrounding the British Isles, John Franklin was called to the captain's cabin. He found Bertram Hayes, famed White Star master in peacetime, looking terrified.

Hayes pushed a telegraph across his desk. It ordered him to turn his big ship around and rendezvous with a destroyer escort many miles astern. "What do you think of that?" Hayes asked the young sergeant, in his mariner's brogue. "I've told 'em to go to hell! I'm not going to turn this ship around out here!"

Hayes kept his ship on her original course, although at a top speed of 22 knots a skillful U-boat commander could still hit the overloaded liner. In the middle of the night, Franklin heard a soldier running up and down the corridor shouting, "All hands on deck to boat stations!" If the ship were hit, Franklin knew that a hundred of his fellow soldiers would drown. The night wore on, the seas grew rough, and the sleep-deprived troops struggled to keep their footing as the great ship rolled from side to side on her zigzag course.

It was not until 7 A.M. that John Franklin and his fellow soldiers saw the hills of the French coast. Just inside the protected confines of Brest harbor, Hayes shut down the engines and *Olympic* glided to a stop.

During the next few months, Sergeant John Franklin would see heavy fighting in France. On September 29, he took part in the breaking of the Hindenburg Line on the Somme Canal. Promoted to lieutenant, Franklin was awarded the British Military Cross for "gallantry and devotion to duty" during the attack upon the canal from Le Catelet to Bellicourt, on September 29, 1918.[10]

After the war, Franklin would stay with the U.S. Army in Paris until his father ordered him back to New York and the shipping business. But Franklin did not start under father Philip's wing at IMM. Instead, as his father had done before him, he worked his own way up in the industry.

The Army experience gave the academically lackluster Harvard dropout a much-needed boost in confidence and street smarts. Eventually John Franklin would succeed his father as America's most prominent shipping executive, one who understood how transatlantic liners could tip the balance of power in another world war.

While young Franklin fought on the battlefields of France, William Francis Gibbs toiled away in a cramped Manhattan office. But despite having steady work as a salaried IMM employee, the thirty-one-year-old Gibbs was frustrated. With America's entry into the war, Morgan and Franklin had put the superliner project on indefinite hold. Although work continued on the superliner, he was distracted by smaller, less interesting wartime conversion projects.

Gibbs wondered what was to happen to his superliner project after the war was over. But he had another interest as well. Scanning the shipping news one day, he came across a story in the *Evening World* suggesting that, if placed in peacetime service and run by an American crew, a refurbished *Leviathan* could capture the Blue Riband from *Mauretania*.[11]

Once the war was over, Gibbs thought he would have his chance to make his mark. Not only would he build one, maybe two 1,000-foot-long ships, but America had seized a fabulous war prize: the biggest ship in the world, along with dozens of fine liners from the German imperial fleet. America, which had neglected its commercial shipping for fifty years, now found itself with a modern fleet that could compete with the British head-on after the war.

For Gibbs, the free German ships were manna from heaven. If operated by a capable American company, *Leviathan* could generate the cash flow and public support needed to finance the construction of his own thousand-foot-long superliner.

As Gibbs dreamed, another man despaired.

It was November 1918, and the head of the mighty HAPAG shipping line was watching imperial Germany collapse around him. Albert Ballin had seen his superb *Vaterland* serving his nation's enemy. Other German ships had been sunk, or trapped in American ports. *Imperator* was docked in Hamburg, rusting and neglected. The last of his three big ships, the incomplete *Bismarck,* was almost certain to be seized as war reparations. And socialist rioters were surging through the streets of Hamburg. As a shipping executive, Ballin described dealing with maritime unions as "the most hateful duty which is connected with my work." And now these same people bayed for the HAPAG chairman's blood.[12]

On November 9, 1918, a broken, depressed Ballin swallowed a massive, fatal dose of sleeping pills and Kaiser Wilhelm II, convinced the fatherland was stabbed in the back, abdicated the German throne. "Better an end with dread, then dread without end," Ballin once said.[13]

Germany surrendered two days later.

A GIANT LIVES AGAIN

In December 1919, a few months after the armistice, William Francis and Frederic Gibbs boarded *Leviathan* at her Hoboken pier with a directive from Philip Franklin and the U.S. Shipping Board: create a set of working plans of that enormous ocean liner from scratch. These blueprints, detailing the ship's current configuration, were required by shipyards putting together renovation bids. Their employers, the International Mercantile Marine, had announced these plans would be used to convert her back into a luxurious passenger liner, the biggest flying the American flag.

Leviathan had just been decommissioned from two years of strenuous trooping duties, and was a total wreck, inside and out. Her gray hull was streaked with rust, her interiors gutted, and her machinery worn out. It was hard to imagine that this ship, only six years old and the biggest afloat, had once been the German imperial flagship *Vaterland,* the apple of Albert Ballin's eye.

On that cold December morning, the Gibbs brothers assembled their team on the liner's upper deck, which was caked with bird droppings. With the ship's three massive funnels towering behind his thin frame, the overall-clad William Francis Gibbs gave a rousing speech—peppered, as his brother remembered in utter amazement, with some "extraordinary cuss words." The young designer then dispatched his men to their work.[1]

To Gibbs, a great ship like *Leviathan* was not just a technical puzzle. One had to understand the ship in the same way one had to get to understand a person's likes, dislikes, and quirks.

"There was nothing to go by but the ship herself," he said. "We knew nothing whatever about her. We did not even know where her center of gravity was, and there was therefore nothing upon which we could base our distribution of weights. To do the work set for us it was necessary to measure every inch of the ship, working from the inside."[2]

Gibbs's team would spend nearly every day between December 1919 and April 1920 working to develop construction drawings and specifications. One hundred draftsmen took over the ship's former Ritz-Carlton restaurant on the promenade deck, setting up tables on the scuffed floor and pinning drawings on the cracked walnut paneling. The cavernous domed room, once the haunt of the imperial German elite, was cold and drafty in winter and stifling hot as spring arrived. Soiled army blankets, ripped drapes, and smashed plumbing fixtures were strewn throughout the passenger areas, which stank of mildew. Bits of plaster and broken glass lay underfoot, and bayonet-mutilated paintings flapped from their gilt frames.

It was worse in the machinery spaces in the lower reaches of the ship. Determining the underwater hull shape of the vessel and its center of gravity was a monumental task because there was no dry dock in America big enough to hold *Leviathan.* Water dripped on the men's faces as they lay on their backs inside the ship's cramped bunkers and double bottom measuring every nook and cranny.

Gibbs, who always wore a stiff derby hat on site, reveled amid the wreckage. He hated the constant requests for tours of the vessel, but for the right audience he became adept at talking about ships in ways that laymen could understand.

He explained to one group of visiting congressmen how to find a ship's center of gravity. "You take the ship, 921 feet—or as it happened to be in that case exactly 921.8 feet on the water line—and you divide that into 20 sections," he began. "Then at each of those sections you go on the inside of the ship, and measure the width of the ship at the water line and at given distances below the water line. Then you lay that out on a drawing. And to make a long story short the result of that is finally you get the shape of the ship on those 20 sections."[3]

The congressmen were astonished by Gibbs's phenomenal, perhaps photographic memory, as well as his immense charisma. They could see that the odd-looking young man in a derby hat had captured the loyalty of his fellow IMM designers and workmen. The sickly child had become a leader of men.

But Gibbs was busy with more than his assigned IMM work. On his own, he continued to refine his superliner design, and he let the congressmen know that he was still hard at work on something even more impressive than this German war prize. In reply to a question about whether it was harder to design a new ship or rebuild an existing one, Gibbs said the work on *Leviathan* was more difficult. "I am in a good position to say as to that," he added. "Because we have designed ships of almost identical size—in fact, a little bit bigger than the *Leviathan*."

"What ships have you designed larger than the *Leviathan*?" a congressman asked.

"These ships designed a thousand feet long that the Shipping Board spoke of some time ago."

"Are the specifications prepared?" Walsh queried.

"Not final and complete," Gibbs replied. "But all the necessary information has been prepared by which the specifications could be finally prepared.

"They were designed by the IMM?"

"Designed by me," Gibbs answered firmly.

"For the Shipping Board?"

"I designed them originally for IMM," Gibbs said. "They have been in process for about four years."[4]

Gibbs had a vision: to reconvert the world's biggest liner into an American-flagged superliner. He would then oversee the construction of two even bigger running mates, built according to his own S-171 designs. Ultimately, his reconstruction of *Leviathan,* the famed World War I troopship, would catapult him to fame and gain him much-needed professional respectability.

Yet as he worked on the reconversion of *Leviathan,* Gibbs realized that his employer, the International Mercantile Marine, was losing interest in building a superliner from his own designs. If IMM was getting cold feet, Gibbs would find someone else, and he was determined to gamble his brief career on it.

The months following the armistice had kept William Francis Gibbs busy. All during the spring and summer of 1919, as *Leviathan* carried thousands of victorious American doughboys home, Gibbs set sail in the other direction, bound for the Paris Peace Conference. His boss had loaned him to Shipping Board chief Edward Hurley, who was impressed by the young man's encyclopedic knowledge of the arcana of the European ocean liner business. Just four years after becoming known as an amateur with a ship plan drawn up in his attic, Gibbs was making a name for himself deciding the fate of America's biggest war prize: the captured German passenger liner fleet.

He also got his first look at the German shipyard of Blohm & Voss in Hamburg, builders of so many of the great Teutonic liners he had read about as a teenager. Gibbs had seen the third and final ship in Ballin's trio, *Bismarck,* which everyone at the shipyard anticipated would be turned over to the British as reparations. German workers, still working at various heights aboard the unfinished giant, saw a knot of British and American shipping executives below them. Gibbs heard a loud clank. An iron wrench slammed onto the pavement not far away. Thinking it was an accident, Gibbs walked on. Moments later, another wrench barely missed him. The workers were not happy about how the war had turned out and even less happy to see Americans in the shipyard.

After nearly having his head bashed in, the American naval architect met with the Blohm & Voss executives and asked them to supply IMM with *Leviathan*'s original construction drawings. The shipyard demanded $1 million for the complete set. Gibbs said no thanks, and sailed back to New York.[5] Meanwhile, Philip Franklin was negotiating with the Shipping Board, which managed government-owned merchant ships, to take control of *Leviathan*. With the troops home, the Navy decommissioned the former German liner and tied her up at the same Hoboken pier where she was laid up from 1914 to 1917. On November 5, 1919, the Shipping Board announced that *Leviathan* and two smaller liners would be assigned to IMM for "management and operation on behalf of the Shipping Board."[6] For their services, the U.S. Shipping Board would pay IMM a handsome $15,000 a month.[7] For Franklin, the government had effectively sold the ships to IMM, creating the core of a new transatlantic service.

But Franklin decided that the reconversion of *Leviathan* was enough of a strain on his company's resources. He announced that the acquisition of *Leviathan* meant the S-171 superliner would be put on indefinite hold. Instead, he would look into rehabbing the medium-size former German liners *George Washington* and *Amerika*, both of which were almost fifteen years old.

Gibbs was furious. Why use three older, slow liners instead of building two modern fast ones?

Nonetheless, the Gibbs brothers set to work on reconditioning *Leviathan*, the former German imperial flagship *Vaterland* and still the largest ship in the world.

As William Francis Gibbs and his team toiled away on the battered troopship, *Leviathan* became the center of an intense public controversy, one that nearly destroyed the ambitious project. On January 17, 1920, IMM president Franklin sent the U.S. Shipping Board a down payment against a total price of $28 million to secure full ownership of the ship and several other former German liners. At that point in stepped the bombastic, populist journalist William Randolph Hearst, who had decided that the IMM was getting a sweetheart deal that needed to be exposed to the American taxpayer, the owners of the ships. The pro-German, pro-Irish, and anti-British Hearst had long hated Woodrow Wilson. For Hearst, a secret, no-bid deal between the Wilson administration's Shipping Board and J. P. Morgan's International Mercantile Marine had all the elements of a damning scandal that would sell a lot of papers and advance his own presidential ambitions. Hearst planned to tell the story as America giving away the captured German fleet, the nation's great prize of war, to IMM, the owner of the British White Star Line and thus a British company in all but name. Eight years earlier, Hearst had bashed IMM after the *Titanic* disaster.

A month after Franklin's down payment arrived in Washington, Hearst's flagship *New York American* charged that the sale of *Leviathan* would cause "great and irreparable harm to the present state of national defense and will destroy the Army transport reserve."[8] Hearst also filed a taxpayer lawsuit against the Shipping Board, claiming that it had no right to sell *Leviathan* and the other twenty-nine seized German vessels for only $28 million.

Franklin fought back. The day Hearst filed suit, IMM reduced its offer to $14 million for *Leviathan* and only a few of the other ex-German liners. "We again agreed to undertake to recondition the steamers and to comply with other terms with regard to their being operated in specified trades and remaining under the American flag," Franklin announced.[9]

Franklin's hopes were dashed when on February 19, the judge sided with Hearst, granting a formal injunction on the sale of the ships to IMM. The next day, President Wilson denied all rumors about selling the ships to Great Britain.[10]

The stalemate continued, and William Francis Gibbs, still on IMM's payroll, continued to work on the *Leviathan* plans. "This situation makes me mad," he fumed to a reporter about how Hearst had caused the Wilson administration to cave. "Here America has the brightest chance she ever will to compete with and excel British shipping in their chief boast: the transatlantic trade. . . . Here is the chance for the United States to run the finest ship on the ocean, and a few million dollars is holding her up."[11] Terrified about fire breaking out or board, he refused to cut maintenance expenses. "Considering the value of this steamer and the fact that it is practically irreplaceable," he wrote one government official, "I feel strongly that the expense for guarding is well justified."[12]

In 1921, however, a new president took office, and Gibbs, frustrated at the impasse his project was facing, decided to reach out to the Harding administration. Although Warren G. Harding assuredly did not possess Woodrow Wilson's intellect, he made at least one smart political appointment: an advertising executive named Albert Lasker, who had almost single-handedly put the small-town Ohio politician in the White House. The founder of an influential school of advertising, Lasker had provided Harding with a simple campaign slogan: "A Return to Normalcy." This was a presumed state of the country before the war, which state had been subverted by the fervent "Make the World Safe for Democracy" idealism of Woodrow Wilson.[13]

A grateful President Harding named Lasker chairman of the United States Shipping Board. Although he wanted to be secretary of commerce, Lasker took the job. He promptly fired the four Wilson appointees on the six-member board and took control. Lasker knew next to nothing about ships, but intrigued by *Leviathan,* he decided not to sell her to IMM. Instead he would keep her under government ownership and lease her to a private operator.

Lasker's decision put IMM's big investment in *Leviathan* in trouble and threatened Gibbs's own work. William Francis decided to do some public relations campaigning of his own. On July 16, 1921, he showed the new chairman all over the big ship, with Philip Franklin and a clutch of newspaper reporters in tow. Lasker saw a row of draftsmen hard at work, as well as a mock-up of a renovated first-class cabin.

The advertising man liked, the way Gibbs sold the project: with firm conviction, utter sincerity, and ardent patriotism. Converted, Lasker felt that he could sell the project to the public. European companies who built big ships like this, Lasker told the press, "did not expect to make money, but considered that owning such fine vessels was the best possible advertisement for the German merchant marine."[14] If restored as an American liner, Lasker said, *Leviathan* would be "the finest vessel ever turned out in the history of the world, both mechanically and from the standpoint of luxury . . . an announcement to the whole wide world as to what can be done in American shipyards and by American mechanics."[15]

"Does this mean that she will be operated under the British flag?" one reporter asked.

"It most emphatically does not," Franklin shot back, adding, "It means that she will be under the American flag and the nucleus of a fast American mail service."[16]

Lasker then returned to the Shipping Board's Manhattan offices to begin his plan to sell *Leviathan* as an America icon.

But relations between Lasker and Franklin were cool. In August 1921, Lasker began to negotiate secretly with shipping men outside IMM. His aim was to create a management team to operate *Leviathan, America, George Washington,* and other Shipping Board–owned vessels as passenger liners. Lasker felt that IMM, thanks to Hearst's attack, carried too much political baggage to be part of the team. Instead, the planned "United States Lines" would be managed by three private shipping companies: Roosevelt Steamship Company, Moore-McCormack, and United American, all controlled by four rich and well-connected young men: Kermit Roosevelt (son of the recently deceased president Theodore); Emmet McCormack and Albert V. Moore (two men who ran a lucrative South American shipping business); and W. Averell Harriman (heir to the Union Pacific fortune).[17]

On October 4, 1921, Philip Franklin was asked to come to the Shipping Board's New York office for an afternoon of questioning by Lasker and six senators, including populist firebrand Robert La Follette of Wisconsin. The president of IMM, who had thought he had the *Leviathan* operating contract locked up, quickly realized that Lasker wanted him and his company out of the project and wanted to keep the Gibbs brothers in.

As William Francis Gibbs wrapped up the *Leviathan* construction drawings to meet the October deadline, he was also planning his and his brother's exit from IMM. Working for a shipping company that seemed to make political enemies at every turn did not seem to be a good use of his hard work. Maybe the best course of action was to start a new firm to move the *Leviathan* renovation forward. But he needed political and financial support to make sure *Leviathan* didn't end up in a scrapyard and his hard work gathering dust on a shelf. His boss Franklin might have been a prudent businessman, but he was no bold visionary. Gibbs decided to throw in his lot with the government. There was also the chance that Lasker could get the money for Gibbs to build one, maybe two, ships of his own design.

The *Leviathan* refurbishment plans, all 1,024 detailed pages, were completed at the end of 1921. They covered not only the specifications for the refurbishment, but the materials and workmanship required. As impressive as the plans were, more impressive was their legal impact. Buried in the massive tome was language giving William Francis Gibbs, as government agent, final say over materials and workmanship. Provision after provision included the phrase "with the intent of these specifications and plans." That "intent" was to be determined by the designer, not the shipyard. "Intent" meant whatever Gibbs decided it meant. The document ensured that any contractor agreeing to use the plans would be subject to Gibbs's oversight. If he felt a piece of work was shoddy, it would have to be ripped out and done over at the contractor's expense.

The plans also included clauses to prevent shipyards from throwing in extra charges for "overlooked" items—an easy-to-abuse practice that let low bidders up their profits. Finally, as part of the ship's final "purification," the shipyard was required to make "necessary changes to eliminate essentially German subjects from their design."[18] These included recarving of the wood mantel and replacing twenty-four stained glass windows in the first-class smoking room, as well as relettering any signs that had been missed during the troopship conversion.[19] By December 29, 1921, there were eight bids from shipyards for the *Leviathan* refurbishment. The lowest bid came in from Newport News Shipbuilding

& Dry Dock Company: $5,595,000 for the refurbishing and restoration work, and another $500,000 to convert her to burn oil instead of coal. Newport News was hurting badly after the cutbacks in naval construction after the war. Yard president Homer Ferguson had underbid to get what he saw as a plum project.[20]

But in order to oversee the bidding process, William Francis Gibbs had to temporarily extricate himself from IMM. With Lasker's blessing, he would be working with a new naval engineering firm, whose sole purpose supposedly was to repurpose the big ship. In February 1922, Lasker announced the formation of Gibbs Brothers Inc., with William Francis Gibbs as president and Frederic Gibbs as vice president. The new company would serve as "owner's agents" for the Shipping Board, making Gibbs the sole government representative in the design and construction process. For their labors, the Gibbs brothers and their staff would receive $182,000.[21]

Up against the government, Franklin had no choice but to let his chief of construction go on loan, and Gibbs walked out of the IMM offices with *Leviathan*'s plans under his arm. A number of top IMM designers also asked for leaves of absence to join him. Most would never come back.

Shortly after the press conference, Franklin relinquished all claims on *Leviathan,* saying that IMM had "decided to comply with Lasker's request and we have consented to the cancellation of our contract."[22]

Months would pass before the Newport News yard was ready to receive the vessel, but William Francis Gibbs was thrilled to be in charge of the biggest postwar marine construction project in America. On April 7, the night before the ship would be moved to the yard, the naval architect gazed at his giant vessel as the Hudson lapped against her sides and a few lights glowed dimly from her upper works. "Human endurance could do no more," he told a reporter. "The ship belongs to the people and our responsibility is very great."[23]

It was only 270 miles to Newport News, little more than a day's sail away, but the Gibbs brothers, worried about problems at sea, had stocked a month's worth of provisions to feed the four-hundred-man crew. They also worried about the ship's frayed single-wire electrical system, which had not been maintained for years. Every single lightbulb, bridge control, and appliance on board was connected to the ship's main electrical switchboard by an individual wire. The return, or ground, wire was then bolted directly onto the ship's structural steel. It was a cumbersome, lethal setup. A short circuit could ignite the flammable insulation placed on top of the wiring.[24] Gibbs's electrical engineer Norman Zippler had designed a safer double-wiring system to end the threat, but it could not be installed until the ship was in the yard. For the voyage down, Gibbs made sure that new fire hoses were installed throughout the vessel, and "fifty streams as large as those of the city fire department can be brought into action at one time if the necessity should arise."[25]

Before the break of dawn on April 8, 1922, smoke billowed from *Leviathan*'s funnels for the first time in nearly three years. Tugs pushed the liner back into the Hudson and *Leviathan,* propellers churning, headed out to sea. The Gibbs brothers had signed on as members of the crew. Two days later, *Leviathan,* having averaged 17 knots, arrived with the dawn in Newport News, Virginia. "Everything was done exactly according to schedule," Gibbs said to the press. "There was not a hitch anywhere. The engines worked beautifully."[26]

William Francis worked at site for the next fifteen months, taking the train back to New York for weekends. Dressed in his black derby and overalls, "Iron Hat," as the workers called him, roamed the ship at all hours, construction drawings tucked under his arm. Meanwhile, Homer Ferguson, the president of the Newport News yard, grew more nervous as every day passed. Gibbs was blocking every proposed change order and revision in the original project specifications. Racking up charges for change orders was how Ferguson had planned to make up for his original below-cost bid. The two men began to absolutely hate each other.[27]

Gibbs didn't care. He had the full backing of the Shipping Board and work progressed rapidly. An army of two thousand workers nailed door frames together, screwed brass light plates in place, and ripped out the substandard electrical wiring, as mountains of supplies were brought on board every day. A set of Yorkshire pudding pans for the galley. Asparagus tongs for the ship's Ritz-Carlton restaurant. A Santa Claus clock for the children's playroom. Twenty typewriters and 160 gramophone records for the passengers.[28] Four seventeenth-century Flemish canvases, plundered from the ship in 1917, were located and rehung in the first-class Social Hall. From the silver library inkwells to the silk-shaded dining table lamps, no expense was spared.

Down in the bowels of the ship, machinists carefully converted each of the ship's forty-six boilers to burn oil instead of coal. This would not only increase the ship's speed, but also eliminate hundreds of stokers from the crew roster. The four great turbine casings were lifted open, and thousands of blades were repaired, replaced, or cleaned. No longer smeared with coal dust and grease, the cathedral-like engine and boiler spaces, crisscrossed by ducts, stairwells, and piping, now gleamed in antiseptic white paint. Standing on scaffolding slung over the side, workers carefully brushed layers of shiny black paint on the hull and white on the superstructure. The smokestacks remained coated in red primer until the spring of 1923, just before *Leviathan*'s trials, when Gibbs gave the order to paint the stacks in the new United States Lines colors: a red base, followed by a white band, and a blue top.

By this time, Franklin and IMM had purchased *Leviathan*'s younger sister *Bismarck* from the British Reparations Board, completed her, and registered her as an English vessel. The new White Star flagship *Majestic* was billed as the largest ocean liner in the world.

In response, Gibbs put together a public relations trick. On April 22, 1923, the Shipping Board announced that when *Leviathan* entered service on July 4, she would top the *Majestic*'s size. Gibbs knew that the White Star flagship had a gross tonnage of 56,551. In ships, as noted earlier, this is a measure of size, not weight, and it is calculated by multiplying ship volume by a numerical constant. As built by HAPAG, *Leviathan* measured in at 2,000 tons smaller than *Majestic*. But using a different, U.S. tonnage multiplier, Gibbs recalculated *Leviathan*'s gross tonnage at 59,956.65.[29] After receiving this news, White Star chairman Harold Sanderson snorted that "there was a ship which it was claimed could blow herself out as with a bicycle pump and then claim to be the largest ship afloat."[30] Gibbs struck back. At a press conference held with Lasker, the naval architect belittled a planned new flagship of IMM's White Star Line. "Even the new *Majestic* . . . will not be in the class of the *Leviathan*."[31]

The U.S. Shipping Board scheduled the maiden transatlantic voyage of the United States Line's flagship for July 4, 1923. But for the first few voyages, the board decided that Gibbs Brothers Inc., not the new United States Lines, would train the crew and operate the vessel. Shipyard management continued to protest Gibbs's tight hand on the contract. When asked by a congressional committee about how he had been able to convince the shipyard to sign the *Leviathan* reconditioning contract, Gibbs replied, "The specifications are drawn so that no contractor can possibly take this contract believing that he can take it at one price and make his profit on the contract out of possible extras."

"Now, if I were a contractor," one congressman asked, "why, it does not seem as though I would be induced to sign a contract like that. I am putting myself absolutely in the hands of the owner."

"That's right," Gibbs said proudly. "I say this, you could not get me to stand between the Government of the United States and a private contractor on work involving the amount of money that this work involves, unless that provision is in the specifications."[32]

Gibbs also faced criticism for his fees. For its time, the $182,000 he had negotiated with Lasker was a large amount for two brothers' role as owner's agent. "The question is very naturally provoked," *Marine News* wrote in May 1923, "as to what charm Gibbs has over the Shipping Board or what influence he controls that brings to him these juicy retainers." The article failed to say that the $182,000 payment was for the entire design team, not just for the Gibbs brothers.[33]

As the ship trials approached, a distraught Homer Ferguson, now facing a $1.25 million loss on the project, met with Collis Huntington, owner of the Newport News yard. Ferguson offered his resignation. But

Huntington refused to accept it. "My wife owns most of the stock in the shipyard," he said, "and she has not been feeling too well recently, so maybe we should say no more about it."[34]

By September 1922, Philip Franklin heard that Gibbs Brothers Inc. was not a one-project firm, but was to become part of a private-public partnership with the United States Lines. Franklin found out in the newspapers that William Francis Gibbs had taken the same plan he had promoted to Jack Morgan six years earlier and sold it to Albert Lasker and the U.S. Shipping Board. The rumored arrangement—*Leviathan,* Gibbs's two planned superliners, and a Montauk terminal all financed at government expense—could easily drive the privately controlled IMM out of business. Lasker hinted that he would hire "a managing staff of experts who had experience in handling the super liners to serve for a fixed fee."[35] It would not be IMM, but Gibbs Brothers Inc.

Furious at being double-crossed, an angry Franklin confronted his absent chief construction manager.

"I understood from you that you have been and are now negotiating with the Shipping Board for the management and/or operation of a steamer or steamers in the transatlantic trade," Franklin wrote William Francis Gibbs, "and I regret to say that I feel such action is in violation of the intent and spirit of our agreement granting you and your brother leave of absence for the specific purpose of supervising the reconditioning of the *Leviathan,* and which we did only at the earnest solicitation of the Chairman of the Shipping Board."

Franklin then added: "If my understanding is correct, I must ask for the resignation of your brother and yourself."[36]

Gibbs replied a few days later. His response was confident, even cheeky. "When you gave your approval to our undertaking the work in connection with the *Leviathan,*" he said, "we did not understand you intended to limit our services to the Government so as to prevent our advising in the solution of the problems facing the Shipping Board and the operation of the *Leviathan.* We realized certainly, and you and your associates made us feel keenly, that any connection with the I.M.M. Company for the time being has ceased. We did not feel that our conversations with the Shipping Board relative to operation were in any way a violation of the intent and spirit of our arrangements with the I.M.M." He then added that if Franklin insisted, they would "terminate any obligation on your part which might be implied."[37]

Yet soon after this falling-out, the Gibbs brothers suffered a major setback, one that certainly influenced his later fear of speaking to the press too soon. Despite the crushing amount of work from the *Leviathan* project, William Francis was still refining his own designs for the Shipping Board. By now, Gibbs's own ships had grown to over 70,000 thousand gross tons (almost a quarter larger than *Leviathan*), with a price tag of $35 million each. These would make them the biggest ships afloat, by far, a statistic that appeared in many newspapers. A revised rendering shows that Gibbs had heavily reworked his original prototype. It was longer, lower, and sleeker, with three squat funnels instead of four tall ones, and a curved superstructure front. Despite its huge size, it had the fine lines and racing profile of a navy destroyer. Gibbs imagined that he could combine the great luxury of *Leviathan* with the speed of *Mauretania* in a ship far safer than anything yet built.

Yet the ambitious young designer had overplayed his hand. On October 17, 1922, Albert Lasker told Gibbs that the Shipping Board would not provide a $25 million operating subsidy for the two ships that were supposedly to operate with *Leviathan* as a threesome. Despite President Harding's urging, Congress refused to spend the money, as Republican members were pressuring the White House to turn over the government-owned fleet to private owners.

Without support from the government, the construction of the new superliners was financially impossible. For his part, Lasker was tired of government work and eager to get back to the advertising business, and had no heart for the political battle to raise the money for two new *Leviathan*-type liners. He decided to resign from the Shipping Board and rejoin his advertising firm once *Leviathan* entered service. However upset Gibbs might have been by the loss of the subsidy and his patron, he did not let it affect the work he had at hand, which was to keep *Leviathan*'s construction on a tight deadline to meet a May 1, 1923, delivery to the Shipping Board. It was met, and on that day, a happy Gibbs gave reporters some classic American philosophy: "When backed by preparation, initiative, and faith, seemingly nothing is impossible in America."[38] The question that remained was whether a single American superliner could compete against European rivals, without the two companion vessels needed for a regular schedule of weekly transatlantic service.

But even if Gibbs's dream of building new ships of his own design was put on hold, the reconstruction of *Leviathan* was still a huge professional triumph, one that made everybody in the shipping world sit up and take notice. Gibbs had taken the lessons learned from engineering publications and his mentorship with Admiral Taylor and successfully applied them to the rebirth of the largest ship in the world. What he lacked in formal training and social grace he made up for in natural charisma and organizational ability. He also had a gift for wooing powerful supporters like Albert Lasker.

Out of his newfound confidence grew a managerial technique that he would use to get what he wanted: being disagreeable. Gibbs realized that if he remained shy and meek, he would be run over roughshod by naysayers. Conflict was part of the job of building great ships. To achieve his purposes, he would have to be good at fighting fights. And winning. And to win, he had to maintain autocratic control over every aspect of a project.

Gibbs also learned the value of secrecy. Walking out of IMM's office with the *Leviathan* plans under his arm would not be the last time he kept his hard work from the prying eyes of rivals, real and imagined. He would also avoid speaking to the press as much as possible. He had no stomach for public embarrassment.

The former recluse knew he was rubbing people the wrong way at times. "Everyone thinks I'm such a mean fellow because I like ships more than people," he joked.[39]

As *Leviathan*'s sea trials approached, Gibbs dreaded the impending publicity. Yet he knew he had to play to the media to convince the American public that it was time for the nation to compete on the North Atlantic sea-lanes.

A great ship proudly flying the American flag, he hoped, would sell itself.

THE PARVENU

The refurbished *Leviathan* left Newport News on May 15, 1923, and sailed up to the Boston Navy Yard for a final inspection at the new Navy dry dock, the first in America large enough to accommodate a superliner. She was greeted by fifty thousand cheering spectators. "So frenzied was the mob," said her captain, forty-eight-year-old Commodore Herbert Hartley, "that the U.S. Marines were called out to hold the crowd in check." Some well-wishers fainted in the intense spring heat.[1]

On June 19, the ship left Boston for the Florida coast with 456 guests and 1,135 crew members on board. The speed trials were to be conducted three days later. In the meantime, guests sang songs around the ship's seven pianos, played shuffleboard, and swam in the ship's Roman swimming pool. The crew conducted extensive tests of the auxiliary mechanical equipment. In keeping with maritime superstition—which had it that females on board were bad luck before a maiden voyage—there were no women aboard.[2]

At 7:17 A.M. on July 22, as the sun rose on the eastern horizon and Florida's Jupiter Lighthouse appeared to the west, Gibbs gave Commodore Hartley a quick pep talk, and then left with his brother for the engine room.

The guests who were awake noticed that the entire ship was now shaking as a result of her increasing speed. The vibrations from the 90,000-horsepower engines were rattling chinaware and bedsteads all over the ship. Those on deck saw a violent wake surging astern and black smoke pouring from her stacks. They waited for some sort of announcement about what was happening, but the public address system was silent.

At 10 A.M., William Francis Gibbs walked down the grand staircase. He took out a piece of paper and tacked it to the ship's main bulletin board outside the Social Hall. It read: "Between 7:17 am and 10 am, the ship had traveled 75 nautical miles."

When asked by reporters what that meant, Gibbs deadpanned: "That means we've made over 28 knots and broken the world's speed record. It means we can trim the best the British have got or anything else afloat in the class of big merchant ships."[3]

Gibbs then sent a cable that was published in the *New York Times* the next day. "If the *Majestic,* the world's next largest liner, was here, the *Leviathan* would pass her by a knot and a half every time." But in regular service, he added, the operators of *Leviathan* would not try to capture the Blue Riband from *Mauretania* "unless the others start some fancy business. Then we will use our untouched reserves."[4]

Gibbs failed to say that the warm water current of the Gulf Stream probably added about two to three knots to her speed. And unlike *Mauretania*'s trials eighteen years earlier, *Leviathan*'s trials did not consist of runs with and against the wind and currents. As for the British, speed and distance meant nothing unless the ship achieved the highest average speed between the British Isles and New York.

Gibbs and his team left nothing to chance to make sure *Leviathan*'s maiden voyage would be a sensation with the public. A brochure featuring glamorous models boasted that "the airiness and spaciousness of *Leviathan* interiors is apparent. And the absence of expensive bad taste is notable."[5] Newspaper ads claimed that *Leviathan* was "the World's Largest and Most Beautiful Ship."[6] The Charleston craze was sweeping the country, and the ship's orchestra released a set of dance records. The maiden voyage was scheduled to begin on the Fourth of July.

When the day arrived and as Captain Hartley and the engine room staff got up steam, ten Army and Navy biplanes whizzed over the ship's three red, white, and blue stacks, even as a hard rain pelted the ten thousand flag-waving spectators crammed onto the pier. Thousands more well-wishers jammed against the street railings. The cheers and shouts were met with three window-rattling booms from *Leviathan*'s whistles as the huge ship backed into the Hudson River.

At the dawn of the Roaring Twenties, Gibbs spent years transforming the big German liner into the finest and biggest passenger ship ever to fly the Stars and Stripes. As an American wartime troopship, *Leviathan* had been well-known to an entire generation of World War I veterans. Now she was, in the eyes of the public, Americanized from stem to stern, and ready to compete. No matter for the time being that she lacked comparable running mates and that her bars were dry.

For her first voyage as the flagship of the United States Lines, *Leviathan* carried more than 1,700 passengers and a crew of 1,100. Taking in $520,000 in passenger fares and carrying mail worth another $20,000, the crossing was going to be a profitable one.[7]

The country that had built her nine years earlier did not share in the jubilation of the American public. After the Treaty of Versailles, ships that the Americans had not taken were seized by the British as war reparations. Germany's once-mighty commercial fleet had been reduced to the mechanically defective *Hansa*—the former Blue Riband holder *Deutschland*—and a few small cargo vessels. Racked by inflation and political unrest, the Fatherland's days of North Atlantic shipbuilding supremacy seemed over. So did its days of industrial and military might.

The first afternoon out, hundreds gathered in the first-class Social Hall, filled with the happy sound of a jazz band and the clink of crystal ware. Alice Longworth, daughter of Theodore Roosevelt, unveiled a life-size portrait of President Harding by Howard Chandler Christy to polite applause. Senator Reed Smoot of Utah was led to say that it was the first time he had not become seasick on a transatlantic crossing. Among those watching was Vincent Astor, traveling to England to oversee the transfer of some family money to his younger sister. As night wore on, people moved to the first-class dining room, where soft lighting showed off a sweeping grand staircase leading down to richly decorated tables, capable of seating seven hundred.[8]

It was the Prohibition era, and Captain Hartley had insisted that the government-owned *Leviathan* would be "as dry as a bone." He said he kept a close eye out for any alcohol that might have been smuggled aboard. A reporter for the *New York Times* was skeptical: "There were some gay little parties in private suites, and if someone produced a bottle at a public meal no official notice was taken of it."[9]

That night, President Harding wired Albert Lasker, who was on board. "I hope the prestige the great ship is giving the merchant marine will prove a compensation to you for your wholly unselfish service rendered to the Government for two years," he wrote. "Accept assurances of warmest personal regards."[10] Less than a month later, the ailing Harding would be dead, his administration embroiled in scandal.

Leviathan dropped anchor in Cherbourg, France, on the morning of July 9, 1923. Captain Hartley had driven her at an average speed of 23.65 knots and made the 3,239 mile voyage in 5 days, 17 hours, and 7 minutes. This was over a day longer and three knots slower than *Mauretania*'s record passage fourteen years earlier.[11] As planned, the United States Lines was not planning to break any records on this trip.

On the other side of the English Channel, residents of Southampton had been awaiting *Leviathan*'s arrival. After tugs nudged *Leviathan* into her dock, stewards set up the first-class dining room for a six-hundred-person luncheon on July 12. The guest of honor was the mayor of Southampton. Ever on the lookout for city revenue, he challenged America to build even more *Leviathans;* the Port of Southampton, he promised, would be ready to receive them.[12]

"We are going to compete with you to the best of our ability," William Francis Gibbs told his audience, "but we will never do a single thing which would not be considered fair on a British football field."[13] This description was not, of course, a fair characterization of how he had recently inflated the size of *Leviathan.*

Though no liquor was served and the summer sun turned the dining room into a steambath, the audience stood and applauded the president of Gibbs Brothers. Yet William Francis chose not to bask in his newfound fame. During the actual reconstruction work, he avoided the press as much as possible. He authored no major articles or papers in the shipbuilding trade journals promoting his achievement. Gibbs thought publicity "treacherous," and he avoided it so as "not to get bitten."[14]

Gibbs did have enemies jealous of his sudden fame: other naval architects, who resented his lack of formal training and knack for forming tight connections with government officials. Those inside that closed world thought of him as an amateur interloper and called him "the Parvenu" behind his back. Proper Philadelphians of his childhood would have said "north of Market."

On the trip back to New York, the liner carried 1,174 passengers—less than a third of her capacity. She did not make notable speed, either. She entered her home port on July 23, having made the return trip in 5 days, 12 hours, and 11 minutes at an average speed of 23.9 knots, still well behind *Mauretania*'s westbound Blue Riband performance.[15]

Gibbs Brothers Inc. would operate *Leviathan* for two more round trips before the U.S. Shipping Board forced the company to relinquish control of what was now regarded as the greatest ship in the world to its government-backed United States Lines. For the three trips run by the Gibbs brothers, during the high season of 1923, *Leviathan* earned a profit of $277,406.01.[16]

But during the next ten years, few crossings made money. The postwar world had brought new challenges to passenger shipping, especially for American vessels. First there was Prohibition, which had proved so troublesome to Captain Hartley to enforce on *Leviathan*'s maiden voyage. America's ban on alcohol had become law on New Year's Day 1920. Not only *Leviathan* but all U.S.-flagged ships had been required to stop serving liquor, a deficiency that pushed many passengers to choose the more spirited accommodations of foreign-flagged shipping lines.

Then there were the moves to restrict American immigration. The refurbished *Leviathan* could carry 4,505 passengers, nearly half in third class and steerage—a class of passenger not welcomed by American anti-immigration forces. In 1924, U.S. laws would slam the country's doors shut to eastern and southern Europeans, dealing a huge blow to the transatlantic liner business. To survive, many lines turned steerage into "Tourist Third Cabin" or "Tourist Class" to appeal to passengers who considered themselves respectable, to travel on the cheap. A White Star advertisement trumpeted the virtues of traveling economically: "The fittings of course, will be somewhat less luxurious, but no less pleasing."[17]

The public still hoped that *Leviathan,* which supposedly had attained over 28 knots on her trial runs, would use those "untapped reserves" of engine power about which William Francis Gibbs had boasted. The editors of the nation's premier naval architectural journal, *Marine Engineering and Shipping News,* wanted to see "the greatest sporting event on the ocean that the world has ever seen." This would be a transatlantic race between *Leviathan* and *Majestic.* "All English-speaking people go in for a fair sport," an editor wrote, "and, instead of engendering any ill-feeling between the two great Anglo-Saxon nations, the reverse, would undoubtedly result. . . . Let's be game!"[18]

That the two ships were German-built didn't seem to bother the people promoting the race.

Nor did the magazine seem to care that it was *Mauretania,* not *Majestic,* that still held the Blue Riband. After her 1920 refit, *Mauretania* was even a faster ship, now capable of reaching 27 knots in spurts. But no competent master would ever think of pitting *Leviathan* against *Mauretania* in a head-to-head race. The German-built hull could not withstand the punishment of moving at 25 knots in rough seas for long stretches. Despite all the improvements Gibbs Brothers had worked into her, she still had poorly designed superstructure expansion joints and a design flaw in the funnel shafts (called uptakes). Rather than have one continuous shaft from boiler to smokestack, *Leviathan*'s German designers split these uptakes in half and moved them to the sides of the ship. The two uptakes then merged into one exhaust near the top deck, and smoke was then ejected from the funnel. This allowed first-class public rooms to be bigger and gave passengers broad, uninterrupted views from one public space to the next.[19] However, this innovation greatly weakened the integrity of the hull.

The design flaw did not become apparent until 1924, when crewmen of *Leviathan*'s sister *Majestic* noted stress fractures in upper decks just forward of the first stack. Management chose to ignore the problem. That December, however, as *Majestic* hurtled through a storm at full speed, passengers heard a terrific noise "like a cannon shot." The main deck had cracked open along its entire width, just beneath the second funnel's twin uptakes, and another crack had opened up along the ship's port side. *Majestic* limped gingerly into port, and underwent expensive repairs at Harland & Wolff. When Gibbs got the news, *Leviathan* was quickly pulled out of service so that the same weak area around her split funnel uptakes and expansion joints could be shored up with additional plating and bracing.[20]

"It was a poor design," Gibbs's electrical engineer said later about *Leviathan*'s funnel design. "Too many sharp corners. We didn't know it until we drove her hard."[21]

One of the officers on *Leviathan*'s maiden voyage would ultimately become the best American skipper on the high seas. He loved the rumbling of the turbines beneath his shoes, how her red, white, and blue funnels towered over every ship in sight. Like William Francis Gibbs, he saw great ocean liners not as mere machines, but as living beings. "You know enough to meet a ship as an individual," he said, "a thing with idiosyncrasies and temperament. Every single ship has personality."[22]

Hot-tempered and cocksure, Harry Manning had no shortage of personality himself. He was just twenty-six when he was appointed second officer of *Leviathan* in the summer of 1923. He had been on the ship for the first time in May 1914, when a group of cadets from the New York Maritime College had toured what was then the *Vaterland*. It was he who overheard German commodore Hans Ruser sneer that America would never have a ship like this.

Now she did, thanks to William Francis Gibbs.

Manning was born Harry Luelker in Germany. However, Harry's mother remarried a British Foreign Service officer on assignment, and the young boy took his stepfather's name. When Harry was ten, he moved with his family to New York City.[23] The neighborhood kids found it easy to bully a brainy, fragile-looking child with a lisp, and Harry often came home with a bloody nose.[24] After high school, Harry chose the New York Maritime College at Fort Schuyler in the Bronx: the Harvard of the sailing world, perched on an outcrop above Long Island Sound.

At five feet, six inches, Manning was the smallest member of his class. By then he was quite capable of defending himself. He rounded out his academic regimen by getting into fistfights with classmates who made fun of him. After graduation and a stint on a small sailing ship, he signed on the aging American Line's liner *St. Paul* as an able seaman for fifteen dollars a month.[25] When the United States Lines was formed in 1922, Manning jumped at the chance to serve as first officer of *George Washington,* which like *Leviathan* had been seized at the outbreak of World War I. He then received an appointment to the biggest and fastest ship flying the American flag, *Leviathan.* As second officer, Manning spent most of his day either on the bridge or in the chart room, and did not socialize with passengers. He was a teetotaler and never smoked, but he taught himself the tango and took boxing lessons. Boxing, he said, was like the North Atlantic in winter. "She will strike you if you get careless," he said. "You must be watching all the time. She lunges in and probes at you."[26]

Second Officer Manning's reputation as a boxer helped him to manage the ship's crew of one thousand, which was packed into a rabbit warren of dark corridors, mess rooms, and machinery spaces in the lowest reaches of *Leviathan*'s belly. Manning kept an emotional distance from the crew. He hated the radical left-wing union activities that were emerging in the shipping business, and he had refused to join the Masters, Mates & Pilots union of the American Federation of Labor. "I'll starve before I join," he said.[27]

Unhappy crewmen could be aggressive. On one trip, Manning put a coffee cup to his lips, gagged, and spat everything out. A steward had dumped roach poison into his coffee. His hands became scarred by knife fights with stowaways and "various obstreperous members of the crew" who hated his authoritarian ways.[28]

But Manning also had his conflicts with the authority of management, specifically his boss on *Leviathan,* Commodore Herbert Hartley. After a few crossings, they despised each other. For Manning, Hartley was a stuffy celebrity worshipper; for Hartley, Manning was an insolent egotist.

One night, First Officer Linder informed Commodore Hartley that a frame section within the foremast mast was rusty and needed inspection. The two men started the long climb to the crow's nest. After inspecting the frame, the commodore and first officer stepped out onto the crow's nest and peered over *Leviathan*'s bow. A cold blast whistled through the rigging, tore at their heavy leather overcoats, and stung their faces.

On the bridge, Second Officer Manning was standing watch.

As he squinted, Hartley saw a large ship barreling toward them. "Must be the *Paris*!" he shouted to Linder, who picked up the crow's nest phone to the bridge and told Manning that the French Line's new flagship was heading right toward them.

"Tell that stupid son of a bitch I can see what I'm doing," Manning snarled into the receiver.

A shocked Linder told Manning that the captain was standing right next to him and could hear every word coming out of the phone.

Fixing his eyes on the crow's nest through the bridge windows, Manning shouted, "I can see the bastard!"

"You know Hartley heard every last word," Linder said.

After coming down from the mast, a silent but fuming Hartley told Linder to report to the bridge and relieve Manning from his watch.

"But Mr. Manning's watch is not over. It's only three-thirty, sir."

"Do as I say," Hartley said.

Linder saluted and climbed up to the bridge, where he saw Manning standing defiantly at his post, his small, slight frame standing against the bridge's polished brass and gleaming woodwork. This little man, who wore his cap pulled low over his eyes, glared back at Linder, as if to remind him that he felt completely at ease commanding a ship with three thousand souls aboard.

"You are relieved," Linder told Manning, adding, "You've done it this time."

In his quarters, Hartley swiveled around in his chair and stared intently at his second officer. "Mr. Manning," Hartley said slowly, "I am not stupid, I am not a son of a bitch. And I am not a bastard."

Manning said nothing.

"Now," Hartley continued, "there are eleven gangways leading off this ship at Pier Eighty-Six. I don't care which one you take, but you take one and *never* come back!"[29]

Manning was unrepentant. "I was an awful son of a bitch in those days," he recalled proudly.

He left *Leviathan* on one of those gangways, but he was simply too valuable a mariner to be left "on the beach," as the sailors would say. By January 1927, he got his first command: the small United States liner *President Roosevelt.*

The day he came aboard, the Hudson was choked with ice as gray clouds dropped sleet on Manhattan piers. But when Manning took his place on the bridge of the small, sturdy vessel, he was a happy man. "Navigation in bad weather was my specialty," he said about his first trip as a master, "and we had no trouble on that score."[30]

Two years later, Captain Manning's death-defying rescue of the crew of a sinking Italian cargo ship made him a national hero. By then chief officer of *America,* he had heard the cargo ship's distress call in the middle of an Atlantic gale. In the storm, it was impossible to bring the large liner close enough for a rescue, so Manning led a small lifeboat crew that fought the storm for an hour, to get to the Italian sailors just as their ship was sinking.

During most of the 1920s, France, Britain, and the United States were all still operating ships built before World War I. The victorious

British quickly absorbed their seized German liners as national symbols. Cunard struggled for a while operating the former HAPAG giant *Imperator*—which they received as reparations for *Lusitania*—losing fifty thousand dollars per voyage. Her new owners stuck with the big liner, and in early 1921 renamed her *Berengaria* in honor of the wife of Richard the Lionhearted. She, along with her big running mates *Mauretania* and *Aquitania,* was converted to burn oil instead of coal, and they became the most profitable and famous trio of liners on the Atlantic, running a weekly service between New York and European ports. IMM's White Star Line also prospered by operating with its German-built *Majestic* in concert with the popular *Olympic.* But as the decade ended and the 1930s began, shipbuilders and nation builders began to look ahead.

Germany, which had lost most of its fleet to war reparations, was left with one choice: to rebuild from scratch. By the middle of the decade, Bremen-based Norddeutscher Lloyd had raised enough private capital and government funding to build two revolutionary ships, whose designs were kept absolutely secret. The British, French, and Italians, all of whom were feeling rich thanks to the booming 1920s economy, also started to make grand plans.

Gibbs guessed that the new ships would make all existing liners, including *Leviathan,* obsolete. In 1927, he wrote of American ships that "there is no vessel that can step into the *Leviathan*'s place, as her obsolescence—I use the term in its strictly technical engineering sense—increases."[31]

The large fee from the *Leviathan* rebuilding did not translate into financial stability for Gibbs Brothers Inc. William Francis and Frederic Gibbs pounded the pavement looking for naval design work, without much success. Rivals such as professionally trained Theodore Ferris and Ernest Rigg snapped up most of the big jobs.

After a roaring start to the decade, America's shipbuilding initiative for the transatlantic trade stalled. Content to sit on its fleet of prewar German vessels, the nation's shipping companies kept sailing older ships rather than build new ones. One way to build his superliner, Gibbs figured, was to get out of the engineering business and find partners to buy the United States Lines from the government.

* * *

On top of his financial worries, Gibbs's family situation was also troubling. His mother, Frances, died in 1921. After her death, an aging William Warren Gibbs and his two daughters moved to the Gladstone, a shabby Philadelphia apartment building once popular with the rich. William Warren later moved to Princeton to live with his youngest daughter, Georgianna, and her family. He gamely continued to look for something new to promote. The *Philadelphia Inquirer* noted in January 1917 that a former sailor named Lester Barlow had perfected an "aerial torpedo" that after being tossed out of a plane could fly up to two hundred miles before exploding at exactly eight feet above a target on the ground. The paper went on to say that "W.W. Gibbs, a well known financier, is interested with Barlow in his invention."[32]

On October 25, 1925, William Warren Gibbs died, aged seventy-nine and possibly senile, at the "Kenwood Sanitorium for the Insane."[33]

When William Francis and Frederic Gibbs went through his papers to settle his affairs, they discovered that their father's estate consisted of one small life insurance policy and the proceeds of a court settlement *Barlow v. McIntyre.* Together they totaled $3,555.02. At his death, William Warren Gibbs did not own a single share of United Gas Improvement Company, which made so many proper Philadelphians wealthy. Nor did he own shares of his other successful venture, the Electric Storage Battery Company, known at his death as Exide.

Worse still, William Warren Gibbs left debts behind. When the estate was finally settled, his children were left nothing.[34]

The once famous and stupendously wealthy Philadelphia promoter was quietly buried in his hometown of Hackettstown, New Jersey.

Two years after his father's death, Gibbs still shared a modest New York apartment with Frederic. Those who knew him well thought this shy, awkward engineer would remain a bachelor. He was terrible at small talk, did not drink or smoke, and was uncomfortable around women. William Francis sometimes even provoked them into leaving him alone.

"If there's anything I can't stand," he once said, "it's sitting between two dames and trying to make conversation out of nothing. They always say, 'What do you do?' I say, 'I'm an engineer.' They say, 'What kind of engineering?' I say, 'I design garbage barges.' As a matter of fact, I do, too. Then they—the dames I mean—drift off. It's all so damn pointless."[35]

Yet in 1927, the forty-year-old William Francis Gibbs made a sudden move that shocked New York society: he got married. Gibbs's unexpected decision gave him a cultured wife who admired his drive and appreciated his eccentricities. The union would bring financial security and social acceptance, but had no effect on his single-minded quest to build a superliner.

MRS. WILLIAM FRANCIS GIBBS

Although groomed to be a New York socialite, Vera Cravath was not one to flit from one fashionable event to another. Born in 1896, Vera never went to school, but was educated at home by private tutors. Fearless in the saddle, she was famous as an equestrian all over the Northeast.[1] She was also fluent in French and distinguished herself as an amateur pianist. A cloud of curly hair framed her long face, and she smiled easily. She loved music, as well as the unconventional, the exotic, and the intellectual. Many years later, music teacher Olga Samaroff Stokowski, ex-wife of the conductor, remarked that Vera "knew her Brahms themes as well as she knew the characters in Hamlet."[2]

She also was daring in her personal life. While visiting Paris in 1925, Vera filed for divorce from her polo-playing husband, James Larkin. Her request was granted, and she took custody of their young son, Adrian.

Her father was New York superlawyer Paul Cravath. The broad-shouldered, hulking Cravath terrorized his staff at Cravath, de Gersdorff, Swaine & Wood, a firm that made its money defending the practices of Wall Street and the Morgan trusts; in fact, many in the Manhattan elite felt Paul Cravath was the J. P. Morgan of American law, one who possessed "sound knowledge of finance and a magnificent ego which swept aside inhibitory obstructions."[3] The Ohio minister's son also was one of New York's major social arbiters: Cravath had helped develop the upscale Long Island community of Locust Valley and its ultra-exclusive Piping Rock Club. Cravath's "magnificent ego" and social striving had also cost him his marriage to Vera's mother, Agnes, a former operatic soprano, in 1926.

Shortly after his separation, Cravath set sail on a friend's yacht for the sunny Mediterranean, leaving his divorcée daughter behind in New York.

William Francis Gibbs met Vera at a party to which he most likely was dragged. In his old-fashioned wing collar and black suit, he looked like a Victorian relic among the smart set around him. He was unsociable and testy in these settings. Society women he found especially irksome.

At first, Vera was put off. "I thought him rather strange," Vera recalled. "But I was fascinated." And they continued talking to each other.

"François," as Vera playfully called him, gradually opened up to her. "It may sound trite but he knew what he was going to do in life," she said. "He wanted to build ships—if that meant being the best naval architect, all right—but building ships was the most important thing in the world to him."

The day after the party, Vera and young Adrian were scheduled to sail for Europe. On arriving in Rome, Vera found a cable waiting for her at the hotel front desk. Gibbs said he was in London. Could he visit her?

"I was impressed, you know, that he'd come all the way down there to see me," she said. "It set me to thinking."

Gibbs arrived a few days later, still wearing his black suit and wing collar, and carrying a bag of dirty laundry. Vera was unperturbed. After dropping off the bag at a laundry, Gibbs ordered a carriage and trotted Vera and Adrian around the sunny streets of Rome. After the ride, Gibbs said a quick good-bye and headed back to America, most likely aboard *Leviathan*. Vera and Adrian followed him not long after.[4]

A few weeks later, a small group of Vera's friends assembled for a dinner party at her father's apartment. The dinner guests waited for their hostess to appear. The time dragged on. The wait grew embarrassing.

Finally a message arrived from their hostess: Vera Cravath Larkin had been secretly married the day before. After dinner, the daughter of New York's most famous attorney invited her startled friends to the Waldorf-Astoria to meet her new husband, naval architect William Francis Gibbs.

Gradually, details of the marriage began to emerge in the society pages. William Francis Gibbs and Vera Cravath Larkin had been married at the Madison Avenue Presbyterian Church, with one witness present on January 6, 1927, and they had been at the New York Municipal Building earlier that day to apply for a license.[5]

Doyennes and debutantes of two cities were wide-eyed at the news of the wedding. "The marriage is of wide interest both in New York and Philadelphia," noted the *New York Times*.[6]

Vera had not told her father about her plans and Paul Cravath was no doubt puzzled when he returned and found his daughter remarried to a forty-one-year-old engineer. But he immediately tried to launch his new son-in-law into New York society. Cravath held a luncheon in the couple's honor at the St. Regis hotel, and a dinner the following night at the exclusive Creek Club on the north shore of Long Island.[7]

Thus began a much more social life for William Francis Gibbs. He would attend the opera, go to restaurants and private dinners, and host dinners in return. He would in time join several elite clubs. Eventually the rich and famous in New York would travel on his ships. But he never truly became a member in good standing of his father-in-law's social world. He shocked too many polite Manhattan dinner partiers with his insulting, acidic wit. But when he was in the mood, he would, in the words of one observer, sit in a chair "like an old rug" and turn out to be "the lion of the party."[8]

Such moments must have pleased Vera when she was with him. The couple appreciated each other's strengths, but aside from gatherings at Olga Samaroff's music appreciation classes, they maintained mostly separate social calendars. The ardor with which he pursued Vera in their whirlwind monthlong courtship seems to have cooled down soon after they were married. Although he had married a rich woman, Gibbs's ambition was unaffected; in fact, it grew more intense.

When he was asked about their relationship many years later, Gibbs's reply had a ring of impatience, as if the answer should be obvious: "My wife is beautiful, intelligent, and considerate. We get along fine. She goes her way and I go mine, so we don't see much of each other—maybe that's why."[9]

Vera did have some influence on her husband's wardrobe. After the marriage, Gibbs swapped his derby hat for a floppy brown fedora, his black bow tie for a batwing of the same color, while giving up his stiff wing collar entirely. For her part, Vera knew she had married a man who lived to work and was set in his ways, and she appreciated him for the driven eccentric he was. "He keeps himself like a trainer would a race horse," she said. "He's up regularly at six thirty . . . he fixes his own breakfast of weak tea with lots of sugar and Uneeda biscuits. He really only eats one meal a day—dinner."[10]

As an independent woman for her era, Vera had no need to live for her husband. Publicly, at least, she described the marriage as a happy one. With William Francis putting in long hours at work, Vera filled her time with cultural and musical activities, sponsoring concerts and holding dinner parties at their apartment at 170 East Seventy-Ninth Street. She actively promoted the careers of African-American artists, at a time of continuing barriers to their rise. She hosted a dinner to honor soprano Dorothy Maynor after her recital at Town Hall; on the guest list were Harlem Renaissance photographer Carl Van Vechten and other notables.[11]

For Gibbs, work always trumped socializing. But he showed a commitment to Vera by adopting her son, Adrian, and on March 24, 1928, Vera gave birth to a son of their own. Vera thought of naming him William Francis Gibbs Jr., but changed her mind. "I thought the die had been broken with the creation of his father," she said.[12] So he received names from both his father and his Cravath grandfather: Francis Cravath Gibbs. Another son, Christopher, followed on September 2, 1930.

Gibbs was not a good family man. While he bent over his drafting board, his two sons spent much of their time at Grandpa Cravath's Still Place in Locust Valley. Like their mother, both Francis and Christopher liked to ride horses through the fields of the estate. The two boys once sent Christmas cards to their father with each on their horses. The photo of young Francis showed that he had his father's long jaw, spindly legs, and high cheekbones.[13] The boys had a privileged childhood, but horses and possessions could never make up for the lack of time with their father. One family member would later describe the Gibbses' parenting style as "rather arms-length"; Vera was "a socialite who lunched with Mrs. Belmont after Metropolitan Opera Guild meetings" and William Francis was "in Manhattan most of the time making sure things were in perfect running order" at the office.[14]

If life in the Gibbs family was less than ideal, father, mother, and grandpa Cravath at least shared a love of music. Cravath was also a cofounder of the Council on Foreign Relations, an organization that promoted Wilsonian internationalism, a connection that could only have helped Gibbs's career. And the exacting Cravath had indeed become fond of his daughter's new husband. He was also apparently fond of ships as well: whenever a liner would pass by his office and blow its whistle, New York's greatest attorney was known to throw open his window and bellow back in response.[15] Despite his intense interest in New York society, Cravath was himself a self-made man from a small town, and recognized talent when he saw it. His law firm, unlike other white-shoe establishments, made its hires based primarily on grades and natural talent, not family background, and gave young associates small parts of complex legal cases to master.[16] This was an approach his lawyer-turned-engineer son-in-law would follow in his own career.

Frederic Gibbs lived alone after his older brother's marriage, but didn't seem to feel left out of the family. In fact, the Gibbs brothers continued to spend most of every day together, sharing a desk as Frederic struggled to balance the books of Gibbs Brothers Inc. Vera also spent time fussing over Frederic, calling her solitary brother-in-law "Fritzy Boy."[17]

Vera's money made little difference to Gibbs's relentless self-marketing, as he was still doing his best to try to drum up business. After the loss of the *Leviathan* management contract, Gibbs Brothers kept itself afloat with minor commissions. Two years after refurbishing and operating the big ship, however, Gibbs finally received a commission to build a liner of his own from the keel up. If completed as specified, the Matson Line's *Malolo* (Hawaiian for flying fish) would be the largest ship yet built in the United States. She would also be constructed in Gibbs's hometown by the firm of William Cramp & Sons, the same shipyard where eight-year-old William Francis saw the launching of the *St. Louis* back in 1894. Carrying 693 first-class passengers, *Malolo* would serve on the Pacific run between San Francisco, Hawaii, and East Asia.

"She will be the finest liner of her type ever built," Gibbs asserted. "It is our intention to incorporate every modern idea that will pass the stern scrutiny of practicability."[18]

As it turned out, the ship nearly cost the Gibbs brothers their lives.

MALOLO

The years after the stunning debut of *Leviathan* were a dismal time for new ship construction in America. The big ship, popular with passengers despite her dry bars, continued to lose money, and her smaller running mates *George Washington* and *America* could hardly compete with the glamorous Cunard, White Star, and French fleets. The government-backed United States Lines was floundering. Even worse, Philip Franklin, now a bitter enemy, had shuttered the American Lines' transatlantic service and made plans to sell White Star back to British interests. Gibbs of course was too proud to ask Franklin to take him back.

The man who had rebuilt America's greatest ship found himself fighting over contracts for coastal vessels and tugboats with more established rivals such as Theodore Ferris. Ferris, a talented engineer, was also an astute businessman who carefully cultivated cozy relationships with shipyards. He also had few qualms about self-promotion. His peers described him as a "naval architect of ability and reputation," and someone who "had enjoyed a highly lucrative practice for years."[1] Ferris and other more established naval architects locked up contracts with the Ward Line and the IMM-controlled Panama Pacific Line, which made freight and mail runs to South America and the Caribbean.

So getting the *Malolo* contract was a major coup, but it was the intervention of Gibbs's old mentor that really kept the company afloat. Admiral David Taylor, who had backed William Francis Gibbs's first superliner in 1914, also found himself looking for work. After retiring as chief of the Navy's Bureau of Construction in 1923, he was serving as the president of the Society of Naval Architects and Marine Engineers, a post that was prestigious but did not provide much satisfaction. He knew as well as any man that it was not a good era for shipbuilding. The Washington Naval Conference, convened in 1921 to end the race in naval armament, resulted in Britain, France, Italy, Japan, and the United States agreeing to scrap most of their older battleships and limit the size and construction of new naval vessels.

But Admiral Taylor could not stay away from the designing of ships; nor did he want to be stuck in a do-nothing job after seeing so much of his previous work undone by the Washington treaty, one that the admiral knew would do little to stop Japan from becoming East Asia's dominant naval power.

Gibbs jumped at the chance to hire Admiral Taylor. "Considering Admiral Taylor's great reputation and standing and our relative inexperience and the difficult periods through which we were passing," he recalled later, "it was with diffidence that we suggested the possible alliance."[2] Taylor, in turn, jumped at the offer to move into commercial ship design. The retired admiral still believed that William Francis Gibbs would one day build the thousand-foot American superliner that the young man came to him with the day they met a decade earlier. For Gibbs's part, having the Navy's former chief constructor on board would help his firm attract contracts and recruit engineering talent in a big way.

Taylor was made a full partner in Gibbs Brothers Inc. in 1925, just as it began construction on *Malolo*. The keel was laid on June 1, 1925, and she was launched almost exactly a year later, with Secretary of Commerce Herbert Hoover attending. Taylor advised Gibbs Brothers on the defense-related features in the design of the vessel, including special mountings for six-inch guns, which allowed her to be converted for Navy use as a troopship or even as an aircraft carrier.

But the project produced one headache after another. Months into construction, a catchboy on one of the riveting teams dropped a molten-hot rivet, setting the scaffolding on fire and burning out most of the ship's nearly completed interiors. What didn't burn was waterlogged and reeked of smoke. Gibbs held the shipyard, Cramps of Philadelphia, responsible. He insisted that it pay the $200,000 (equal to about $2.5 million in 2012) needed to repair the ship.

Cramp did so, unhappily. Like Newport News, the shipyard was starved for work after World War I, and saw the $7 million *Malolo* job as a ticket back to profitability. The yard was wrong. As he had done when he oversaw the construction of *Leviathan,* Gibbs refused to allow any change orders and demanded that any deficiencies in material quality be fixed at Cramp's expense. In the end, $4 million of unanticipated costs drove the shipyard into bankruptcy just as *Malolo* was completed.[3]

On May 26, 1927, the year of Gibbs's marriage, his new liner arrived off Nantucket for sea trials. But the runs along the measured course were postponed. The fog was too dense for the ship to be brought up to full speed in a heavily trafficked shipping channel. Gibbs and Frederic stayed on the bridge with the ship's captain, while Admiral Taylor went to rest in his stateroom. A light rain pattered on the decks.

Suddenly, Taylor heard the thumping of boots and frantic shouts from the bridge above. When he looked out his porthole, he saw a set of lights emerge out of the fog. A black object then revealed itself, barreling directly toward *Malolo*'s port side.

Admiral Taylor's throat went dry. "They say she was a small boat," he recalled, "but let me tell you she looked like 20,000 tons. She came up with a 'bone in her teeth,' and she looked as big as a snow bank."[4]

A few decks up, the Gibbs brothers also saw a freighter's prow aimed at a point right beneath the port bridge wing.

Taylor braced himself for a hit. He heard two sharp blasts from horns of the oncoming ship. The bow of the Norwegian freighter *Jacob Christiansen* then sank into *Malolo*'s flank like an axe blade. The impact knocked many on board off their feet. Down below, a switchboard operator was busy calculating his overtime pay when green seawater burst through a wall and swept him from his desk. He leveraged himself against his office door, forced it open, and managed to escape.

Gibbs heard a cacophony of sounds from below: tearing steel plates, popping rivets, ripping electrical wires, and shattering glass. Keeping cool, he walked over to the watertight door controls and flicked the alarm switch. Bells jangled up and down the length of the ship. Near the waterline, the scuppers—openings in the hull where the ship's pumps ejected bilgewater—snapped shut.

But the icy Atlantic had already doused the boilers. With electricity gone, lights across the ship went out. Then the emergency generator kicked in, bringing the bridge's red and green running lights back on. In the corridors and public rooms of the now-darkened ship, a few emergency lamps flickered dimly.

The Norwegian freighter had rammed *Malolo* directly against the bulkhead separating her two boiler rooms, ripping a vertical gash two feet wide and fifteen feet high into her hull. Taylor thought *Malolo* was doomed. This seasoned ship expert had never seen a ship be "rammed square amidships and still remain afloat."[5]

Taylor and Gibbs knew what had happened to the liner *Empress of Ireland* when she was struck amidships eleven years earlier. On May 30, 1914, the midsize luxury liner—owned by the Canadian Pacific Railway—was on the first leg of a voyage between Quebec City and Liverpool. On board were nearly 1,400 passengers and crew, including a large contingent from the British Salvation Army. As she headed into the traffic-clogged St. Lawrence River, the fog closed in. At 2 A.M., as he poked his ship through the murk, Captain Henry Kendall peered off the starboard bow and saw a set of lights twinkling in the distance, lights that were moving directly toward his ship. Before Kendall could take evasive action, the Norwegian cargo ship *Storstad*'s bow gouged through *Empress of Ireland*'s side, piercing the bulkhead that separated her engine room from one of her boiler rooms. Both compartments rapidly flooded and the ship lost power, leaving passengers scrambling for the exits in total darkness. The watertight doors failed to close, and the sea poured in through open scupper valves and portholes. Within ten minutes, *Empress of Ireland* careened onto her side. Hundreds of people clung to the hull; hundreds more were trapped within her flooding interior. Finally, the vessel shook and slid into the icy St. Lawrence. Some 1,014 of the *Empress*'s passengers and crew perished.

Now, with water flooding *Malolo*'s boiler rooms, William Francis knew there was a possibility that his brand-new ship would flop onto her side and sink like *Empress of Ireland* before her.

Leaving the bridge, William Francis and Frederic scampered down several flights of stairs until their shoes hit cold seawater, soaking their trousers. They finally reached one of the watertight doors. It was wide open—they had designed a deliberate delay in the automatic closing to let crewmen escape to the upper decks.

Gibbs grabbed for one of the levers that would close the doors manually. "Just at that moment," Frederic recalled, "the watertight door slowly began to close." The timed delay was over and all the automatic doors between *Malolo*'s twelve watertight compartments automatically clanked shut. Frederic found the wait for the watertight doors to close agonizing.[6]

Water nevertheless filled the *Malolo*'s boiler rooms and other spaces. But the decks held, keeping the water from flooding the rest of the ship. Unlike *Titanic,* whose watertight bulkheads extended only partway up the hull, *Malolo*'s reached all the way to the superstructure. The bulkheads separating the two boiler rooms from the engine room and cargo holds strained under the pressure, but held firm.

Slowly, *Malolo* righted herself, but she was still sinking. Next to one of the breached boiler rooms, a drain cover in the indoor swimming pool had broken, allowing seawater to rapidly flood the two-deck-high space. If it flooded, the ship might lose whatever margin of stability she had left and sink. One of the Cramp workers stripped to the waist and plunged into the pool. After several lung-bursting dives, he managed to lock the drain tight.

Finally, the *Malolo* stopped sinking. She sat upright, weighed down with six thousand tons of seawater and drawing thirty-six feet. But for now, she was safe.

No evacuation was necessary, but *Malolo*'s captain sent out a distress call, asking for a tow. Tankers *Gulfland* and *City of Pretoria* raced to the scene, their captains eager for salvage money. With her whistles silent, *Malolo*'s captain guided approaching rescue ships through the fog by ringing the liner's brass bell. When the two tankers arrived, they each launched a lifeboat full of their strongest sailors. "Both crews," the *New York Times* reported, "pulled with as much vim as a Harvard crew." When one of the rescue ships tried to tow the liner through now-choppy seas, the line grew taught and snapped. Finally, three powerful tugs came alongside and began to pull the injured *Malolo* toward New York.[7]

On the morning of May 29, 1927, *Malolo* dropped anchor off Staten Island. She was so low in the water that she looked more like a barge than a luxury liner. Curious bystanders lined the Battery as tugs towed her to the Brooklyn Navy Yard, where she would be pumped dry and patched. At the yard, photographers took pictures of the design team on the bridge. The Gibbs brothers stand next to each other, their hats off. William Francis, in a dark wool coat with his thinning hair blown by the wind, is looking away from the camera, an irritated expression on his face.[8]

But even the bankrupted Cramp Shipyard praised the ship's designer. When asked by reporters if he was satisfied with the ship's design, J. Harry Mull, president of Cramp, responded, "Satisfied? I doubt if the hole in the *Lusitania* was as large as this one we have. We are here in New York harbor. That speaks for itself."

The only injury was a crew member's sprained ankle.

For William Francis Gibbs, the publicity generated by the *Malolo*'s near sinking would link his name with an obsession with safety at sea. For him there was no worse sin than cutting corners to save the client money, especially at the expense of passenger safety. It was the duty of a ship designer not just to meet regulations, but to exceed them, no matter how much the client or shipyard complained about the cost of extra compartments, automatic scupper closers, and lifesaving gear. In July 1927, he published his first piece in a trade journal: "Collision Vindicates Safety Measures," in *Marine Engineering and Shipping News.* Gibbs proudly pointed out that his design team had followed not merely

U.S. requirements, but the standards set in 1914 at the International Convention for the Safety of Life at Sea, which America had refused to ratify even after Senator Smith's findings on the *Titanic* disaster. Spending money in the short term to create the safest ship possible was "good business, aside from ethics and morality, to make such an investment."[9] Rectifying the flaws in *Titanic*'s design, in fact, had been Gibbs's mantra throughout the entire design process. Small wonder *Malolo* survived a collision with a freighter that would have sunk the bigger ship.

It took another disaster, however, for the world to finally act on those promises made back after *Titanic*'s sinking. In 1928, the British passenger ship *Vestris* capsized in a storm off the Virgina coast, killing 127 people. Shifting cargo caused her to list, and then seawater poured in through an open coal port that could not be closed. After the tragedy, seafaring nations convened in London to strengthen the international Safety of Life at Sea regulations, first set down in 1914 in response to the *Titanic* disaster. The SOLAS convention ratified the rules on stability and compartmentalization that had been raised at the 1914 meeting but ignored after World War I broke out. However, it was not until 1933 that the safety standards William Francis Gibbs worked into *Malolo* became international law.

Malolo proved to be a popular and successful transpacific liner. But as fine and safe a ship as *Malolo* was, she was comparatively small and no international record breaker. By the late 1920s, the public was much more interested in a new generation of transatlantic giants emerging from the drawing boards of Britain, France, Italy, and Germany. With its ships clearly obsolete and the United States Lines about to be sold into private hands, it seemed clear that as long as the 1920s economic boom continued, America would build a new superliner.

The question was, who would design her?

Gibbs knew he had some stiff competition from another naval architect, a favorite of shipyards and lines alike for turning out attractive ships for the American coastal trade on time and on budget. Theodore

E. Ferris did not make many headlines during the 1920s, but he had a steadier and more lucrative practice than Gibbs. And he had made fewer enemies.

Gibbs decided that the best course of action was to find sympathetic partners and attempt to buy the ships of the United States Lines outright. Once the company was secured, he would find financing to build his own transatlantic superliner.

A GERMAN SEA MONSTER

The 1920s was a boom time for the transatlantic business. The big ships had nicknames: *Berengaria* was the "Berry," *Majestic* the "Magic Stick," *Olympic* the "Old Reliable." During the high summer season, men in straw boaters and women in cloche hats crammed the pier heads, waving handkerchiefs as the giants backed away from their piers into the Hudson, blasting their whistles and belching black smoke from their stacks.

The five- or six-day trip was filled with masquerades, passenger talent shows, shuffleboard tournaments, and smoking room bridge games. After dark, the booze flowed and hot jazz bands blared from the ballroom stage. The Prince of Wales and Queen Marie of Romania joined Cornelius Vanderbilt III and Vincent Astor at captain's tables overflowing with grilled antelope, quail eggs, and caviar. New York's corrupt mayor, Jimmy Walker, set a new standard for shipboard dandyism: he packed his steamer trunks with forty-four suits, twenty pique vests (to go with his tailcoats), twelve pairs of trousers, and a hundred cravats. At the captain's table, Commodore Sir James Charles of the Cunard Line demanded that male guests don evening dress, complete with military decorations and hereditary medals. A legendary gourmand, he dropped dead during one of his gargantuan feasts aboard *Aquitania* in 1928.[1]

As the stock market soared and more people embarked on the European "Grand Tour," some expressed disapproval of the party atmosphere that prevailed aboard the big liners. "If you are a lover of the seas and the ships that sail them," one passenger wrote his friend Franklin Roosevelt, "then the *Majestic* is nothing more than a gorgeous hotel filled with the usual obnoxious crowd . . . as the steward will tell you, the best class of people travel second class at third class rates."[2] Many old salts felt that the big ships had lost their vitality and purpose, degenerating into floating pleasure palaces catering to the whim of the rich American tourist. "Everybody on the *Berengaria,* even the dogs," one officer sniffed, "were 'socially prominent.' "[3]

Of all the big liners, only the United States Lines' *Leviathan* consistently lost money. The reason was simple: her bars were dry, and Americans flocked to the foreign ships instead. "One thing I learned," joked her captain, Herbert Hartley, "was that in passing the Volstead Act, Americans had voted dry but they traveled wet."

As *Leviathan* approached New York harbor at the end of each voyage, liquor found in steamer trunks was seized by the master-at-arms and dumped overboard.

"I am sure no less than a million dollars worth of the world's choicest whiskies are resting at that hallowed spot on the floor of the Atlantic," Hartley remembered.[4]

Some wondered whether this era of extravagance was sustainable, but by the late 1920s, the European shipping companies, flush with profits from the economic boom, were making big plans to build a new generation of liners—with heavy government support.

The French Line took the first step. Among rich Americans, the company had a reputation for elegant service and superb cuisine. Its president, Jean dal Piaz, had a strict policy of not building sister ships: "Vivre, ce n'est pas copier, c'est créer (To live is to create, not to copy)," he declared.[5] By 1927, dal Piaz unveiled the 43,000-ton *Ile de France.* On the outside, she was boxy and traditional, and she was no record breaker. It was her interiors that stunned the world: she was the first big liner to be decorated in the so-called Art Deco style. There was no mimicking of European palaces aboard the new French flagship. She was a ship of the Jazz Age, full of bright colors, indirect lighting, and streamlined furniture. The chic *Ile de France* quickly stole the cream of the American tourists away from the aging Cunard and White Star liners.

At the same time, the Germans started construction on two giants that *would* be contenders for the Blue Riband. They would be named *Bremen* and *Europa.*

Even Washington was taking notice. "There is no question," declared one New York congressman at a Shipping Board hearing in early 1927, "that the United States Lines is hampered by reason of their failing to have enough ships of the *Leviathan* type, or the large type."[6]

A new race was shaping up, and William Francis Gibbs had no intention of sitting on the sidelines.

Now in his early forties, he surveyed the transatlantic passenger fleet and realized that the big liners built before 1914 were showing their age. Cunard's *Mauretania,* still the holder of the Blue Riband, had just turned twenty. Although still the fastest ship on the seas, she was small and cramped compared to bigger rivals, lacked private bathrooms and other luxuries demanded by modern travelers, and rode rough in heavy seas. Booking a ticket on *Mauretania* seemed more like a necessary inconvenience if one wanted to cross the Atlantic in less than five days. None of the other prewar giants—*Olympic, Majestic, Leviathan, Aquitania, Berengaria*—could approach her speed. And they too were starting to have mechanical and structural problems as they aged.

Gibbs knew that a new generation of European liners—the two German ships in particular—would push aside the dowagers, while the hapless, government-controlled United States Lines was in no financial position to compete against what was coming. By the late 1920s, the management team Lasker had set up to run the government-owned United States Lines was coming apart. Under intense political pressure from Republicans, the Shipping Board decided to sell *Leviathan* and other passenger ships to a private operator.

Given their track record, William Francis and Frederic Gibbs assumed they had the best chance of buying the United States Lines. The Gibbs brothers teamed up with J. H. Winchester & Company, a New York–based shipping firm, to put in a bid. In it, they argued that to compete with the weekly sailing schedules of the foreign lines, *Leviathan* had to have two fast running mates. Constructing two such ships was impossible, they pointed out, because the government would not put up the subsidies needed to do it. Instead the brothers offered to buy *Leviathan* from United States Lines for $10 million, and at the same time modernize two seized German liners: *Mount Vernon* and *Monticello,* both of which sat idle after finishing World War I trooping duties. With new engines, the brothers concluded, the two old four-stackers could become a duo of medium-sized liners that could make 23 knots and keep pace with the bigger flagship. Gibbs had admired these two vessels since his college days—the former *Kaiser Wilhelm II* and *Kronprinzessin Cecilie* were sleek, rakish ocean greyhounds that he felt deserved a new lease on life. Once the three liners were producing steady cash flow, the brothers felt they could successfully lobby for the construction of one or more superliners.

Although the Gibbs proposal was initially well received, a new player with better political connections but no experience in shipping entered the bidding. Paul Wadsworth Chapman was a New York promoter who had recently financed the construction of the largest passenger airplane yet built, but he had little knowledge of the shipping business. By buying up all the ships of the United States Lines, he hoped to create a sea and air transportation network between America and Europe.

Chapman offered just over $16 million for *Leviathan* and her smaller fleet mates, and promised an initial payment of $4 million in cash, with the remainder to follow. His proposal also called for a government loan of $42 million to build two new 25-knot superliners, the same size as *Leviathan* but able to compete with the two much-anticipated German superliners then under construction. In a few years, Chapman promised, his privatized United States Lines would have a three-ship, American-flagged weekly transatlantic service that could take on the Europeans.[7]

In February 1929, Gibbs testified before the Senate Commerce

Committee that Chapman's two *Leviathan*-class superliners, to be built with a $42 million loan from the government, would lose millions of dollars annually. Chapman's offer, Gibbs said, is "analogous to a man offering to pay another $4,000,000 on condition that the latter lend him $42,000,000."[8]

What's more, Gibbs argued, a Chapman default would do serious damage to the growth of America's merchant marine, by weakening investor confidence in marine securities.[9]

Confident in his analysis, Gibbs held firm to his $10 million offer. But he found himself hard against the chairman of the Shipping Board: Thomas V. O'Connor, the former head of the International Longshoremen's Association, who had become a favorite of ship owners. O'Connor allied himself with Chapman and threw him the sale. A day later, O'Connor savaged Gibbs in the press. Gibbs's statements, the chairman said, "are the sort of thing that gives aid and comfort to our foreign competitors."[10]

Chapman, now in possession of *Leviathan* and the rest of the United States Lines fleet, then hired as his naval architect a man for whom Gibbs had no respect: Theodore Ferris, the former chief designer for a failed World War I project, the Emergency Shipbuilding Program, set up to build hundreds of wooden freighters to ship men and supplies to the front. The expensive enterprise was a complete disaster, and Ferris was accused of corruption. An outraged Ferris resigned under a cloud and returned to private practice. When appointed as head of the Emergency Fleet Corporation, he was paid the handsome sum of $2,500 per month, and according to a Senate report, the "reputation which the position entailed would insure Mr. Ferris of one of the most profitable practices throughout the entire world and this practice would have continued as long as he chose to remain in business after he severed his connection with the government."[11]

Lusting to design his own superliner, Ferris saw Gibbs not just as a competitor, but an enemy as well. The feeling was mutual. Unlike Gibbs, Ferris was a highly trained engineer who had spent his early years paying his dues in shipyards and drafting rooms rather than the grass courts of

Merion Cricket and the halls of Harvard. The so-called Dean of Ameri can Naval Architects had designed nearly 1,800 ships, including yachts for European royalty. Tall and broad-shouldered, Ferris radiated a confidence and poise that his volatile, eccentric rival Gibbs lacked.

In addition to snapping up the United States Lines superliner project, Ferris picked up more good press notice by designing two new coastal steamers for the Ward Line, *Morro Castle* and *Oriente.* Both were intended for passenger and freight service between New York and Havana. The ships were relatively small, 10,000 tons each and 508 feet long, but they showcased many of the innovations that Ferris included in his proposed superliners. With a sustained speed of 21 knots, they were among the fastest vessels under the American flag.

The two "Havana ferryboats" debuted in late 1930 and were an immediate commercial success; a relaxation of Prohibition aboard American-flagged ships meant that her five hundred passengers could enjoy a nonstop drinking party. Both ships were luxurious, with public rooms adorned with exotic woods and plush fabrics. During hot tropical voyages, an elaborate system of vents and shafts funneled cool air into the passenger quarters.

The press reported that "every possible safety feature has been considered in the construction of these vessels to make them the safest ships afloat at this time."[12] *Morro Castle* was equipped with ten fireproof steel lifeboats and two fire detection systems. Fire doors, controlled from the bridge and placed at key intervals, would stop flames from spreading.[13]

But there were to be no heat or smoke sensors in the public rooms; watchmen making rounds were expected to see or smell any evidence of fire.[14]

Ferris's triumphs must have been galling to Gibbs, who had lost his opportunity to move from being a ship designer to a ship owner. Not only that, but Germany had finally unveiled its two new contenders for the Blue Riband, ships that made Gibbs's predictions of obsolescence come to pass. Other great ships, each faster and more advanced than the last, would follow in their wake.

* * *

On July 28, 1929, thousands of well-wishers lined the banks of Germany's Weser River as the new Norddeutscher Lloyd flagship *Bremen* left her home port of Bremerhaven and sped toward Cherbourg, France. The company's general manager boasted that *Bremen* was "the pulsating symbol of Germany's comeback as a world economic factor."[15] After leaving France and steering a course toward New York with more than two thousand excited passengers under his care, Captain Leopold Ziegenbein was determined to take the Blue Riband away from Cunard's *Mauretania,* holder of the prize for more than twenty years.

Looking more like a navy cruiser than a luxury liner, the 51,000-ton, 950-foot-long *Bremen* and her slightly smaller sister *Europa* were streamlined from bow to stern: two squat smokestacks, a low superstructure, and a rounded bridge front. Along with the world's most powerful turbines yet built, *Bremen* incorporated an innovation pioneered by Admiral Taylor a decade earlier. For the first time, the drag-reducing bulbous bow first used on American naval vessels was employed on a commercial transatlantic liner.[16] Her engines were also much more efficient than older ships'. *Bremen* needed only 20 boilers to achieve nearly 28 knots, using 20 percent less fuel and putting out nearly 50 percent more horsepower than *Leviathan,* which needed 45 boilers to make 23 knots.[17]

Even her whistles blared Teutonic superiority. Most liners of the time had three whistles. *Bremen* had five. "The only time the *Bremen* gets any real fun out of its five whistles," a *New Yorker* writer commented, was when she was in her home port, when her captain would "let loose with all of them simultaneously in three prolonged blasts: their rough version of saying 'Aloha,' or 'Farewell.' They never do that on this side, however. Afraid the West Side folk wouldn't understand."[18]

Inside and out, *Bremen* was unabashedly modern. As described by writer John Malcolm Brinnin, "her designers took care to follow the lines of her structure and to make their interiors conform to all its sweeps and nuances . . . a marine look, a 'shippiness' that ran directly counter to the cozy house-in-the-country look of British ships. . . ."[19]

Her designers took their cue from the French Line's popular *Ile de France,* with her modern, sleek Art Deco interiors. Down in the engine room, *New York Times* reporter Ferdinand Kuhn stared at the whirring turbines and marveled at how there "was no outwards sign that she was making the greatest speed ever reached by an ocean liner."[20] *Bremen*'s mighty power plant, the seagoing equivalent of a supercharged Mercedes-Benz engine, packed over 100,000 horsepower, or one-third more power than the old *Mauretania.*

Bremen also carried a harbinger of the future between her smokestacks. At eight o'clock on July 23, as the German greyhound tore past the Nantucket lightship, passengers taking an early morning stroll heard the roar of an airplane engine, followed by a loud hissing sound. Looking up, they watched a Heinkel seaplane, thrust forward by a burst of compressed air, bank toward New York. The plane was carrying a portion of the ship's mail. "If all is well," Kuhn reported, "the cargo of mail should be landed at Quarantine about 11 o'clock, and 1,000 New Yorkers will receive letters and postcards before the ship docks. The letters will carry 65 pfennings postage and bear a vivid blue marker 'Luftpost Avion.' "[21]

Four days, 17 hours, and 47 minutes after leaving France, *Bremen* swept past New York's Ambrose Lightship, black smoke flying from her low buff funnels and a great wave shooting up and away from her bulbous bow. During the 3,164-nautical-mile passage, she had averaged 27.83 knots, soundly beating *Mauretania*'s 26.06-knot record of 1909. She had also bettered *Mauretania*'s sailing time by seven hours.

The next year, *Bremen*'s sister *Europa* averaged 27.92 knots along the same course, shaving half an hour off *Bremen*'s record time. For the first time in nearly two decades, the coveted Blue Riband was firmly back in German hands.[22]

Captivated by the speed and modernity of *Bremen* and *Europa,* the American public hugely admired the German achievement. "Behind those few hours lies a narrative of international rivalry for speed, of stupendous engineering effort, of scientific research and of courageous businessmen willing to spend $15,000,000 to build a ship capable of beating the *Mauretania* by a knot," *New York Times* science writer

Waldemar Kaempffert wrote.[23] Avant-garde designers like Modernist Swiss architect Le Corbusier looked to the ships of the late 1920s as paradigms of a new, technology-driven aesthetic, in which function swept away all ornamentation.

Others were less than enthralled of this elegant yet sinister-looking liner. When *Bremen* swept past *Leviathan* at full speed in the English Channel, the crew of the aging American flagship realized that indeed a new era had dawned. "We consoled ourselves in the thought that, after all there was a 20-year development in marine propulsion," recalled one of *Leviathan*'s officers, "and 'just wait, some day' but it was hard to take, nonetheless."[24]

The great British and French shipping lines were quick to act after the debuts of *Bremen* and *Europa*. In 1929, the French Line began developing a new contender for the Blue Riband. The ship featured a radical new hull design by Russian-born naval architect Vladimir Yourkevitch. His liner, initially called T-6, featured a bow with a big bulb at the base like *Bremen* and *Europa*. Above the waterline, the prow curved sensuously forward like the bows of nineteenth-century clipper ships. And unlike the full, rounded hulls of older vessels, the hull would be fine and tapered. The ship would weigh 79,000 gross tons, stretch 1,029 feet long and 117 feet wide, and could maintain a service speed of 30 knots. She would be powered not by geared turbines, but by a turbo-electric plant that men like William Emmet had pushed for use in large liners since the early 1900s. Her designers aimed for an output of 150,000 horsepower.

Yourkevitch had brought his design to Cunard first, only to be turned down by the conservative company. But the French loved Yourkevitch's vision. In 1931, workers at Penhoët–St.-Nazaire on the Loire River laid the keel of what would become the great liner *Normandie*.

Word of the new French ship shook self-satisfied Cunard out of its lethargy. Its grand trio of *Mauretania, Aquitania,* and *Berengaria,* all built before World War I, was fast becoming obsolete. After *Mauretania* lost the Blue Riband, Cunard's passenger revenues declined rapidly. Cunard started a snobby advertising campaign: "Whole families of America's highest type would rather miss Ascot, or the first day of grouse shooting, than cross in any other ship afloat, but the *Mauretania*."[25] But to no avail. The old ship, with her stuffy Edwardian décor and lack of private bathrooms, was commercially doomed.

So Cunard too began to plan a superliner, one that would sail with a sister ship to create a two-ship weekly service between Southampton and New York. Like *Normandie,* she would have three funnels. At 1,019 feet long, 118 feet wide, and at 81,000 tons, she would be larger than the French vessel. Her engines would be traditional Parsons steam turbines, not turbo-electric.

William Francis Gibbs could only watch in frustration as the German, French, and British lines forged ahead, while the United States Lines floundered under Chapman's bad management. The promise of a Blue Riband–winning American superliner seemed to be slipping away.

And then, in late October 1929, the American stock market crashed. The catastrophe wiped out the fortunes of millionaires and the savings of ordinary tourists. Ticket sales slipped that winter, and then spiraled rapidly downward as the economy soured. Companies fought over a shrinking pie of passengers, and many of the older ships began to struggle when faced against the modern *Bremen, Europa,* and *Ile de France.*

As the Great Depression paralyzed the nation, a chastened Paul Chapman discovered that operating a big ship like *Leviathan* required not only money, but also expertise and judgment. On December 11, 1929, six months after *Bremen*'s Blue Riband triumph and six weeks after the stock market crash, *Leviathan* was traveling eastbound with 845 passengers aboard. As the ship approached the British coast, a northwestern gale began viciously pounding her. The old liner creaked and groaned ominously. From his observation perch ninety feet above the ocean, Second Officer Sherman Reed grew anxious about the ship's speed. Reed telephoned Commodore Harold Cunningham requesting that he slow *Leviathan* down.

"Maintain your speed," the captain snapped at Reed. "This ship will get in on time."

A few minutes later, a forty-foot wave came barreling straight toward

Leviathan's bow. The officers on the bridge braced themselves for impact. With a roar, the bow buried itself in the swells, tossing up a great cloud of white spray. The entire ship resounded first with a gigantic boom, then a loud crack.

Leviathan's upper decks had split open just in front of the midship expansion joint, a vulnerable area where a double bank of elevator shafts and the ship's split funnel uptakes cut through the superstructure. Her hull took the blow and remained intact. Though the ship was in no immediate danger of sinking, a large part of her superstructure was now open to the elements.

At the ship's forward end, the crack had skewed walls and jammed stateroom doors shut. "Great confusion!" all over the ship, Reed recalled. "Bridge telephones ringing like mad."[26]

Somehow *Leviathan* stayed in one piece and limped into Southampton, where workers plated over the yawning gap. She returned to New York on December 16 and spent the rest of the winter undergoing permanent repairs at the Boston Navy Yard. The accident cost Paul Chapman nearly half a million dollars.[27]

Aware of this structural flaw, Gibbs knew that *Leviathan*'s structural integrity had been put at risk for the sake of grand passenger spaces. His chief electrical engineer, Norman Zippler, said that after learning of this accident, his boss made it a rule never to use split funnel uptakes in any future passenger ship design.[28]

Almost as soon as *Bremen* won the Blue Riband with her four-and-a-half-day crossing, others were looking beyond the achievement. Waldemar Kaempffert of the *New York Times* posed the question Gibbs was also asking: "Is a three-and-a-half day ship possible?"[29] Gibbs believed it was—if designers could advance the mechanics of propulsion. Efficiencies would not be enough. Revolutionary technologies were required.

The deciding factor for speed, many engineers thought, lay in the untried domain of something known as high-pressure, high-temperature steam. This could drive small turbines at tremendous speed. *Bremen,* which used British-licensed Parsons turbines, was driven by steam at 338 pounds per square inch putting out 100,000 horsepower, the highest ratings yet achieved in a passenger liner, but some argued that steam pressure could be doubled or even tripled, resulting in smaller, more efficient engines. "By increasing the steam pressure to, say 735 pounds per square inch," a German naval architect wrote, "and by further increasing the turbines revolutions to, say 2,300 to 2,500 . . . it might be possible to install from 150,000 to 170,000 horsepower in a ship even smaller than the *Europa* or *Bremen* and attain a service speed of more than 30 knots."[30] With such engine efficiencies, the space not taken up by engines and boilers could be used to carry revenue-generating items: mail, expensive cargo, and passengers.

High steam pressure was tried in experimental power plants. One small engine, produced by Germany's Wagner Hochdruck turbine company, produced 400 horsepower at 21,000 rpm with a boiler pressure of 735 pounds per square inch.[31] But no naval architect dared to use it in a merchant vessel. It was untested and thought to be much too dangerous. It also meant that shipyards could no longer simply build engines according to British designs licensed by Parsons, the inventors of the marine steam turbines. New engines would have to be relicensed.

But Gibbs had based his very first superliner plans on another technology untested in ships, the Curtis electro-turbine engine. He had made up his mind to use high-pressure, high-temperature steam in his superliner design, freeing American shipyards from dependence on British technology. But the way things were going, it appeared he would never get the chance to build the ship. Among naval architects, Theodore Ferris's reputation was growing, while Gibbs's was treading water at best.

Looking for more company revenue, the brothers brought in Daniel Cox, one of the nation's most respected yacht designers, as a partner. But the Depression was also killing the yacht market. By 1931, Gibbs & Cox was down to twelve men on staff. Frederic paid single men $25 a week, married men $50.[32]

If things didn't get better soon, William Francis Gibbs's firm would fail. Despite the dark times, Gibbs refused to actively advertise his firm's service in the press, preferring word of mouth. "If you employ a publicity man you will begin to believe what he says about you and then you are lost," he would tell his employees.[33]

In 1931, Theodore Ferris unveiled plans for his duo of transatlantic superliners for the United States Lines at a dinner hosted by the Society of Naval Architects and Marine Engineer. Before a large group of naval architects and Navy brass, Ferris laid out drawings for two 970-foot-long, 108-foot-wide liners able to carry almost 3,000 in three classes, and a crew of 1,050. Ferris promised a service speed would be 30 knots, about two knots faster than *Bremen* and *Europa,* and a top speed of almost 32 knots in smooth seas.[34] The interiors would be stunning, befitting a pair of floating Waldorf-Astorias. And Ferris intended his superliners to be the safest ships in the world, able to stay afloat with any four of their eighteen watertight compartments breeched. Unlike Gibbs's superliner design, Ferris's ships had fuller hulls and larger bulbous bows, features that would supposedly add to their speed and stability.

The liners would bear names that not just resonated with the American public, but also were closely connected with the career of William Francis Gibbs: *Leviathan II* and *Leviathan III.*[35]

During the question-and-answer session, Gibbs's business partner Admiral Taylor spoke up, asserting that the planned liners were already obsolete compared to the new ships being constructed in France and Great Britain. Taylor added that the proposed ships might be too wide to clear the Panama Canal, greatly reducing their military value. "If we are not to have a blue ribbon contender," Admiral Taylor concluded, "it would seem that a smaller and cheaper vessel would be good enough for a second flight."[36]

Ferris's reply was both caustic and defensive. The admiral, he said, "must be in possession of information concerning the probable speed of the British and French superliners which no one else has." The speed of his ships, sniffed Ferris, will be "second-to-none as regards any now building or existing."[37]

As the meeting neared its end, a member of the Society of Naval Architects and Marine Engineers' council rose for a rare encomium.* "I move that a rising vote of thanks be given to him as appreciation of his splendid effort," he declared, "and of his valuable contribution to the art of shipbuilding and the annals of our Society."

The entire audience stood and gave Ferris an ovation.[38]

Yet America's shipping business was still in turmoil. Ravaged by the Depression, Paul W. Chapman had defaulted on his loan payments to the government, just as William Francis had predicted. The U.S. was threatening to repossess his ships even as Theodore Ferris was unveiling his superliner designs to the public in 1931. While Ferris promoted his new ships, a chastened Chapman got together with Vincent Astor and IMM representatives to make a deal for United States Lines.

On the IMM team was thirty-six-year-old John Franklin, the son of the company's president, Philip Franklin. The elder Franklin, who had backed Gibbs's first superliner designs, was now ailing. John, the lackluster Harvard student and heroic soldier, had proved himself to his father in the shipping business. After returning from the front, John Franklin cut his teeth at the Norton Lilly shipping company for twenty-five dollars a week.[39] Here he began earning respect from leading members of the industry. John Franklin's experience with the tank regiment also gave him a common touch that allowed him to work with the maritime unions, a talent few maritime executives had. But it wasn't until John succeeded at Norton Lilly that Philip brought him aboard IMM.

As 1931 ended, Kermit Roosevelt (who had started his own shipping line), John Franklin, and their colleague Basil Harris teamed up to create a new and preeminent American shipping line. According to the *New York Times,* their proposed 181-vessel fleet would be the "greatest union of steamship companies in the history of this country and will form the most formidable shipping combinations in the world."[40]

---

* A rising vote of thanks was a very rare occurrence at a Society of Naval Architects and Marine Engineers meeting, according to member William duBarry Thomas in a letter to the author dated October 22, 2008.

Roosevelt-IMM made it clear that it wanted to divest itself of its foreign flag companies, and that passenger service, a consistent money loser, would be a minor part of their operations.

It would be the young John Franklin's first major business coup. With Astor's influence and Franklin's business smarts, the United States Lines would start building new passenger ships. After prolonged negotiations, United States Lines fleet was sold to the Roosevelt-IMM group on October 30, 1931, for just over $3 million—a steal compared to the $15 million Chapman paid for the fleet back in 1929.[41] Some old liners were returned to the U.S. Shipping Board and laid up; two smaller liners still in construction—*Manhattan* and *Washington*—were finished and put in service.

Yet it was not the financial problems of the United States Lines that captured the public's imagination. As his dreams of building *Leviathan II* and *Leviathan III* slipped from Theodore Ferris's grasp, one of the worst maritime disasters of the twentieth century would forever tarnish his reputation. And Gibbs's fascination with new technology would change the fortunes of his firm and his life.

DEATH BY FIRE

On the early morning of September 8, 1934, people who lived along the north New Jersey shore woke up to find a nor'easter gale sending waves crashing onto their wide sandy beaches. It was after Labor Day and the boardwalks were empty, the beaches largely deserted, the summer cottages closed up. But local residents looking out to sea could see in the distance a pillar of black smoke. Under its shadow, ten miles from shore, frightened passengers huddled at the stern of an ocean liner on fire. The lettering on her bow read MORRO CASTLE.

News of the disaster was radioed to authorities onshore, and the story hit the news. Soon an airplane carrying a newsreel crew swooped over the ship. Their film would later be shown in movie houses across the world. "As we approach the burning inferno, her sides white with heat," the newsreel announcer intones, "we see many who've jumped overboard struggling in the water, whilst the floating furnace casts a pall of smoke over the rain swept sea."[1]

The crew of British cruise liner *Monarch of Bermuda* picked up survivors adrift in the lifeboats and bobbing in the windswept waters. *Monarch*'s Captain Jeffries tried to secure a towline from his ship to *Morro Castle,* but rough seas broke the line and the floating funeral pyre drifted toward the Jersey shore, smoke still billowing from her upper decks. "It was a ghastly sight," a shaken Jeffries told reporters.[2]

At seven-thirty that evening, *Morro Castle* ran aground on the sandy beach of Asbury Park, New Jersey.

The following morning, crowds swarmed around the still-smoldering wreck, which reeked of burning wood and death. Coast Guard officials who boarded the wreck saw that the fire had spared nothing, burning *Morro Castle* right down to the steel decking and bulkheads. Furniture, paneling, and electrical wires were reduced to heaps of ash. On the fantail, along with discarded clothing and shoes, they also found the charred body of a small boy.[3]

One reporter got aboard the vessel with a movie camera. As heat from the decks blistered the soles of his feet, he captured the horrific scene: "Gaping portholes with the glass melted by heat, a charred hulk . . . girders and beams everywhere twisted by the terrific heat of the all devouring fire, every bit of planking was burned off the decks amidships, and the blasting flames had even buckled the rails, twisting everything that did not melt in the blistering heat."[4]

He then pointed his camera to a row of scorched lifeboats still in their davits. The fire had spread so quickly that most of them could not be launched.

The cause of the *Morro Castle* fire has never been conclusively determined, but a sequence of events came out in the inquiries. Several hours before the fire started, Captain Roger Wilmott was found dead in his cabin after complaining of stomach cramps. First Officer William F. Warms took over as acting captain for the rest of the trip from Havana to New York. At 3 A.M., a steward smelled smoke coming from the first-class writing room. He opened a storage locker full of linen, wood varnish, and cleaning chemicals. Blue flames burst from the locker, spread across the ceiling, and engulfed the entire room in seconds.[5]

Some pointed to negligent maintenance. Others suspected radio operator George Rogers set the fire deliberately to cover up his murder of the captain. Many crew members whispered that he was a psychopath with a criminal past. As panic swept through the ship, Rogers remained at his post as the flames engulfed the radio cabin, and he later enjoyed promoting himself as one of the heroes of the disaster. But twenty years later after the fire, Rogers was convicted of two unrelated murders. He died in prison.

When Theodore Ferris designed *Morro Castle,* he had built fire doors that would close automatically when a space reached a certain temperature, to seal off the burning area. But Ward Line management had ordered the tripwires disabled, probably to prevent the doors from closing accidentally. Even if the doors had been closed, there was still a six-inch gap between the combustible ceiling and the steel decking. Traveling through the gap, the flames bypassed the doors and set adjacent compartments ablaze.[6]

In keeping with regulations, Ferris had also put heat sensors in the cargo holds and cabins but not the public rooms. Acting captain William Warms saw no indicator lights flash on the bridge's fire detection panel when the flames burst out of the writing room's storage locker. By the time Warms found out about the fire, it was burning out of control, having spread to the two-deck-high lounge and engulfing the ship's main staircase and elevators.

Moreover, above the lounge sat the ship's Lyle gun, a device designed to fire a line and rescue bucket to another ship for an evacuation. Inexplicably, stored next to the Lyle gun was more than a hundred pounds of gunpowder. Its explosion had sent flames blasting high into the night sky, blowing out the entire center section of the ship. Wind blew through broken windows and portholes, whipping the flames into greater fury.

It was the exploding gunpowder, not the muffled fire alarms, that roused most of the ship's 318 passengers, many of whom had been drinking heavily to celebrate their last night at sea. The flames found more and more fuel: wood paneling, cotton sheets and linens, damask drapes. Fire zigzagged through the ventilation shafts that "sea cooled" every single stateroom. The power then failed, plunging the ship's smoke-filled corridors into darkness. Some panicking crew members made a run for the lifeboats not yet consumed by the inferno, leaving passengers stranded on the blazing vessel. The remaining crew tried valiantly to turn on the hydrants and spray water on the advancing flames. Following the standards of the time, Ferris estimated that only six of the ship's forty-two hydrants would ever be used at one time. When all were used at once, the trickling hoses had virtually no pressure in them and were useless.[7]

Two days after the burnt-out hulk beached itself, the *New York Times* reported that 183 passengers and crew were either dead or missing. Four hundred ninety-one survived. The death toll was later revised to 124, less than a tenth of *Titanic*'s. But the fire on the ship was front-page copy for weeks. On September 18, U.S. Steamboat Inspection chief Dickerson N. Hoover noted that "not a single fire door [was] closed on the ship," and that there was no evidence remaining to determine whether or not acting captain Warms had shut down the ship's ventilation system.[8]

Questions were also raised in the design community about the use of ornate, period furnishings aboard ships. *Marine Engineering*, the nation's premier naval architectural publication, showed a picture of the ship's beautiful wood-paneled main lounge. Although such a space was an "excellent example of the modern ship decorator's art," wrote the article, they were the kind of decorative schemes that "presented a fire hazard that must be eliminated" in new American ships.[9] Wood furniture and plush draperies, even if they created a luxurious ambience, were lethal if they ever caught fire aboard a passenger ship. The answer was to treat flammable fittings with fire-retardant chemicals and to eliminate wood from ship construction to the maximum extent possible.

The owners of the *Morro Castle* scrambled to defend themselves. "Nothing was overlooked by the Ward Line to prevent, as far as human foresight could, the happening of such a disaster," said Franklin Mooney, president of the Ward Line's parent company, Atlantic Gulf &

West Indies Steamship Lines. "We note with great interest the suggestion made that future construction should be of steel largely in place of wood," he suggested, "and this is already having our consideration."[10]

Franklin Mooney was also sweating from a Senate inquiry investigating the business practices not just of his own company, but of the entire American merchant marine. He might have hoped that the conviction of acting captain Warms and two officers for criminal negligence would have satisfied the public's cry for justice. Warms had refused to send out an SOS until it was too late, and had failed to order an evacuation until the fire had driven passengers away from the lifeboats and toward the stern. Worse still, the lifeboats that were launched were filled mostly with crew members. The hundreds of passengers marooned on board had two choices: stay on a burning ship or jump into the sea. Warms's conviction was eventually overturned, and within a couple of years, he was back at sea.[11]

Six months after it washed ashore, tugs yanked what was left of *Morro Castle* from the beach and towed it to the ship breakers in Baltimore.[12]

By 1935, Congress began looking into how government policy had contributed to the disaster. Ferris's ships, both the two coastal vessels for the Ward Line and the superliners planned for the United States Lines, were all paid for with government money. Critics now pounced on evidence that the shipping companies had fattened their profits with mail contracts and abused construction subsidies.[13] Even worse, the Ward Line managed to make a $263,000 profit on the disaster by insuring *Morro Castle* for more than her actual value, even as the company tapped the government for $500,000 in mail contract collections.[14]

Besides the misuse of government funds, the *Morro Castle* disaster also brought to light horrific working conditions aboard American ships. *Morro Castle* and her sister *Oriente* in particular had a reputation among sailors as "unhappy ships," driven hard and fast by their owners. Sailors who had crewed in them took the lessons of that experience far into the future of American shipping.

* * *

Only two years after being lionized for his superliner designs by the Society of Naval Architects and Marine Engineers, Theodore Ferris found that his name was forever tied to death on *Morro Castle*. Public outrage crushed him. By the time Congress wrapped up its investigation, his plans to build two superliners was dead, and his career was finished. He would never design another passenger ship.

"I put everything I had into the ship," Ferris told a journalist bitterly. "She had every known scheme of fire prevention and was adequately compartmented. A good, strong ship, broad of beam, and with everything modern. That took something out of me. Her total loss staggered me. I somehow felt as if I were to blame for the casualties." Then he added, "Maybe it was my fault. . . ."[15]

But as Ferris's reputation collapsed, William Francis Gibbs's name was rising anew. Obsessed with fire and fire safety since his childhood—when the family coachman would take the Gibbs boys into the night to watch lumberyard fires—Gibbs never forgot the *Morro Castle* disaster. Compared to a sinking or collision, fire at sea was profoundly more terrifying and lethal. One day, William Francis hoped to build a completely fireproof ship, one that protected her passengers from the danger of fire at sea, even if it meant that their accommodations would be less plush than those of his competitors.

After the *Morro Castle* disaster, William Francis decided to hedge his bets in the volatile passenger ship business and try his hand at something steadier. As the Roosevelt administration pumped more and more money into the rebuilding an outdated U.S. Navy, Gibbs slowly maneuvered himself to the center of the transformation. By the mid-1930s, his once-struggling firm was poised to become one of the military's most successful contractors. Gibbs also realized that the best way to get support for his superliner project was to win the support of the expanding Navy, not the fickle shipping companies, especially if a fast superliner could be a key military asset in a major future conflict.

FDR'S NEW NAVY

As the American shipping industry floundered in the early 1930s, European companies forged ahead. In 1932, Fascist Italy stunned the world by entering the transatlantic game with two new superliners: *Rex* and *Conte di Savoia*. The two ships would leave from Genoa, call at Mediterranean ports, and then sail the South Atlantic to New York. Both were 50,000 tons, both sported rakish, two-funneled profiles, and both brought Italianate luxury to the high seas. *Conte di Savoia* boasted a vaulted first-class lounge that would have not been out of place in Florence's Palazzo Borghese.

To build the liners, Il Duce Mussolini had forced the merger of Italy's three big shipping companies into one government-backed entity called the Italian Line. He was all but certain *Rex* would capture the Blue Riband on its maiden voyage, leaving Genoa in September 1932. But by the time *Rex* reached Gibraltar, the engines had broken down. When repairs were completed, most of her A-list celebrity passengers hopped the train to France, to catch the westbound *Europa*.

After a year of disappointing passenger loads, the Italian government gave *Rex*'s captain permission to burn twice as much fuel as normally allotted to get as much speed out of his ship's engines as possible. When *Rex* steamed into New York harbor, an exhausted Captain Tarabotto ordered the posting of a notice in the passenger areas: "Notwithstanding great part of crossing hindered by strong opposite winds and heavy fog, *Rex* beats all preceding records as to speed as well as to time spent in crossing Atlantic Ocean. . . . Such result entitles the *Rex* to the blue ribbon."[1]

Rex had made the 3,181-mile westbound trip from Gibraltar to Ambrose Lightship in 4 days, 13 hours, and 58 minutes at an average speed of 28.92 knots, beating *Europa*'s average speed by exactly one knot and a little over three hours.

On the captain's desk was a cablegram from Il Duce: "Good! Very good!"[2]

After *Rex*'s 1933 Blue Riband win, a British MP named Sir Harold Keates Hales commissioned an actual trophy—a baroque pile of silver, gold, and onyx—to be displayed aboard victorious vessels. Around the same time, the European lines agreed upon an official course for the Blue Riband: from Bishop Rock lighthouse off the southwest tip of England to Ambrose Lightship at the entrance to New York harbor. During the summer months, the contending vessels would use the northern track of 2,886 miles. During the winter, liners would take the slightly longer southern track of 2,958 miles to avoid icebergs. Because liners sailing from Europe to America had to fight the Gulf Stream and prevailing winds, the westbound average speed would be the official metric of the Blue Riband contest.[3]

Mussolini's ship enjoyed only a brief moment in the limelight. In 1935, *Rex* was completely outclassed by the new ship from France: *Normandie.* At 79,000 tons, she was the largest ship in the world, able to carry 828 in first class, 670 in tourist, 454 in third class, and a crew of 1,345.[4] Like *Leviathan* and *Bremen, Normandie* was designed with split uptakes for the ship's two working funnels. The resulting open space produced a first-class dining room stretching more than three hundred feet—longer, the French Line claimed, than the Hall of Mirrors at Versailles. Crowned by a coffered ceiling, the room could seat seven hundred diners at a time. On the promenade deck, three of the ship's first-class public rooms—the Grand Salon, the Smoking Room, and the Café-Grille—were linked by an unbroken line of grand staircases and arched openings. The lounge's walls were covered with painted Dupas glass panels and its chairs in red Aubusson floral upholstery. In addition to an indoor swimming pool, first-class passengers had the use of a winter garden with caged birds and exotic plants, and the first permanent floating cinema, complete with a modest stage and dressing rooms.

Under the command of Captain René Pugnet, *Normandie* set sail on May 29, 1935, from Le Havre—among her one thousand passengers was French president Albert François Lebrun. The new liner's prow cut through the ocean so smoothly that the bow barely kicked up a wave, but as she approached her cruising speed of 29 knots, passengers experienced a problem that had plagued the ship during her trials: bone-shaking vibration. While the first-class passengers lounged and danced in relative quiet, people occupying cramped tourist and third-class staterooms found it very hard to sleep.

The crowds that lined the Hudson by the thousands to greet what had been hailed as the most beautiful ship in the world knew nothing about the vibration. The crowd did know that *Normandie* had captured the Blue Riband from *Rex* on her first voyage, sailing the now-established course between Bishop Rock and Ambrose Lightship in 4 days, 3 hours, and 2 minutes at an average speed of 29.98 knots, a full knot faster than her Italian rival. Flying from her mainmast was a thirty-foot-long blue pennant symbolizing the speed prize.

When *Normandie* docked at Pier 88, crowds of spectators pushed and shoved to get a look at a world unimaginable to all but royalty and movie stars. For most of them, even a $91 tourist class stateroom (about $800 in 2012 terms) on *Normandie* was completely out of reach.[5] But despite Depression era poverty, unemployment, and bad news—or perhaps because of those things—the American public was ocean-liner crazy. On Broadway, Cole Porter's 1934 hit high-society musical *Anything Goes* had been set aboard a mythical transatlantic superliner named *American*. Expecting big crowds to greet *Normandie*, the French

Line charged people fifty cents for a tour of the first-class salons, still redolent with the aroma of French cigarettes, cognac, and perfume.

Among those lined up to pay to get on the ship was a nondescript man in a shabby black suit and floppy fedora. William Francis Gibbs, accompanied by his bespectacled, professorial chief electrical engineer, Norman Zippler, had come to check out the competition. When their group's guide looked away, the two men bolted for a door to the lower decks, which were strictly off-limits to visitors.

Of principal interest to them were the engine rooms, and *Normandie*'s four unique turbo-electric generators, the largest, most powerful ever placed in a commercial vessel. At cruising speed, the engines could deliver 160,000 horsepower to the four propellers, nearly 50 percent more power than the steam turbines of *Normandie*'s closest rivals.

After hours of playing cat-and-mouse with *Normandie*'s crew, an exhausted Zippler asked his boss if they could sit down. He was tired and afraid of getting caught and thrown off the ship by angry French Line officials.

Gibbs agreed, and they found a hidden corner.

"Take out your book and I will give you some dictation," Gibbs said quietly. For the next three and a half hours, the two men huddled in *Normandie*'s belly as Gibbs rattled off from memory all he had seen—gauge readings, measurements, and devices.[6]

Gibbs greatly admired *Normandie,* mainly because she was so revolutionary, blending advanced engineering with glamour, grace, and charisma. The first ship he ever designed, as a raw amateur working with his brother in 1916, had reached for many of the same elements: over 1,000 feet long, scooped-out hull, fine lines and a flared bow, and powered by turbo-electric engines. He was impressed with *Normandie*'s power plant. The steam generated by the boilers spun four turbo-alternators that generated 33,400 kilowatts of power to the four electric engines.[7]

But even before his escapade into *Normandie*'s belly, Gibbs had already concluded that American firms such as General Electric, Babcock & Wilcox, and Foster Wheeler could produce better engines than those on European ships. "The great power plants of the United States," he said, "were way ahead of the power plants of any great nation . . . this knowledge and experience of engineering that had been gained by these great concerns at tremendous cost could be adopted and made practical for use in a naval vessel."[8] However, the big American shipyards such as Newport News, New York Shipbuilding, and Bethlehem refused to use them. To save on costs, they preferred to purchase licenses from the Parsons Company of England—whose namesake had invented the steam turbine in the 1890s—and use these older designs in their commercial and naval vessels.

Gibbs was already well ahead of the curve. Five years before *Normandie* entered service, he had grasped a new chance to use American-designed turbines in American-designed ships, when the Grace Line approached him with a request that seemed too good to be true: to design four new, high-quality passenger-cargo steamers.

Grace Line operated steamships between New York, Caribbean ports, and the West Coast. They wanted four 10,000-gross-ton, combination passenger-cargo liners. Because of their long voyages, the ships had to combine comfort and speed—the classic passenger liner requirements—with economy of operation.

Before giving Gibbs & Cox the job, Grace Line president D. Stewart Inglehart asked Gibbs to make a presentation before the company's board of directors. When Gibbs appeared, the directors peppered him with questions until one of them stumped him.

"I don't know the answer to that one," he responded calmly. "I can find it out."

Inglehart later called on Gibbs, to tell him how the board had voted, and why. "We gave you the job because you didn't bluff on that question," Inglehart said. "It was a catch question. We had the answer to it. After you left that day we said, 'Here is a man we can rely on to tell the truth,' and that decided the issue."[9]

The four ships would be named *Santa Rosa, Santa Paula, Santa*

Elena, and *Santa Lucia,* and were to be ready in 1932. With *Malolo*'s profile in mind, Gibbs streamlined the superstructures, used two low *Bremen*-style smokestacks, flared the bow forward, and added a horizontal fin to the back of the first stack to deflect smoke away from the upper decks. Compared to the conservative style of Theodore Ferris, the Gibbs & Cox look was moving steadily toward the sculpted and sleek.

The four *Santa*s also introduced a new collaborator to the Gibbs & Cox design team, a sophisticated and confident New York interior designer named Dorothy Marckwald. Grace Line and Gibbs wanted a simple and elegant look that would appeal to the 250 all-first-class passengers. A successful, unmarried professional woman born into New York society, Marckwald understood what the Grace Line wanted and she delivered with restrained, colonial-inspired décor. The most spectacular public room was a columned, two-deck-high dining salon, painted in cream and pastel tones. On warm tropical nights, the ceiling could be retracted, allowing passengers to dine and dance under the stars.

One of Marckwald's great strengths was her insight into the special needs of shipboard design. She refused to follow trendy resort styles, whether from the "cottages" of the rich at Newport or the era's grand Adirondacks hunting lodges. "We knew the elk-horn style would soon be dated," she said. She also had an eye for color. "One thing we don't do on a ship," she once said, "is use color that is at all yellowish green. You know—anything that will remind a traveler of the condition of his stomach."[10]

Despite his guarded manner around women, Gibbs was pleased with what Marckwald produced and decided to employ her services for future contracts. He also respected her intelligence and her willingness to speak her mind. His confidence in her allowed him to focus on the area where Gibbs & Cox was making great strides, the engine room.

His treatment of the wife of a Grace Line executive, however, was a strange mixture of gallantry and misogyny. After a long business luncheon with the couple, William Francis remarked: "It's not possible. No woman in the world can keep her mouth shut that long." "I don't see why you just don't do exactly everything Mr. Gibbs suggests," she told her husband. "He's the most wonderful man in the world."

William Francis then sent her a dozen red roses and a note praising "the only woman in the world who can keep her mouth closed."[11]

Back at his office, Gibbs had carefully studied the innovative new engine design pioneered by German engineers during the late 1920s: high-pressure, high-temperature turbines. Gibbs believed enlarged versions of this improved turbine engine, built and designed by American manufacturers, would work nicely in the four small *Santa*s, and he found a shipyard that could do the work. The boilers could superheat the steam to 743 degrees and pressurize it to 375 pounds per square inch, giving the ships a service speed of about 20 knots. The compact American turbines would not just free up space for cargo, but also be more fuel efficient than the turbo-electric drive Ferris had used on *Morro Castle.*[12]

News of the well-designed *Santa*s had spread, and the U.S. Navy took notice. For the naval brass, the appearance of the Grace Line ships could not have been more perfectly timed. By 1933, as Germany installed a Nazi government and Japan marched into Manchuria, the need for a new American Navy had become increasing evident. The Navy was eager to expand and update its aging fleet. And in the White House was a president whose understanding of naval requirements could be counted on. Franklin Delano Roosevelt believed that, like dams and bridges, new naval vessels provided both a clear public benefit—national defense—and the jobs Americans so urgently needed.

Among "New Navy" advocates, there was no more fervent advocate for the abandonment of the Parsons design than the combative Vice Admiral Harold Bowen, the guiding light of the Navy's Bureau of Engineering. Early in his career, Bowen had experienced faulty naval technology when he was chief engineer of the battleship USS *Arizona,* built in 1916 with geared turbines based on the British model. Bowen was furious that the battleship could not make a single voyage without the engines breaking down. He also joked that the ship's engines "should have been put in a museum to show young engineers what turbines looked like before the development of the Curtis (American) turbine and before we knew how to cut reduction gears."[13]

For Bowen, it was almost criminal that American naval development was held hostage by conservative shipyards and their reluctance to risk profit margins to work with American suppliers. These Navy contractors were, in his words, "totally dependent upon Parsons, Ltd., for research in engineering, which at this time reflected the inbred conservatism of British engineering."[14]

Like William Francis Gibbs, Admiral Bowen thought that America's big industrial corporations could design and build much better propulsion systems for American naval vessels. General Electric and Westinghouse had developed turbines well suited for high-pressure steam, "characterized by their high speed, small size, and small number of parts . . . adaptable to the highest steam temperatures contemplated, [they] were very rugged, and free from distortion." When paired with the now-perfected double reduction gears, like those Gibbs & Cox used in the four *Santa* ships, Bowen believed, the resulting naval vessels could outrun and outcruise anything afloat.[15] When he saw the *Santa*s, he knew their designer offered something America needed. "These vessels," Bowen said, "embodied engineering practices far in advance of anything that had ever been attempted in the U.S. Navy."[16]

President Roosevelt's massive Navy rebuilding program became a reality. Bowen's friend Admiral Samuel M. Robinson, chief of the Navy's Bureau of Engineering, pushed for the use of the new engines, in a new class of destroyers. Going with tried-and-true Navy relationships, he selected United Dry Docks of Staten Island to design and build them. But the company balked at the immense and unfamiliar technical work involved. They approached Gibbs & Cox.

"United Drydocks came to us and asked us whether we would undertake the technical work, supervision of the technical work, for them if they obtained a destroyer to build," Gibbs recalled later. "We told them we were glad to do that on condition that the Navy would welcome an outside firm in the industry."[17]

Destroyers were among the leanest and meanest ships in the fleet. A typical destroyer was only about 350 feet long and of just 1,500 tons (one-thirtieth the gross tonnage of a big transatlantic liner). At sea, they protected battleships, cruisers, and troopships from enemy vessels, usually submarines. Because its job was to circle slower vessels to protect them, a destroyer needed a short turning radius, along with speed and maneuverability.

Naval work was unchartered waters for Gibbs & Cox. The firm had just lost their most valued naval advisor. Shortly after the completion of the Grace liners in 1932, Admiral Taylor was struck down by a stroke at the age of sixty-eight and rendered a complete invalid.[18] Gibbs was devastated at the loss of his surrogate father, and the greatest supporter of his yet-unrealized superliner project. "During our years of association and after his illness," he wrote, "he had endeared himself to us all." Gibbs saw in Taylor not only vast technical skill, but intuition, courtesy, and consideration for others. "My brother and I counted on him as our best friend and wisest counsel."[19] Taylor would die eight years later.

Gibbs had no choice but to carry on without his mentor's help. "I only knew one naval officer at the time, I think, Admiral Robinson, who was then Chief of the Bureau of Construction and Repair," Gibbs recalled. As it turned out, Robinson was just the right person to know, and his subordinate Admiral Bowen was eager to work with the Gibbs brothers because of their experience with the new engines. Robinson's decision meant that Gibbs & Cox was going to have a future.[20]

The final contract called for Gibbs & Cox to design a distinct new class of destroyers, to be named for Alfred Thayer Mahan, the father of American sea-power strategy. Sixteen vessels were to be built, a massive amount of work, and Gibbs & Cox's Depression-era cutbacks had left it short of staff. "Our office was small," Gibbs recalled; "the only thing we could do was to give our technical aid in guiding such projects."[21] The Navy agreed that its own staff would prepare basic plans of the ships, which Gibbs & Cox's technical staff would turn into "working plans" to guide construction at United Dry Docks.

At Admiral Bowen's invitation, Gibbs took a train to Washington to make a presentation on the principles of high-pressure, high-temperature steam at the Bureau of Construction, headed by Vice Admiral Emory S. Land. Gibbs talked for five hours about how the new propulsion system would shrink engine room space, improve fuel economy, and increase a destroyer's cruising radius.

Key to these advantages was the rapid revolution of the turbines made possible by the very hot, pressurized steam. As the revolutions increase, Gibbs pointed out, the size of the machine can come down.

If a large Parsons turbine is subjected to such pressure and temperature, Gibbs explained, "the whole thing expanded, causing the rotor blades to rub against the casing and the bearings to wear out." But the small, high-speed turbine Gibbs favored "has very little expansion. The result is that it is extraordinarily reliable."[22]

Gibbs almost certainly discussed his designs with President Roosevelt. Few written communications survive between the two men. But the commonalities between their lives and interests are remarkable, not least their love of ships. FDR was four years older than Gibbs, but both went to Harvard, both attended Columbia Law School, both were part of New York social and business circles, and both had an intense interest in naval architecture and maritime history. An avid sailor since his youth, the president was also a big reader of naval publications and a passionate model ship builder. At least one Gibbs & Cox employee, William Garzke Jr., remembered that there had been numerous telephone conversations between the two men. "I am positive that the two of them knew each other and talked about their interests in ships," he recalled. "It remained off the record for fear of favoritism."[23]

With Admiral Land's approval and perhaps a nudge from President Roosevelt, Gibbs set his staff to work on the *Mahan*-class destroyers. Like the Grace Line's *Santa* class, his engineers specified double-reduction gears—the first time such equipment was to be used in an American naval vessel. Gibbs also called for four boilers capable of funneling 650-degree steam at 700 pounds per square inch into three turbines. That would generate 49,000 horsepower, enough to give the 341-foot vessels a cruising speed of 36 knots.[24]

Bowen pored over the *Mahan*-class plans and concluded that "this machinery was the most rugged and reliable of any main drive installation ever installed in all respects." More good news followed in May 1935, when Vice Admiral Harold Bowen replaced Admiral Robinson as chief of the Navy's Bureau of Engineering and engineer in chief. Around the same time, Gibbs & Cox moved into an expanded suite of offices at

21 West Street, a nondescript skyscraper in lower Manhattan that faced the Hudson River.

When the first *Mahan*-class destroyer began sea trials, Chief of Naval Operations Admiral William D. Leahy found himself bombarded with crew complaints about the cramped engine compartments of the new, smaller turbines. An exasperated Leahy called the Navy chief constructor into his office and asked him what was going on with these new ships.

"Admiral," Bowen replied, "I'm not going to build an obsolete navy."[25]

Meanwhile, William Francis traveled between New York and the various shipyards, making sure that the destroyers were built to his precise specifications. As a new method of quality control, Gibbs & Cox built a model shop, where workers built large models according to the blueprints churned out by the design department. "No piece of equipment or member of the hull was installed until the drawings were approved by the appropriate Bureau of the Navy Department," Bowen noted after making the rounds at Gibbs & Cox. "It was much easier for the pipe shop to lay out its piping from the model than to construct it by laying templates in the actual vessel."[26]

For Gibbs, the model shop became the Holy of Holies at 21 West Street, guarded by a maze of security doors and locks. One of the rare non-Navy visitors, a bishop, wondered whether he would be let in. "You know how bishops talk," he said to the naval architect. Gibbs, a practicing Episcopalian, smiled and replied, "I have great confidence in your character, but even more in your abysmal ignorance in the things you will see."[27] He was proud of the veil he wrapped around the "radical technical developments" that were "without precedent" in naval history. "The people did not know what was going on," he bragged years later, "and it's doubtful if the layman now realizes what we finally came up with. Government leaders had no desire to tell them, and that was right in line with my expectations. . . . What a fortunate thing!"[28]

The destroyers exceeded everyone's expectations. After *Mahan*'s trials, United Dry Docks chairman of the board James H. Davidson wrote Gibbs an assessment of the vessel's performance: "Congratulations on your designing . . . I never before witnessed so smooth a performance as she gave us. Estimated HP 44,000, with five burners not in use and two nozzles cut out. I feel she will approach 40 knots if crowded."[29]

Following the successful completion of the *Mahan* class, the Navy made good on its word and offered Gibbs & Cox additional destroyer work, this time from the Bath Iron Works in Maine and Federal Shipyards in New Jersey. The next group of destroyers, the *Somers* and *Benham* classes, began entering service in 1937. This time Gibbs & Cox's engineers upped the steam temperature to 850 degrees and pressure to 600 pounds per square inch for both classes of vessels.

Still under attack from the conservatives in the Navy Department, Bowen decided to take on the skeptics on the Board of Inspection. Instructing the new commanding officer of the *Somers* before its first runs, he told him to "do everything he could think to bust up the machinery of his ship."

The commander tried, and afterward reported that it couldn't be done. As far as design went, the ships were "unbustable."[30]

Gibbs had pushed naval design where it had never gone before. By the late 1930s American naval vessels had surpassed the British in steaming radius, operating efficiency, speed, and fuel economy. No longer would U.S. ships have to rely on the old British Parsons engines. The new destroyers, Bowen wrote, finally gave America technological independence.[31]

For Gibbs & Cox, the success of the destroyers was important in two ways. First, in designing for the fastest ships in the Navy, the firm turned away from designs for individualized commercial construction and toward designs for multiple, mass-produced vessels. The move would pay unexpected private-sector benefits, spurring new design and construction techniques that increased the efficiency of commercial shipbuilding.

Second, the Navy contract put the firm of Gibbs & Cox on solid footing. Memories of the lean Depression years began to fade. William Francis found that he did not get along with his yacht-designing partner Daniel Cox, and so he bought out Cox's stake in the company. As president of a firm whose ranks would swell to 1,200 employees by the 1940s, Gibbs was established fully on his own.[32]

Gibbs savored the absolute control he now had. When Stalin's Soviet Union—finally recognized by the United States in 1933—sent a private business agent to court Gibbs' firm, the naval architect was willing to hear him out. Told that Moscow sought plans for a moderate-sized, combination battleship-destroyer, Gibbs presented the agent with plans for a colossus of a ship able to carry sixty planes on its flight deck. In the words of a journalist, the battleship "would dominate any naval engagement as easily as a shark dominates a school of mackerel."[33] The Soviets declined. "Gibbs is a very peculiar man, you see," the somewhat bewildered agent said. "I mean, he is like an artist. He likes his work."[34]

Gibbs & Cox also developed a reputation for being demanding but fair employers. The Gibbs brothers' Philadelphia childhood may have been rigidly defined by religion, education, and family background, but in an age when Jews and other minorities were excluded from all too many white-shoe firms, William Francis hired "anyone who qualified, regardless of race or national origin."[35]

As Gibbs & Cox began to grow and prosper, the firm was aware of the threat of war. In fact, as Gibbs saw it, "world conflict was inevitable." He knew that when it came, "an enormous expansion of shipbuilding, both naval and merchant, would be required."[36]

The prospect of war brought one of Gibbs's professional qualities front and center: his obsession with secrecy and security. From his own excursions to check out competitors' ships, he knew how much could be learned from even a short but expert inspection. Opening his firm's own work to other eyes was a mistake he would not allow Gibbs & Cox to make. Even the innocent public was not to know specifics. "It's doubtful if the layman now realizes what we finally came up with," he recalled years later. "Government leaders had no desire to tell them, and that was right in line with my expectations."[37]

Gibbs remained a meticulous and tough manager. Armed with reams of notes and his photographic memory, he would compile records of every meeting and phone conversation, along with his firm's research, into a series of bound volumes. When his advice was ignored and something went wrong, his so-called black booking made the lives of his critics miserable.

The strengthening relationship between Gibbs and the Navy elite began to raise some eyebrows. His abrasive management style and thirst for innovation upset established ways in a notoriously traditional military branch. But his ally, Admiral Bowen, couldn't have been happier with the routing of those conservative Navy men he called "mossbacks."

"The Navy isn't hiring an entertainer," Bowen said. "It's hiring a man for his brains."[38]

The shipyards, which had always resented Gibbs for insisting on strict compliance with contract terms, hated him even more for usurping the role of their in-house naval architects and ruining their old established contracts with Parsons. "Gibbs has a great way of getting what the owner wants," complained one executive at Newport News, "and then steering things around to what Willie wants."[39]

Despite the Navy work, the innovations, and his own growing personal wealth, William Francis Gibbs was still not getting what he really wanted: a crack at designing a superliner. If fitted with a high-pressure, high-temperature engine larger than the one in the destroyers, he reasoned, an American superliner would be unbeatable, a sea champion that would surely claim the Blue Riband.

Meanwhile, a new transatlantic liner from Great Britain had taken center stage in the public's imagination. This ship, subsidized by a massive government loan, would become a British national icon.

For Gibbs, who felt that most British naval architects were "condescending, supercilious bastards," the new Cunard flagship, *Queen Mary,* would be the ship to beat.[40]

THE *QUEEN* AND THE *AMERICA*

On May 27, 1936, all of England cheered for *Queen Mary* as she left Southampton on her maiden voyage. On the ship's bridge, a cluster of reporters circled Sir Edgar Britten, her first captain. "Are you going to try for the Blue Riband, Captain?" an American reporter asked.

"Well, naturally that's what we're out for," Britten replied flatly. "What did we build her for?"

The Cunard Line was eager to prove to the world that the *Queen* was indeed the fastest liner on the seas.

As a publicity stunt, British Olympic runner Lord Burghley sprinted around the four-hundred-yard enclosed promenade deck in only sixty seconds . . . while wearing evening dress and patent leather shoes.[1]

At 81,000 tons, *Queen Mary* was slightly larger than *Normandie.* Her conservative British designers had not given her any of the French ship's swoopy lines. Although her prow flared forward, she had no bulbous bow. Her hull was full, her superstructure stacked and ponderous. Ugly, square-mouthed ventilators clustered around her three stovepipe funnels.

Inside, the color scheme was a somber palette of deep reds, rich browns, and brass accents. Artisans from the Bromsgrove Guild adorned her public rooms and staterooms with tooled leather and acres of rare wood veneers from every colony in the Empire. Big alabaster urns, which glowed yellow at night, lined her first-class main lounge. Overstuffed sofas and chairs cluttered the Smoking Room and "Long Gallery." The Verandah Grille, the reservations-only nightclub overlooking the stern, boasted a sunken dance floor. Not everyone was pleased with the bizarre blend of Radio City Music Hall and Buckingham Palace. "The general effect is one of mild but expensive vulgarity," sniffed one design critic. "The workmanship is magnificent, the materials used splendid, the result is appalling."[2]

If *Queen Mary* lacked the French vessel's flair, her engine room packed a real wallop. John Brown's naval architects installed the most powerful geared turbines ever used on a commercial vessel, able to generate 158,000 horsepower. But because most of the design work was done in the late 1920s, the engines were not nearly as advanced as those Gibbs designed for the new American destroyers.

Gibbs excelled as a designer of naval vessels but he still craved the spotlight of the transatlantic passenger trade. He followed the construction of *Queen Mary* with intense interest, even if he disliked Cunard's stodgy design. Even more, he resented what he perceived as the British nation's sense of entitlement to maritime supremacy. He also guessed that *Queen Mary*'s great bulk would cause her to behave badly in rough seas.

Yet rather than building ships to match the new Cunarder in size and speed, American shipping companies wanted smaller, economical liners that could pay their own way and not depend on government money. The American public seemed to like it that way.

But *Queen Mary* would never have made it out of the builder's yard if it were not for a massive infusion of cash from the British government. In December 1931, only a year after the keel was laid, Cunard was hurting so badly that Chairman Sir Percy Bates called off construction on the ship known only as Hull 534. For three years, the hulk sat half-completed on the Clydebank slipway, its upper decks scarred with rust and festooned with bird nests.

Figure 1. 1733 Walnut Street, the Gibbs family home in Philadelphia. It was one of the grandest homes in the city, occupying the northeast corner of prestigious Rittenhouse Square.

Figure 2. William Warren Gibbs, father of William Francis and Frederic Gibbs. A self-made man from the small town of Hope, New Jersey, William Warren Gibbs was said to sit on more boards of directors than any other man in America. His son William Francis would later say that he would have never had amounted to anything if his father had not gone bankrupt.

Figure 3. The liner *St. Louis* slides down the ways at Cramp Shipyard, Philadelphia, on November 12, 1894. In the crowd was an eight-year-old William Francis Gibbs. "That was my first view of a great ship and from that day forward I dedicated my life to ships," Gibbs later recalled. "I have never regretted it." *The Mariners' Museum.*

Figure 4. William Francis Gibbs at Harvard, 1910. Gibbs disliked Harvard and its social world, and lived in constant fear that his peers would mock his hobby of redesigning British battleships. *Courtesy of the Harvard University Archives* (HUD 310.870).

Figure 5. The Cunard liner *Mauretania,* holder of the Blue Riband from 1907 to 1929, being nudged into dry dock. William Francis Gibbs and his brother, Frederic, traveled on this legendary ship's maiden voyage. Long, lean, and designed to be converted into naval auxiliaries, *Mauretania* and her sister *Lusitania* were the first express liners outfitted with steam turbine engines. *Corbis Images.*

Figure 6. The White Star Line's *Olympic*, built in 1911 and the sister to the ill-fated *Titanic*. Bigger and slower than *Lusitania* and *Mauretania*, these White Star liners were built for luxury and smoothness over raw speed. Converted to a troopship during World War I, she carried John Franklin and his tank battalion to the front in 1917. Known as "Old Reliable," *Olympic* would remain a popular liner until she was sold for scrap in 1935, following the merger of White Star and Cunard. *The Mariners' Museum.*

Figure 7. William Francis Gibbs in his study, circa 1915, showing his preliminary blueprints and sketches. *The Mariners' Museum.*

Figure 8. William Francis Gibbs's prototype superliner design, circa 1919. One thousand feet long, weighing in at 70,000 tons, and having an estimated service speed of 30 knots, if built this ship would have been the largest and fastest liner in the world. Compared to previous ships, with their tall funnels and stacked upper decks, Gibbs's design had a low, sleek profile that conveyed speed and power. This basic concept would evolve into the SS *United States*. *The Frank O. Braynard Collection.*

Figure 9. Admiral David W. Taylor, who launched William Francis Gibbs's naval architecture career and provided him the formal training he never received in college. *Photo courtesy of Naval Surface Warfare Center, Carderock Division.*

Figure 10. J. P. Morgan (center) arrives at the 1912 Pujo hearings in Washington, D.C., accompanied by his daughter Louisa (left) and son J. P. "Jack" Morgan Jr. Jack Morgan put up the financing for William Francis Gibbs's first ocean liner prototype. *The J. P. Morgan Library.*

Figure 11. Philip Franklin, president of the International Mercantile Marine as well as Jack Morgan's right-hand man in the shipping world. The split between him and William Francis Gibbs led to the founding of Gibbs Brothers Inc., later Gibbs & Cox. *Courtesy of Laura Franklin Dunn.*

John M. Franklin beside one of his tanks in World War I. His battalion was the only unit of American Heavy Tanks at the front and was longer in the line than any British battalion of Heavy Tanks had ever been.

Figure 12. Lieutenant John M. Franklin, son of Philip Franklin and future president of the United States Lines, standing by one of his battalion's tanks during World War I. *Courtesy of Laura Franklin Dunn.*

Figure 13. A young Vincent Astor, jauntily attired for mid-ocean promenading, aboard *Mauretania*, 1922. Deeply affected by the loss of his father in the *Titanic* disaster, Astor had a strong interest in naval design and used his vast wealth to make the United States Lines a powerful force in the shipping world. *Corbis Images.*

Figure 14. Rebuilding *Leviathan,* the former German imperial flagship *Vaterland,* from a battered troopship to luxury liner in Newport News, Virginia, 1922. His masterful management of the building of *Leviathan* into America's flagship luxury liner launched William Francis Gibbs's career as a naval architect. *The Frank O. Braynard Collection.*

Figure 15. A restored *Leviathan* enters the Boston Navy Yard dry dock, May 1923. *The Frank O. Braynard Collection.*

Figure 16. *Leviathan*'s first class smoking room, typical of the lavish period interiors of liners built before World War I. *The Frank O. Braynard Collection.*

Figure 17. Paul Cravath and his teenage daughter Vera, circa 1913. Her surprise marriage to William Francis Gibbs in 1927 shocked New York society. *Encore Editions.*

Figure 18. William Francis Gibbs (right, in his signature "Iron Hat" derby) supervising the construction of the Matson liner *Malolo* at Philadelphia's Cramp Shipyard, 1926. *The Mariners' Museum.*

Figure 19. Theodore Ferris, the "Dean of American Naval Architects" as well as William Francis Gibbs's archrival in the American superliner competition in the late 1920s. *The Mariners' Museum.*

Figure 20. *Bremen*, the revolutionary German liner that snatched the Blue Riband from *Mauretania* in 1929. The success of the sisters *Bremen* and *Europa* led to the construction of bigger, faster ships by the Italians, the French, and the British. Theodore Ferris came up with designs for two American superliners very similar-looking to these German ships, but they were never built. *The Mariners' Museum.*

Figure 21. The burning cruise ship *Morro Castle* aground off Asbury Park, New Jersey, with thousands of curious spectators crowding the beach, on September 8, 1934. One hundred thirty-four passengers and crew perished when the ship caught fire on the final leg of a trip between Havana and New York. The disaster ended Theodore Ferris's career and reinforced William Francis Gibbs's obsessive fear of fire. *The Mariners' Museum.*

Figure 22. The French liner *Normandie* in New York harbor. The most luxurious transatlantic liner ever built, she boasted a revolutionary, sleek hull design as well as turbo-electric propulsion. Although acclaimed as the most beautiful ship of the 1930s, *Normandie* rarely sailed more than half full. William Francis Gibbs and his assistant Norman Zippler conducted some industrial espionage aboard *Normandie* following her 1935 maiden arrival in New York. She lost the Blue Riband to *Queen Mary* in 1938. *The Mariners' Museum.*

Figure 23. The first class lounge aboard *Normandie*. Three decks high, the room boasted light fixtures by Lalique and painted glass panels by Dupas. The split funnel uptakes allowed for an unbroken view from the lounge to the smoking room and its sweeping grand staircase. Such luxury both awed and intimidated the public. *Normandie*'s cavernous public rooms proved hazardous, as they allowed for the rapid spread of fire. *Corbis Images.*

Figure 24. *Queen Mary* at full steam ahead, with a great wave cresting up from her bow. Compared to her rakish rival *Normandie,* her hull and superstructure were blocky, her interiors traditional. Yet she was the only big liner of the 1930s to earn a profit. During World War II, she and her running mate *Queen Elizabeth* were painted gray and converted into troop carriers able to carry fifteen thousand soldiers each. The Cunard flagship held the Blue Riband of the Atlantic from 1938 to 1952. *The Mariners' Museum.*

Figure 25. The first-class lounge aboard *Queen Mary,* decorated in a British interpretation of Art Deco. The *Queens* boasted wood paneling from every colony in the British Empire. Such spaces were criticized for their "mild but expensive vulgarity." *Corbis Images.*

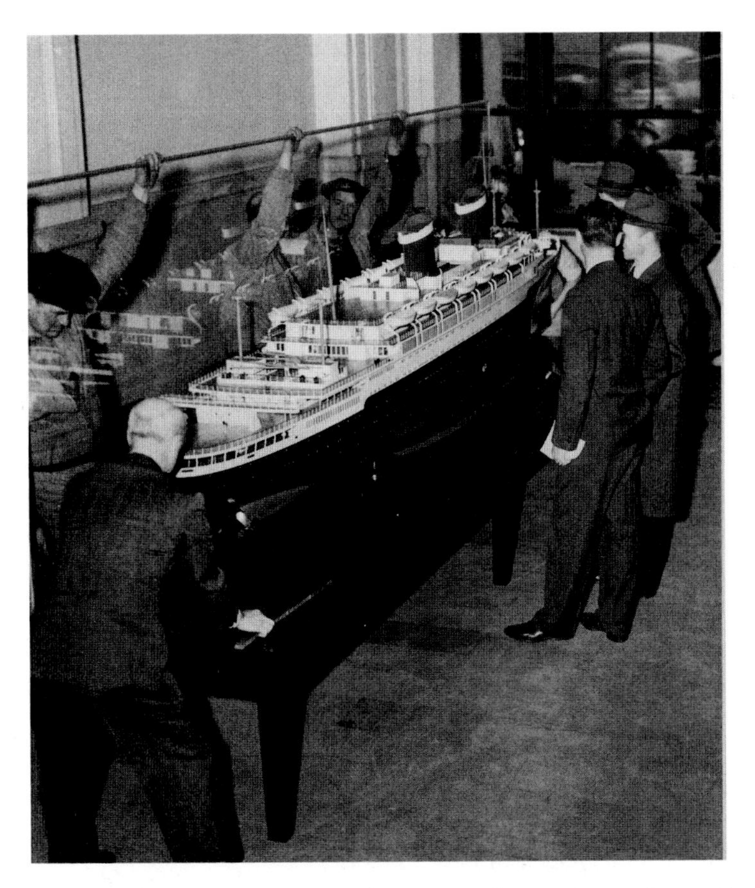

Figure 26. The delivery of a model of the United States Lines' new flagship, *America*, completed in 1940. Neither large nor fast enough to be in the same class as the European giants, this medium-sized liner was Gibbs's practice run for the much larger SS *United States*. *The Frank O. Braynard Collection.*

Figure 27. The Liberty ship *William A. Richardson*. These "ugly duckling" cargo steamers, with plans adapted for mass production by William Francis Gibbs, played a pivotal role in victory at sea. A total of 2,700 Liberties were produced during World War II. *The Frank O. Braynard Collection.*

Figure 28. The burned-out, capsized hulk of the troopship USS *Lafayette,* formerly the French Line's *Normandie,* on February 10, 1942. The fire cast a dark pall of smoke over all of Manhattan Island. She would be righted and then sold for scrap. *The Mariners' Museum.*

Figure 29. William Francis Gibbs, hailed as a "technological revolutionist," on the cover of *Time*, 1942. By then he had swapped his derby hat for a floppy brown fedora. Two years later, he would be exonerated of wartime profiteering charges. *Courtesy of Time Inc.*

Figure 30. William Francis Gibbs (left) and Frederic Gibbs in the "Glass Menagerie" at 21 West Street. *The Frank O. Braynard Collection.*

Figure 31. Pine models of Gibbs's superliner designers. The top and middle models are from the 1940s hull design for *United States*. The bottom model is of his S-129 prototype from 1916. *The Frank O. Braynard Collection.*

Figure 32. Brigadier General John M. Franklin returned to the presidency of the United States Lines in 1946, prepared to transform the company into a world-class passenger and freight company, crowned by the world's fastest and most modern superliner. *Courtesy of Laura Franklin Dunn.*

Figure 33. Elaine Scholley Kaplan, the only female engineer on the *United States* design team. A brilliant mathematician, Kaplan was in charge of designing the ship's propellers. *Courtesy of Susan Caccavale.*

Figure 34. Decorator Dorothy Marckwald aboard the Grace Lines' *Santa Rosa,* next to company president Lewis Lapham. Marckwald believed that the interior of *United States* should be crisp and modern, yet she found Gibbs's obsession with fireproofing to be trying at times. *Courtesy of Gordon Ghareeb.*

Figure 35. Laying the keel of *United States,* February 8, 1950. William Francis Gibbs is sixth from the right, standing along the railing. *The Mariners' Museum.*

Figure 36. Workers assembling the ship's double bottom. *The Frank O. Braynard Collection.*

Figure 37. Cars lined up against one of the two aluminum funnels, before it is hoisted onto the top deck of the ship. These were the two largest stacks ever placed on a passenger ship. *The Mariners' Museum.*

Figure 38. The installation of *United States'* top-secret rudder. *The Mariners' Museum.*

Figure 39. *United States'* two starboard propellers. The five-bladed configuration for the inboard propellers was meant to reduce cavitation and vibration at high speed. *The Mariners' Museum.*

Cunard's rival White Star was faring worse. IMM had repossessed the company after a scandal following its resale to British owners, and the American company received only a fraction of the $34 million still owed to them for the purchase. Saddled with a debt-ridden company that he had already tried to get rid of once, Philip Franklin panicked at the news of a potential merger between Cunard and White Star. He knew that a tie-up between the two would wipe out IMM's stake in the troubled British line. Franklin sailed to England to try to block the deal and prevent 534's completion.

He failed. In April 1934, the two rivals combined to form Cunard White Star Line Limited, with Cunard chairman Sir Percy Bates running the business. The British government swiftly loaned the new company £9 million to complete Hull 534 and start work on a running mate of comparable size and speed.[3]

Shortly before her launch in 1936, Bates approached King George V with the idea of naming the ship after the latter's grandmother, Queen Victoria. Traditionally, Cunard ships had names ending in *ia*.

"Your Majesty," Bates said respectfully, "the Cunard line is building the best, biggest, and speediest ship in the world, and requests your gracious permission to name her after the most illustrious and remarkable woman who has ever been Queen of England."

The king replied, "My wife will be delighted."[4]

And so the ship would be named not *Victoria,* but *Queen Mary.*

When the ship was launched, King George V addressed the two hundred thousand spectators and millions more who listened on the radio around the world. "It has been this nation's will that she should be completed," he declared, "and today we can send her forth, no longer a number on the books, but a ship with a name in the world, alive with beauty, energy and strength."[5]

His wife released the champagne against the bow, and with a roar, the great ship slid into the Clyde River.

Tugs then towed *Queen Mary* to the fitting-out basin where she would be, in the words of the king, transformed into the "stateliest now in being."

* * *

As *Queen Mary* approached New York harbor on her 1936 maiden voyage, planes bearing photographers and newsreel cameramen swooped around her. The cameramen got dramatic shots of the enormous wave kicked up by her prow.

But *Queen Mary* failed to take the Blue Riband away from *Normandie* on her first voyage. It was not until August 1936 that Sir Edgar Britten pushed his ship to what he thought was the limit, taking the record with an average speed of 30.14 knots and a westbound sailing time of 4 days and 27 minutes.

His triumph was the first in a two-year back-and-forth battle between the *Queen* and *Normandie.*

Finally, in 1938, *Queen Mary* raced across the Atlantic in 3 days, 21 hours, and 48 minutes, beating *Normandie*'s 1937 time by an hour and 19 minutes. She averaged 31.69 knots.

And there the Blue Riband stayed.

Queen Mary beat *Normandie* in more than just speed. Of all the big liners on the Atlantic in the 1930s, the British *Queen* was the only one to make her owners a profit, and established itself as the A-list ship for celebrities, who found her atmosphere much more welcoming than the more opulent French liner. Rich Americans appreciated Cunard service, redolent of both a grand hotel and a British country house. She also appealed to ordinary travelers, who found her tourist- and third-class cabins vastly superior to those aboard *Normandie.*[6] The British public, most of who never sailed on a liner, saw her as a trophy winner that vanquished foreign rivals. Despite complaints about her "Odeon cinema" décor, *Queen Mary* seemed to be a ship with a soul. Grand and fast as she was, she was cozier and statelier than the other European giants, whose nationalistic opulence could border on intimidating. While other superliners glittered, *Queen Mary* glowed. One could feel at ease aboard the *Queen.* Her regal name gave this grand machine a true feminine quality and linked her firmly to the nation whose flag she flew.

But not everything about *Queen Mary* was admired, especially her tendency to roll. Cunard had hoped that *Queen Mary*'s size would dampen her motion, and chose not to bolt down heavy furniture or install handrails in the corridors. It was a mistake. During one storm, with the liner tilted almost 45 degrees, a piano broke from its moorings in the tourist-class lounge. In the words of one crew member, "after two or three days of this the piano was reduced to its iron frame plus strings, and as it cartwheeled its way around the devastated room it uttered the most weird cacophony of noises."[7]

Cunard eventually stripped *Queen Mary*'s interiors and found out that her engine machinery was too heavy and her superstructure too high, making her a bottom-heavy, or "cranky," ship. The shipyard fixed the problem, at least partially, by placing ballast and water tanks higher up in the ship's hull.[8]

William Francis followed *Queen Mary*'s problems with an odd sense of delight, saving newspaper articles that mentioned the ambulances greeting the Cunard flagship at her Manhattan pier to pick up passengers who had lost their footing and broken an arm or a leg. His designer mind wondered: how would a big ocean liner behave in rough seas if powered by the smaller, lighter turbines now employed in his newest destroyers?

The competition between the British and French liners captured the headlines, but they were not the only vessels on the seas. Both German and Italian liners held regular schedules throughout the 1930s to New York, although their visits were becoming increasingly controversial as fascism in both countries intensified. Italy's *Rex* and *Conte di Savoia* sailed under the flag of Mussolini's regime. By the mid-1930s, the specter of Nazism loomed over Germany's *Bremen* and *Europa,* to the point where Americans, Jew and Gentile alike, were unnerved by her swastika-draped salons. (Kosher kitchens were shrunk, then shut down completely.) On July 27, 1935, a group of anti-Nazi rioters stormed aboard *Europa* as she sat docked at her Manhattan pier and cut the ropes to the swastika flag. When the demonstrators tried to stomp on it, protestors and crew members came to blows. The New York police came aboard and broke up the fracas.[9] After the incident, an enraged Adolf Hitler made clear that the swastika flag would never be dishonored again: he decreed it the official flag of Germany.

* * *

As the European ships made headlines with speed records and anti-Nazi riots, the United States Lines still struggled to survive, even though it was now backed by Vincent Astor's vast wealth. The company was chafing under an aggravating provision in their government contract: the aging liner *Leviathan* would have to remain in Atlantic service until 1936, making a minimum of five round trips per year. Her revenues were the worst of all the big liners—she was averaging only one-quarter full—even though Congress had repealed Prohibition and liquor could once again flow on American-flagged ships.[10] Vincent Astor and the Franklins decided they had to take a drastic step to avoid being financially sunk by a seagoing dinosaur.

In 1933, IMM simply defied the terms of the contract and tied up the big liner at her New York pier, where she remained all year. Government pressure led to renewed service the following year, but *Leviathan* would make only five more crossings, losing half a million dollars, before IMM laid her up again.

The United States Lines was not the only company disposing of its old ships. To complete *Queen Mary,* Cunard White Star had to rid itself of many of its finest liners, most of which were built before World War I. After having been reduced to making budget booze cruises to the Bahamas for most of the early 1930s, the rust-streaked *Mauretania* sailed to the Scottish scrappers in 1935. President Franklin Roosevelt, a longtime admirer of the Cunard speedster, lamented that "the steel of her is being recast into shells, guns and other machines of destruction to human life."[11] Other prewar liners such as White Star's *Olympic—Titanic*'s sister ship—and Cunard's *Berengaria* quickly followed her to the breakers. White Star's *Majestic* was given a brief reprieve when British Admiralty converted her into a training ship, but she was destroyed by fire shortly after her conversion.

As the older European ships disappeared, IMM president Philip Franklin had had enough with the superliner business. Paying the Shipping Board half a million dollars to be released from the *Leviathan* obligations, he ordered the ship out of service. In 1935, tugs towed the dead ship over to the same Hoboken pier where she had rusted after World War I. *Leviathan* would never carry another paying passenger.

But the breached contract drew the political scandalmongers, who charged that the *Leviathan* retirement was a corrupt bargain among cronies, reaching all the way to the president of the United States. Assistant Secretary of Commerce Ewing Y. Mitchell called the half-million-dollar deal "an example of official corruption," a scandal that did not go away when Roosevelt axed him. Influential Washington journalist Drew Pearson, in his column "Washington Merry-Go-Round," wrote that "the friendship between Franklin Roosevelt, his cousin Kermit, and Astor—the latter two chief owners of I.M.M.—the trips on the Astor yacht *Nourmahal* and the laying up of the *Leviathan* by I.M.M.—with Roosevelt's permission—undoubtedly will figure in the next campaign."[12]

Pearson was wrong. FDR, overseeing a recovering economy, easily won the 1936 election, and the Franklins and Vincent Astor tightened their grip on the United States Lines. Yet the shipping industry had not escaped public scrutiny and wrath. Investigations after the *Morro Castle* disaster, only two years earlier, opened a window on the industry-government nexus. Critics attacked the U.S. Shipping Board and the 1928 Jones-White Act, which threw tax dollars in the form of mail subsidies and construction loans to American-based shipping operations. "A saturnalia of waste," Senator Hugo Black of Alabama found after a congressional inquiry. "Inefficiency, unearned exorbitant salaries and bonuses and other forms of 'compensation,' corrupting expense accounts, exploitation of the public by the sale and manipulation of stocks . . . the 'values' of which are largely based on the hope of profit from robbing the taxpayer."[13]

The result of it all was the Merchant Marine Act of 1936, which did away with the old Shipping Board and government subsidies. In their place came a new U.S. Maritime Commission and new, targeted subsidies. Roosevelt's government would still be supporting U.S. shipbuilding and U.S.-flagged ships, both for the union jobs they could create and the merchant marine fleet's potential to meet military needs in time of war.

Roosevelt appointed campaign-contributor Joseph P. Kennedy the commission's first chairman. Kennedy was used to spearheading such government operations. As the first chairman of the Securities and Exchange Commission in 1933, Kennedy had vigorously prosecuted Wall Street malefactors after the 1929 crash (after having been one the worst, of course, before his appointment). His new job was to make sure that shipping companies no longer helped themselves to taxpayer money. Among those who would work with Kennedy was Admiral Land, chief of the Navy's Bureau of Construction, the powerful position once held by William Francis Gibbs's first patron, Admiral David W. Taylor.

One of the Maritime Commission's first actions was to take a thumbs-down position on the building of a superliner. "This type of vessel," its report read, "is believed to be economically unsound. . . . The building of these vessels, at the expense of other, more economical ships, cannot be justified by the United States." The future of American passenger liners, the commission argued, lay in "fireproof, vibrationless, attractive, and economical vessels of reasonable size and speed."[14]

At IMM, John Franklin—who moved up to replace his retiring father Philip as president in 1937—did not disagree. Because *Leviathan,* IMM's largest U.S.-flagged ship, ended up a money loser, Franklin was loath to let his United States Lines build another superliner. Indeed, many Americans felt that the nation should leave the superliner race to the Europeans. One newspaper complained that *Leviathan* and ships like her were "a sample of the waste of taxpayers' money" and that the merchant marine should "rest on its own keel and rent such ships if they are needed in wartime."[15]

Thus, with the Maritime Commission's report in mind, the United States Lines went in another direction: formally requesting a construction subsidy for a moderate-sized ship as a replacement for *Leviathan.* Although not in the same size league as *Queen Mary* or *Normandie,* the proposed American flagship would be one of the most modern and technologically advanced ships in the world. But the United States Lines needed the government's help, and she would have to be more practical and cost-efficient than the European giants.

John Franklin called on William Francis Gibbs to design the new medium-sized liner, known in shipping jargon as a "cabin liner." Although increasingly busy with his work for the U.S. Navy, Gibbs threw himself into the ship's design. Gibbs & Cox created plans for a liner 725 feet long, 95 feet wide, and of 33,000 gross tons. Judging by the hull shape—flared bow with a slight bulb, a concave forward section, and overhanging spoon stern—it appears that Gibbs put his notes from his secret exploration of the *Normandie* to good use. Above, two low, raked, tear-shaped "sampan" funnels, both with fins to deflect smoke from the passenger decks, crowned her sweeping superstructure. The ship would have a patriotic name: *America.*

Incorporated into *America* were features Gibbs was building into his destroyers. Steam, superheated to 850 degrees, would power two steam turbines, putting out a total of 20,000 horsepower. The vessel could cruise comfortably at the economical speed of 22 knots. If pushed, she might reach 25. Her passenger capacity was 1,438 in three classes. Unlike big liners of the past, *America* also had no expansion joints in her superstructure—the ship's relatively short length, the hull engineers calculated, eliminated the need for that risky design feature that allowed a ship to flex slightly in heavy seas.[16]

Navy secretary Claude Swanson, happy about *America*'s design, said the liner was "an extremely valuable adjunct to the Navy in time of national emergency."[17] But the U.S. Maritime Commission found itself torn between the Gibbs & Cox design and one submitted by the naval architects at the Newport News Shipbuilding and Dry Dock Company. The shipyard in its presentation put forth its design ideas, and doubtless also argued that the United States Lines—and therefore the government funders—could save money by keeping all the designing in-house.

William Francis and his assistant Matthew Forrest attended to hear the shipyard's presentation. When it ended, Gibbs asked for a long lunch break and went back to his hotel room with Forrest. On the spot, just as he had covertly copied data from *Normandie,* Gibbs worked into the Gibbs & Cox plans every possible advantage of the shipyard's design. When the meeting at the commission resumed at 3 P.M., Gibbs unveiled a set of reworked plans.[18]

With support from John Franklin, Gibbs & Cox got the design contract, and Newport News got the contract to build the ship. It is unclear how happy the shipyard was about that result. Newport News management remembered how the perfectionist naval architect had destroyed shipyard profit margins in the past.

Workers at the yard laid *America*'s first keel plates in 1938. Gibbs & Cox once again retained decorator Dorothy Marckwald and her assistant Anne Urquhart to design the interior. Marckwald created a total of twenty-three public rooms for three classes, all designed with an eye for restrained luxury. The first-class ballroom made use of "light tones of wood and fabric with metal and metal leaf ornamentation."[19] Artwork for the public areas would be made out of lightweight metals and gesso. Among the most fetching pieces was the mural in the bar, depicting stylized Art Deco scenes of shipboard life, complete with white-jacketed stewards, bartenders holding cocktail shakers, and elegant passengers in evening dress.

Gibbs had been fanatical about fire prevention since childhood, and the *Morro Castle* disaster had only reinforced his obsession. In his designs for *America,* he partitioned the interiors not with wood, but with a new, fireproof particleboard called Marinite. To test the product, a sample stateroom was built, moved outdoors, filled with wood, and set on fire. The only effect, according one press report, was "slight warping and scorching of the bulkheads." The wood was gone, but the rest of the unit was left completely intact.[20] The secret to Marinite's fire resistance was simple: it was infused with the miracle material known as asbestos. In addition, *America*'s fire safety system was completely automated, with the vessel also divided into several fire zones with fireproof doors. Unlike *Morro Castle,* there were no gaps in the bulkheads and decks that allowed fire to bypass closed fire doors and jump from one room to another.

As *America* was being born, another great ship was being put to rest. "As the maintenance of this ship is a useless burden and its continuous maintenance is of no value to the company," a 1937 IMM report read, "it is our desire to immediately dispose of the *Leviathan* by selling her for scrap."[21]

On January 25, 1938, tugs pulled the dirt-smeared *Leviathan* away from her Hoboken pier. Workers had already sliced ten feet from her three stacks and cut down her two masts—she had to pass underneath the Firth of Forth Bridge on her way to the breaking yards of Rosyth, Scotland.

Just before her final departure, *Leviathan* received one last well-wisher. A feeble, aged Philip Franklin stood on a tugboat that drifted around *Leviathan*'s stern. His son John Franklin, new president of the United States Lines, stood atop their headquarters at One Broadway, glad to be ridding his company of this "damn wreck."[22] The ship was dark except for her mast and running and bridge lights. A single tugboat bearing Philip Franklin followed *Leviathan* all the way to the Battery before turning back.

Sometime after the Scottish wreckers started to rip *Leviathan* apart, William Francis Gibbs pasted one last article into the scrapbook he had started nearly two decades earlier. "Now they have ripped out her fittings—the tapestries which the craftsmen of Gobelins took two years to weave," the piece read, "the paneling made especially for her adornment by a famous artist. Aching, scarred, her gutted timbers gape at a pitiless heaven."[23]

As *Leviathan* was reduced to scrap and *Normandie* and *Queen Mary* battled for the Blue Riband, propeller-driven airplanes were making their first scheduled flights across the Atlantic. Even Harry Manning, now commodore of the United States Lines, bought himself an airplane. He and millions of others read about Charles Lindbergh's solo Atlantic flight in 1927, and Manning began to wonder if someday, sea travel and the Blue Riband would be part of history.

Soon after he became president of the United States Lines, John Franklin was appointed a director of the fledgling Pan Am Airlines. Its president, Juan Trippe, asked all new directors to make a transatlantic trip aboard one of the new Clipper planes to Europe. Franklin was terrified by the prospect. "The airplanes in those days didn't have such impressive safety statistics then," he recalled.[24] He made the flight, and then decided to stay in the shipping business.

Airships seemed to open a new, safe door to passenger air travel. On October 11, 1928, a 776-foot-long dirigible, *Graf Zeppelin,* left Friedrichshafen, Germany, for the United States. The airship had space for twenty passengers, housed in jewel-box-like living quarters (known as a gondola) slung beneath its belly. Four and a half days later, *Graf Zeppelin* touched down at Lakehurst, New Jersey, creating a sensation. A bigger, faster, more luxurious version of the *Graf,* the *Hindenburg,* was commissioned in 1936. Able to carry about sixty passengers, it was still travel only for the superwealthy. But if costs came down and more passenger space were added, airships might be able to compete more directly with ocean liners. That dream ended when the hydrogen-filled *Hindenburg* exploded over Lakehurst on May 6, 1937.

Other wildly popular aviators followed Lindbergh. One of them was Amelia Earhart, who, after a celebrated transatlantic flight in 1928, sailed back to America aboard the United States Lines' *President Roosevelt,* commanded by Captain Harry Manning. En route Earhart and Manning became friends. The sailor's interest in flying and his knowledge of navigation kept them bonded. Nine years after they met, Manning was Earhart's first choice to be the navigator on her attempt to be the first person—man or woman—to circumnavigate the globe along the equator.

Manning left his ship to join her and they took off from California in March 1937. But on the second leg of the flight, Earhart's Lockheed L10 Electra blew a tire and crashed on takeoff from Honolulu. The plane had to be shipped back to California for repairs. During the wait, Manning resumed command of his ship.[25]

When Earhart took off again, on June 1, it was not Manning but his replacement, navigator Fred Noonan, who accompanied her. A month later, July 2, their flight famously vanished over the western Pacific. Its fate is still a mystery.

Still, despite his fascination with flying, Manning believed that the safest and best way to cross the Atlantic was by ship. He, like most people, did not see the propeller-driven airplane as any threat to the supremacy of the huge transatlantic liners. Naval architecture was on the cutting edge of technology—and its greatest advances were still to come. Or so Manning thought.

On August 31, 1939, thousands of reporters and spectators packed the shipyard of Newport News for the christening of *America*. Amid great fanfare, First Lady Eleanor Roosevelt smashed the bottle of champagne against her bow and the new flagship of the American merchant marine slid down the ways. A reporter wrote: "Officials looked everywhere among the gathered celebrities for Gibbs. He had vanished. They finally spotted him. Bored by speechmaking, he was perched like a bald eagle at the top of a scaffolding to get a better view of the ship when she hit the water."[26]

The following day, September 1, the Nazis invaded Poland and World War II began. The United States would not be part of it for more than two years. But *America*'s maiden voyage to Europe was canceled by the passing of the Neutrality Act, which forbade American liners from engaging in transatlantic trade.

John Franklin's new flagship had cost his company and the government more than $17 million. He could not afford to simply tie her up. To skirt the Neutrality Act, Franklin thought about transferring all ships of the United States Lines to Panamanian registry. So did other shipping companies in international trade.

The possibility of the move sparked outrage from the National Maritime Union. Its new president was Big Joe Curran, a sailor who had been galvanized into union activity at the time of the *Morro Castle* disaster. He was a born organizer and leader. As a young man, Curran claimed to have a recurring dream that he later recalled for a *Time* magazine reporter: "Little Joe—now Big Joe—waved his hairy paw, whereupon the great ports of New York, Boston, Marseilles, San Francisco, and Antwerp were paralyzed. All over the world shipping was paralyzed. Then the President of the U.S. called Joe and said: 'Joe, you have paralyzed the world by a wave of your hand. What do you want?' Said Joe (in the dream): 'More pork chops.' "[27]

As a young sailor, the square-jawed Curran despised the rich men who controlled America's shipping industry. To Curran, they were oppressors of the workingman. The owners of the shipping lines drove their crews like slaves, and skimped on safety and comfort all in the name of profit. Described as a "loudmouthed, naive and slightly bewildered young man with a genius for organization," Curran was one of many vocal seamen who complained about the long hours, abysmal pay, and horrible food on American-flagged vessels.[28]

When asked about the *Morro Castle* disaster, Curran blamed it on a system that overworked and poorly trained the crew. "No wonder the *Morro Castle* was a mess," Joe Curran said. "They just lined us all up on the pier, glanced at our papers, and said 'You, you, and you.' No questioning, nothing."[29]

The disaster galvanized Curran to become a leader in the fledging National Maritime Union. With the cooperation of a labor-friendly White House, Curran hoped that the new union would bring better pay, benefits, and hours to American sailors.

The last thing Curran wanted was *America* to be operated like the ill-fated *Morro Castle,* or for American shipping companies to switch to flags of convenience. He gave a fiery speech to three thousand of his followers at the New York union hall. "Even if [the shippers] . . . do change flags and their ships are sunk they will squawk for American protection," he shouted as he shook his fist in the air. "Flag-swapping will put 10,000 seamen out of work." He then threatened to show Roosevelt and Admiral Land of the Maritime Commission that his men meant business. "If they put alien seamen on these ships we are going to take 10,000 sailors onto the White House porch, and picket the Maritime Commission . . . until they drop."[30] Many in Congress also dismissed the move to Panamanian registry as a "subterfuge" to rid American ships of expensive American crews.

John Franklin responded by saying that the move was not a matter of cheating American sailors, but an act of desperation to keep his

American company in business. "If we go out of business now," Franklin asserted, "we can't expect to come back later and say, 'Gentlemen, we're sorry, but we've been out to lunch.' "[31]

For Big Joe Curran there was no let-up in his demand for "more pork chops," or for his indignant, near hatred of ship owners. But Curran deeply admired John Franklin, the no-nonsense Harvard dropout. According to Franklin's daughter Laura, the two would frequently meet at John Franklin's Upper East Side apartment to settle disputes over a scotch and soda.[32] Here, in private, union boss Curran was "Joe," and John Franklin was "Jack."

Even though she was barred from the North Atlantic run, Franklin's new flagship *America* met with near-universal praise. But John Franklin's father Philip, who had hired William Francis Gibbs to build a thousand-foot superliner back in 1916, did not live to see *America*'s launching. The former president of IMM died on August 14, 1939, at his family's Maryland estate.

In the end, United States Lines vessels stayed under the American flag. *America* was allowed to make a few cruises to the Carribean during the 1940 season, while the United States was still at peace, and other United States Lines ships sailed emergency runs to Europe to bring American nationals home.

By then the North Atlantic had once again become a killing field. The ocean liners of all nations were now key wartime assets, as well as valuable targets.

During World War II, William Francis Gibbs's organizational genius would help change the course of the conflict, as well as earn him a level of respect he never enjoyed in peacetime. Roosevelt's secretary of the Navy, Charles Edison (son of the inventor), would later say: "If I ever made a contribution to the Navy, it was keeping William Francis Gibbs in the picture."[33]

And America's wartime industrial machine—the Arsenal of Democracy—would finally make Gibbs's dream of a great American superliner a reality.

A FULL MEASURE OF TOIL

From the first days of the war, the world's great peacetime liners had been gradually disappearing, whether into wartime service or destruction.

Two days before Hitler invaded Poland in 1939, the Norddeutscher Lloyd flagship *Bremen* was docked in New York alongside the *Queen Mary, Ile de France, Aquitania,* and *Normandie.* Rather than risk seizure by the United States, Berlin ordered *Bremen*'s Captain Ahrens to leave New York without passengers and sail at full speed back to Germany.

As the blacked-out *Bremen* slipped out of New York under cover of darkness on August 30, 1939, nine hundred crewmen lining her decks gave the Nazi salute to the Statue of Liberty.[1] Should the valuable potential war prize be intercepted, the German high command ordered her captain to set her on fire. Painted battle gray and strapped from stem to stern with fuel drums, *Bremen* dashed across the Atlantic at 30 knots, her top speed, calling first at the Soviet port of Murmansk and then arriving at Bremerhaven, docking near her sister *Europa,* which had been in port when war broke out.

Hitler planned on using both vessels as troopships in his invasion of Great Britain. But in March 1941, Bremerhaven's citizens awoke to see a towering blaze roaring on the waterfront. One of the ship's officers had boxed the ears of an insubordinate cabin boy. Enraged, the young man set *Bremen* on fire. The gutted, twisted *Bremen* heeled against the dock and smoldered for days.[2] The wreckers then fed the ship's remains into the Nazi munitions factories. *Europa* continued to sit neglected at her berth until the end of the war, although two huge doors were cut in her side should she be needed to transport troops for an invasion of Great Britain.

The British liners docked in New York in September 1939 stayed in their relative position of safety in the then-neutral United States, their masters waiting for the British Admiralty to assign them war duties. The biggest one, *Queen Mary*'s new running mate, was born in the crucible of conflict. On March 3, 1940, a gray-painted *Queen Elizabeth* slipped out of her fitting-out berth at Clydebank, Scotland. Her cavernous interiors were bare, her new furniture and artwork left in storage. Her skeleton crew of four hundred officers and men thought that they were bound for Southampton for sea trials. At 83,673 gross tons, 1,031 feet long, and 118 feet wide, *Queen Elizabeth* was the largest passenger ship in the world and an easy target for Nazi bombers. As his ship cleared the harbor, Captain John Townley opened a sealed letter with his instructions: set a course to New York City, at full speed.

To divert German intelligence, the British leaked a rumor that *Queen Elizabeth* would dock in Southampton the next day. Sure enough the Luftwaffe made a sweep over the city the day of her presumed arrival. Southampton went up in flames, but *Queen Elizabeth* was already several hundred miles out to sea. Run at her top speed of nearly 30 knots, black smoke streaming from her two stovepipe funnels, *Queen Elizabeth* successfully evaded German submarines and bombers. After a harrowing five-day voyage, *Queen Elizabeth* arrived in New York City on March 7, 1940. There were no cheering crowds or gushing fireboats to meet the new Cunard flagship. Tugs nudged the massive liner to her berth at Pier 90, where she and *Queen Mary,* also painted gray, would await war duty.

Also taking refuge in New York was the great French liner *Normandie* and her smaller running mate, *Ile de France.* But on June 25, 1940, France fell to the Nazis. Free French ally Britain swiftly took command of *Ile de France.* The liner prepared for war duty in British service, while *Normandie* remained marooned at Pier 88.

Although burdened with work for the Navy, William Francis Gibbs and his firm made time to help supervise the reconditioning of *Ile de France* as a troopship before she departed for service in Asia from Todd Shipyards on Staten Island. He found that by American standards, the glamorous *Ile de France* was a horrific firetrap. In the words of one official, to refit *Ile de France* and other French liners for trooping duties, "bold stripping and redesigning was the only course open to making the ship safe," and "no firm in Britain could have done the work in less time."[3]

Within a few months, both *Queens* departed New York for wartime trooping duties in the Pacific and Indian Oceans. Designed to carry 2,500 passengers in peacetime, the two Cunard superliners were each refitted to carry up to 16,000 soldiers per trip. The liners traveled alone because they were too fast for escorts to keep up with them. Their only defense against bombers was antiaircraft guns and other light armament mounted on their upper decks. Their only defense against U-boats was their 30-knot speed. This was too fast for a German U-boat commander either to set his torpedo sights on or intercept, because his submarine could only make about 10 knots.

By this time, only a few American ships were braving the North Atlantic passenger routes. A dark wartime pall had fallen over Manhattan even though the United States remained neutral—the gray-painted passenger liners at the piers, the news about German atrocities against the Jews and Poles, and the streams of refugees arriving every week were sinister signs that this was a war that America probably could not avoid.

As the war escalated, even clearly marked American ships found themselves on U-boat targets.

In the early morning of July 11, 1940, the United States Lines flagship *Washington* was stuffed to the gunnels with more than one thousand people fleeing the Nazi onslaught. After only six months of fighting, the Nazis now occupied or controlled most of Europe. France had fallen two weeks earlier. Russia and Germany had signed a nonaggression pact and carved up Eastern Europe. In preparation for a future land invasion, Hitler ordered the Luftwaffe to gear up for a vicious bombing of Great Britain that he thought would pummel her cities, destroy her industries, and break her people's will. At sea, German U-boat commanders struck savagely at any Allied shipping bound for British ports—except ships flying the American flag.

For passengers anxious to reach the United States, the options were shrinking. The Italian Line had just laid up its superliners *Rex* and *Conte di Savoia*—both ships would become casualties of war, bombed and sunk at their anchorages. Because Italy was still technically neutral, the United States Lines had shifted its main European port to Genoa—though many worried that it was only a matter of time before company president John Franklin canceled all future commercial sailings.

Ships that did sail were packed. On one voyage, conductor Arturo Toscanini bunked up with the ship's chief surgeon, and cosmetics tycoon Helena Rubinstein slept on a smoking room sofa.[4] Few passengers carried many worldly possessions with them. Most felt lucky to be on board at all. A westbound transatlantic ticket had become one of the most valuable commodities in Europe.

As Captain Harry Manning raced *Washington* at full speed along the Portuguese coast on the morning of July 11, he felt that his liner was safe from the U-boat threat. To make absolutely clear that *Washington* was a neutral ship, the company had emblazoned two enormous banners across the ship's hull. Bracketed by two painted American flags, it read: WASHINGTON: UNITED STATES LINES. Yet the markings did little to quell the tension on board. The ship was jammed with anxious American expatriates fleeing Paris and London, as well as a few fortunate Jewish families.

America's best skipper had more than his ship to live for now; he had recently married Florence Isabella Trowbridge Heaton, a passenger he had wooed during a recent trip. The marriage, which produced one daughter, would be a failure. "I couldn't serve two masters," Manning retorted when asked why.[5]

Manning also knew that his overloaded vessel could only make about 20 knots, too slow for her to avoid a Nazi torpedo.

At six in the morning, just before the dawn, Manning scanned the horizon with his binoculars and saw a dark object spring to the surface.

A Morse lamp blinked from the U-boat's control tower. "Stop ship. Ease to ship. Torpedo ship!"

The instructions were in English.

Manning barked out, "All stop."

Washington slowly drifted to a halt and bobbed gently in the swells.

"Abandon ship," the U-boat commander then blinked.

Manning prepared for a general evacuation and sounded the alarm calling all hands to boat stations. In response to the screeching sirens, passengers jumped out of their beds and trooped to the open decks. The crew swung the boats out and loaded them with women and children first. "Watertight doors were closed," Manning recalled, "the general alarm was sounded. . . . Not a passenger showed signs of hysteria or confusion. The crew behaved well, obeying orders without question or criticism. . . . We maintained radio silence."

All this time, one of the officer cadets pounded furiously away at the switch of *Washington*'s Morse lamp. "American. *Washington*."

The U-boat commander was unmoved. "Ten minutes," he signaled back.

Manning could not believe what was happening. The U-boat commander wanted not only to sink a neutral ship, but a passenger liner loaded with women and children.

He ordered that the cadet keep signaling, "American. *Washington*."

"I wanted to convince the submarine commander by blinker that he was in error in assuming that we were a belligerent craft or to keep talking until the break of dawn revealed it to his own eyes," Manning recalled. "It was ticklish. I know how 'trigger itch' will work in such a case and how an overenthusiastic young officer might be anxious to sink such a fat prize, as we indubitably appeared magnified by the gloom."[6]

The charade continued for ten minutes, as hundreds of bleary-eyed passengers shivered on the boat deck, lifebelts strapped around their chests. The lifeboats dangled from their davits, swinging above the waves.

Manning and his officers waited. "We held our breath on the bridge," he said, "and awaited the blast that would announce the doom of the finest ship under the American Flag."

Suddenly, the signal blinking from the U-boat's conning tower changed.

"Thought you were another ship; please go on, go on!"

Manning ordered "Full ahead," and *Washington* left the U-boat in her wake. Yet Manning insisted that passengers stay at their boat stations. Sure enough, at 6:53 A.M., another U-boat surfaced.

Manning knew his passengers were already traumatized, and he did not want to have yet another showdown. "Rather than risk another encounter," Manning said, "I swung the ship into the sun again which brought the unwelcome neighbor dead astern and steamed away. The submarine made no move, possibly the sun blinded him, and it soon disappeared astern."[7]

Washington called the next day at Galway, Ireland, where Manning picked up another few hundred refugees. He then pushed his ship at full speed ahead to New York City, where the overloaded ship arrived safe and sound on June 21.

The president of the United States Lines was furious. "Goddammit, Captain," John Franklin roared when Manning proudly told him about his second U-boat escape. "Will you please stop trying to fool those damn submarines!"[8]

A year later, the United States Lines ceased all commercial sailings. President Roosevelt ordered the Navy to seize *Washington, Manhattan,* and the brand-new *America* and prepare them for war duty. America's entry into the war seemed imminent.

As the United States Lines shut down its passenger operations, William Francis Gibbs was in the midst of designing a humble cargo vessel, called an "ugly duckling" by Franklin Roosevelt, a ship that would help tip the balance of power in World War II in both Europe and East Asia by delivering crucial American supplies to the front.

* * *

William Francis Gibbs could not have been happier when a group of British diplomats came to 21 West Street begging for help. He never much liked the English, nor what he thought of as their condescending politeness, and he relished the chance to be blunt in their presence.

It was the summer of 1940, and America remained officially neutral, but the Todd Shipyards—builders of many of Gibbs's destroyers—and the U.S. Maritime Commission had brokered this secret meeting. The British delegation was looking for a naval architect to come up with a template for twenty new cargo vessels, all of the same design, that would carry provisions and war materials to a blockaded Great Britain as it continued its lonely fight against the Nazis. Until the Royal Navy developed better methods of finding and chasing down U-boats, the only way to keep supplies flowing to Britain was to build ships faster than the Germans could sink them. And the only way to do that was by negotiating a contract with America's big shipyards, which could turn out ships in huge quantity thanks to mass-production methods developed by Gibbs & Cox.

Yet William Francis Gibbs did not seem eager to design these ships. "You don't need them," he snapped.

The British were stunned by Gibbs's refusal to prepare plans, and demanded an explanation. Was he turning down a contract?

"If England is within twenty ships of winning the war," Gibbs said with a wry smile, "she has won the war already."

"How many would you suggest?" one of the delegates asked.

Gibbs thought for a moment, and said, "Sixty would be a start."[9]

Relieved, the British delegation presented Gibbs & Cox with specifications for a basic tramp steamer developed about thirty years earlier. The challenge for Gibbs was to take the design for the tublike vessel and translate it into construction drawings that American shipyards could use. And the answer was a prototype for a simple-to-assemble vessel with interchangeable parts. The hull would be welded rather than riveted. To save on machinery costs and construction time, the 10,000-ton, 442-foot vessels would be powered not by steam turbines, which were too complicated and expensive to mass-produce, but by old-fashioned, reciprocating engines that would drive the ship at only 11 knots—fast enough to make headway in the Atlantic, but slow enough to make them easy pickings for U-boat commanders. The boilers were to burn not oil, which Britain had to import, but coal, which was plentiful in the British Isles.

The new U.S. Maritime Commission chairman, Admiral Emory Land, approved the prototype, and a contract was signed on December 10, 1940. But even with a staff of a thousand Gibbs & Cox quickly ran into trouble. "The lack of detail in the plans they brought from England presented a big problem for American shipbuilders," naval historian and eyewitness Frederic Lane recalled. As a result, the American staff had to do "considerable interpretation and amplification."[10] For someone who insisted on complete control like William Francis Gibbs, adapting the plans must have been absolutely frustrating, further confirming his dislike of British shipyard practices, with all their bench work and hand fitting. Working long hours, his staff expanded the prototype's eighty basic working plans into the 550 that the shipyards would need.[11]

The sixty vessels were built by Todd Shipyards—half of them in California, half in Maine. The keel of the first ship, *Ocean Vanguard,* was laid down in April 1941; on October 27, she was launched by Admiral Land's wife.

When the British ships were being planned, Admiral Land also asked that Gibbs slightly refine the design for American use, if—and it looked increasingly likely—the United States entered the war. Gibbs made a few technical improvements and presented a model of this cargo ship to President Roosevelt for approval.

FDR looked at the vessel and called it an "Ugly Duckling." But he gave his blessing, and American shipyard began cranking them out in preparation for American entry into the conflict.

The public would call it the "Liberty ship."

For the Liberty ships, Gibbs built on the efficient methods of mass production he had created for the fast construction of his destroyers. Here he sharply reduced the number of parts used, and increased the prefabrication of major sections of the ship. Deck ventilators, funnels, boilers, engine components, even entire hull sections were assembled by subcontractors off-site, hauled to the shipyard, and welded into place. Shipyards became less like high-end couturiers, building each ship to suit an individual customer's tastes, and more like garment shops, assembling one-size-fits-all products from precut fabrics. The first Liberty ship took 147 days to build; as the learning curve steepened, Liberties began rolling off the lines at an average construction time of 42 days.

The first Liberty ship built for American use was christened *Patrick Henry,* for the Revolutionary War orator who declared: "Give me liberty or give me death!" Gibbs gave America Liberties . . . 2,620 of them over the next four years.

Little more than two months after the first Liberty was launched on September 27, 1941, the vital role of Gibbs's production speed became glaringly apparent following an attack on America's largest Pacific naval base. During the course of World War II, the humble Liberty ship would carry three-quarters of all of America's war supplies to destinations in Europe and Asia.[12]

As the Liberty ships were being put together, Admiral Land found Gibbs hard to work with. He was upset that Gibbs & Cox "pretty generally" ignored most inquiries from the Maritime Commission's Audit Branch.[13] For a time, the admiral decided to let the irregularities pass because of the urgency of the project. But in June 1941, William Francis Gibbs and Admiral Land locked horns over how much money Gibbs & Cox was being paid.

The original agreement with the U.S. Maritime Commission to build the "British Liberties" came to about $600,000 for all sixty ships. When asked to modify the British Liberty design into the EC2 "American Liberty," Gibbs agreed to drop the price to $750,000 in total. The commission then increased their order to 312 ships, for a total of $1.1 million for the lot.[14]

The proposition infuriated Gibbs and no doubt business-savvy Frederic. "We have no precedents for low compensation," Gibbs said to Land on June 24, 1941. He capitulated, however, and told his staff to finish the working plans. But he decided he was through dealing with the Maritime Commission and its admiral. He would concentrate on work for the Navy, which continued to be a loyal client.

Admiral Land was glad to get rid of William Francis Gibbs. "We felt that we paid plenty for it and they felt that we did not pay enough," Land testified to Congress, "so we just split business with them and said, 'Good bye boys,' and we kissed each other on both cheeks and we are running our own show now."[15]

However, by the end of that year, Gibbs would be overseeing an even bigger effort than building Liberty ships: the construction of a naval fleet that would battle the Axis in the Atlantic and Pacific oceans.

On December 8, 1941, William Francis dictated a memorandum to the 1,200 people who worked for him at 21 West Street:

To: The Staff
War having started with Japan it is fitting that I should remind you of the vital position of this organization in National Defense.

Each of you well know that I appreciate the high technical and ethical standards and esprit de corps which you have helped maintain.

We now face the supreme test for which we have been trained and organized during the years we have toiled together. I am confident each of us will vie in giving the full measure of toil, cooperation and loyalty which will insure that this organization serves the Nation and the Navy to the limit of its latent capacity.

I have always been proud of you, but never more than in this hour when we face together this solemn responsibility. I know I can count on each of you to do his utmost.[16]

The Japanese attack on Pearl Harbor immediately transformed America into a wartime nation, with all of its industrial might now focused on producing enough bullets, bombs, tanks, and airplanes to overwhelm the enemy. All work on civilian ships ceased: the nation needed destroyers, cruisers, aircraft carriers, and troop transports, and needed them fast. This was what William Francis Gibbs had been waiting for, and he was ready to step into the center of the war machine.

On December 12, President Roosevelt formally seized the laid-up *Normandie* for use as a military transport. Her name was changed to USS *Lafayette,* and Secretary of the Navy Frank Knox ordered that the French flagship be converted into a troopship right at her pier—not by Gibbs & Cox, but by Robins Dry Dock & Repair Company. During the next two months, workmen dismantled and carted away the ship's lavish Art Deco furnishings and artwork. In their place came thousands of standee bunks and slabs of heavy linoleum flooring that could take the punishment of army boots.

USS *Lafayette* was scheduled to depart for Boston for a final dry docking and inspection on February 12, 1942, before loading her first troops.[17]

But on the afternoon of February 9, 1942, the workers at nearby Gibbs & Cox noticed a column of orange smoke rising from the Hudson River liner piers. Looking to the north, they could see the enormous troopship afire.

Two workmen had been removing a steel pillar from USS *Lafayette*'s first-class Grand Salon with acetylene torches. Sparks landed on a pile of kapok lifejackets, sending flames shooting up to the ceiling. Before fire hoses and extinguishers could be trained on the conflagration, the fire had found its way into the cork insulation used in *Normandie*'s construction. The fire spread from one huge salon to the next, gorging itself on paint, rags, and wood scraps. Like *Morro Castle,* the ship had no automatic sprinklers. Within half an hour, USS *Lafayette* was ablaze from bow to stern, casting a pall of smoke over Manhattan.[18]

William Francis Gibbs found the fire appalling—and familiar. It seemed that fire had been a constant scourge for passenger vessels, and it seemed that ships of the French Line especially had a tendency to go up in flames. In 1939, *Normandie* had actually been trapped in its dry dock at Le Havre by the burning of the French liner *Paris* when she heeled over onto her side.

The burning of *Normandie/Lafayette* in 1942 was one more reminder to Gibbs that he was right to be so fanatical about fire safety measures in the ships he designed. He also took a special pleasure in having designed the world's most powerful fireboat, *Fire Fighter,* which had a retractable, fifty-five-foot-tall water tower that could shoot three thousand gallons per minute.[19] *Fire Fighter,* which worked in New York City waters, was among the fireboats that rushed to the scene and doused the burning ship's decks with river water.

Admiral Adolphus Andrews of the U.S. Navy directed the fireboats to direct their huge jets to the stricken liner's upper works. As the liner listed farther away from the pier, gangplanks snapped from their moorings and plummeted into the Hudson. The New York Fire Department raised a sixty-foot ladder against the ship's prow, and one by one, 2,500 trapped workers inched down to the street. The inferno injured ninety-three men and killed two.[20]

In New York that day was *Normandie*'s designer, naval architect Vladimir Yourkevich, who had escaped from the Nazi invasion of France. As he watched the ghastly spectacle unfold, Yourkevitch asked Admiral Andrews to tell his workers to open *Lafayette*'s sea cocks. She would then sink only a few feet and settle on the river bottom, upright and secure.

"This is a Navy job," the admiral snapped, and ignored the suggestion.[21] Yourkevitch turned his back on the burning superliner and left. He knew what would happen next. In the early morning hours of February 10, the burned-out USS *Lafayette* drunkenly rolled over her port side and sank into the Hudson River mud.

Before the Navy erected a huge fence to shield the capsized giant from the gawking public, the Hearst-controlled *New York Journal American* wrote an editorial capturing the tragic demise of the French flagship. "Get a little closer—as close as a triple line of police, Army and Navy men will permit—and the chill of desolation that creeps along your spine is almost overpowering. For the finality of death lies like a clammy hand across the prostrate *Normandie* as it tilts unhappily in its shallow watershed waiting for whatever fate the future may bring."[22]

The military saw a draft of the article, called it demoralizing, and censored it.

The Navy asked William Francis Gibbs to assess what should be done with the former French flagship, which was now blocking two piers and tying up valuable waterfront space. Should it be broken up on the spot or salvaged and refitted? Gibbs & Cox responded quickly, urging that she should be rebuilt as a troopship, able to carry nearly 19,000 soldiers.[23]

Yet it was not to be. Over the next two years, the charred *Lafayette* was stripped of its upper decks and slowly righted. By January 1944, the hulk was moved to the Todd Shipyards in Brooklyn. The following year, the Navy declared the former *Normandie* surplus, and the wreck was towed to Bayonne, New Jersey, and cut up for scrap.

Yourkevitch rushed to the site. Over the hiss of acetylene torches and wrenching of steel, he pleaded to the owners of his beloved ship to stop demolition. Even the hard-bitten scrap yard owner was moved by the Russian naval architect's despair. "I think it broke his heart," he recalled.[24]

Gibbs never forgot the *Normandie* fire. Like the *Morro Castle* disaster, it was another tragic reminder of how quickly fire could destroy a passenger ship.

Less than a year after Pearl Harbor, the war propelled Gibbs to meteoric fame, and his firm to extraordinary prosperity. Some of the publicity was unwelcomed, and he regretted letting at least one reporter into his office. On September 28, 1942, *Time* magazine placed an image of William Francis Gibbs, complete with battered brown fedora, steel-rimmed glasses, and deadpan, dour visage, on its cover. The backdrop was a bustling shipyard. Hailing him as a "technological revolutionist," *Time* noted that the Liberty ship program boosted commercial ship production by more than 70 percent. War mobilization had reached a fever pitch, and his team designed not just destroyers, but also cruisers and landing craft.

"The job that William Francis Gibbs' firm does is titanic," the *Time* editors wrote. "On one multiple order of ships, they may issue 6,700 purchase orders daily. Not a day goes by that the company does not contract for at least $1,000,000 worth of materials."[25]

William Francis Gibbs was somewhat unnerved at being hailed as a "technological revolutionist" by the nation's biggest news magazine. He felt that too much publicity could backfire and cause the entire enterprise to collapse. Even so, Gibbs & Cox had come a long way. The company offices now occupied thirteen floors of 21 West Street. Two thousand engineers and draftsmen sat at rows of desks under the glow of harsh fluorescent lights for hours at a stretch. The wartime blackout of lower Manhattan meant that all of the windows had to be covered at night with curtains or cardboard, further adding to the claustrophobic atmosphere. Some windows were painted out entirely. But the measure saved lives: without it, prowling U-boats could pick out silhouettes of American ships against New York City's bright lights.

During these long days, William Francis would startle his employees by sneaking soundlessly up behind them while they sat bent over their drafting tables, then pointing out something he didn't like.

"Take it away," he would growl when shown work that did not meet his standards. "Bring me the best."[26]

Many of his staff resented his overbearing management style. "Sir Francis," some called him under their breaths.[27]

The firm's Spartan austerity extended into William Francis's own corner office, and the demands he placed on others he placed on himself. He worked just like his junior engineers, with a simple drawing board in front of him. His drafting table was topped with cheap wood and had legs of green metal. The drab gray walls of his office were covered with models and photographs of his ships. He cracked down on any extravagance in office décor, claiming that it was "the first sign that your company is getting muscle-bound."[28] For ten hours a day, or more, three secretaries struggled to take everything the boss said verbatim, including his spectacular cursing streaks.[29]

And his work style, if autocratic, was far from remote. His

office—glass enclosed on three sides so he could observe the goings-on in the main drafting room—had a sign on the door identifying it as "The Glass Menagerie." Rather than checking with his secretaries to see if the boss was free, employees could simply walk in and interrupt him, even in mid-dictation. Gibbs would listen to what the employee had to say, respond, and then resume whatever he was doing as if nothing had happened. "The place sometimes looks like the box office of a Broadway hit because of the line of men and women waiting to talk to him," a visitor noted.[30]

His brother, Frederic Gibbs, maintained a separate office, at One Broadway, where he could quietly handle the firm's business affairs. Unlike the "Glass Menagerie," Frederic's office was quiet, neat, and uncluttered, its walls almost bare. A massive black adding machine sat on the left side of his desk. A visitor thought that like the potted plants he grew from seed, the painfully shy, introverted Frederic did not feel isolated at all, but "thrived in his special spot—he is looked on around the shop as an anchor to windward."[31] Frederic was so tight with money that he even charged office furniture to the client.[32] Gibbs could not have been more pleased with his brother's skinflint ways. Frederic's financial acumen, which first began to show itself forty years ago in their parent's attic, was now really bearing fruit.

As a master organizer, purchaser, and procurer, William Francis crisscrossed the country, using his Liberty ship method technique with hundreds of small contractors. Many of them, such as the American Bridge Company, had never before built a ship component. Dressed in work overalls and holding blueprints under his arms, he would take a group of midwestern contractors to a docked ship and shout, "Look her over, find out what you can make, ask our guides what it is called and come back with your order."

The results were remarkable. As his friend Frank Braynard observed later, "Countless small plants all over the nation began making air funnels, life rafts, hatch covers, and you name it."[33] Gibbs's techniques pulled their skills and productive capacity together, making a national army out of individual plants and offices across the country.

All the while, Gibbs's political allies continued to lobby on his behalf. Just before America entered the war, William Francis Gibbs and Secretary of the Navy Charles Edison had presented President Roosevelt with drawings of a fearsome "battleship-carrier-cruiser combination," a ship that in Edison's words would have "more speed, more guns and more protection than anything afloat" and could act as a "lone wolf."[34] (It was similar to what Gibbs had shown to the Soviets a few years before.) A few weeks after Pearl Harbor, Edison, now governor of New Jersey, wrote Roosevelt urging that if Gibbs was "given a free hand, two such ships would float before the war ends." Edison also added, "Gibbs knows nothing of this letter."[35]

The president agreed that such ships would be worthwhile, but ultimately nothing came of the proposal. It was judged to be impractical and too expensive.[36]

Yet as business boomed, the resentment of the Maritime Commission for William Francis Gibbs continued to fester. The boiling point came in early 1943, when Admiral Land let it be known that he was going to replace the Liberty ship with a new design. Around the same time, the War Production Board asked Gibbs to become its comptroller of shipbuilding.

The War Production Board, created by presidential executive order, was the nation's highest authority over wartime distribution of goods and services; everything from consumer rationing to the allocation of industrial supplies fell under its purview. Gibbs's new position gave him supervision of the production of naval vessels such as destroyers and cruisers. But he hoped to extend his influence into merchant ships, his greatest passion, and an obviously important factor in the war effort.

Even the president and his top economic advisors could not help but notice that William Francis Gibbs, although a brilliant organizer and designer, was incorrigible when he operated outside the controlled world of his firm. After ten months in Washington, Gibbs was summoned to the White House for a meeting with Roosevelt and Land. At issue was Gibbs's refusal to accept a replacement for the Liberty ship, a stance that threatened to disrupt wartime production.[37]

After the meeting, Roosevelt sent Bernard Baruch, famed financier and FDR economic advisor, to meet with Gibbs. Baruch advised the naval architect to leave his government post and return to New York. An angry Gibbs quit, resigning as comptroller of shipbuilding on September 11, 1943.[38]

Despite the humiliation, Gibbs could look with pride on his triumphs, namely the ships he had designed so well. The destroyers and cruisers with the Gibbs & Cox stamp could outsteam, outmaneuver, and outrun all else on the seas. "When our fleet went to war no one had the faintest idea what the possibilities were for superb logistics," he boasted about his ships' performance in the Pacific Theater. "The Japanese to their amazement found [Admiral] Halsey's ships doing things they did not dream possible. It was a complete tactical surprise to the enemy."[39]

Yet William Francis could not stop thinking about how the really big ships, the great liners, were contributing to the war effort. He knew that thanks to their size and speed, the British *Queens* and the United States Lines' *America* had assumed heroic stature in the North Atlantic. After losing the USS *Lafayette,* the United States Lines owned only one large troopship—the Gibbs-designed *America,* requisitioned by the U.S. Navy in 1941 after only a few months of commercial service. The ship was renamed USS *West Point* and refitted to carry nearly 9,000 troops. But it was the big British superliners *Queen Elizabeth* and *Queen Mary* that were really important to the war effort, shuttling 16,000 soldiers per voyage to Great Britain in preparation for the D-Day invasion of Normandy.

Hitler, of course, knew that the sinking of one of the *Queens* would deliver a major, perhaps even fatal, blow to the invasion of Nazi-occupied Europe. As D-Day approached, he offered to award the Iron Cross and $250,000 in cash to any U-boat commander who could send one to the bottom. It was a prize that could never be claimed. The *Queens* could steam away from any submarine trying to take aim with torpedoes.[40]

William Francis was soon to confront problems in Washington that threatened to undo everything he had achieved. His time holding dual positions in government and the private sector had produced some powerful enemies. In the spring of 1944, the House Naval Affairs Committee announced it would investigate the naval architecture firm of Gibbs & Cox. The charge was war profiteering.

His superhuman drive and wartime achievements had attracted the attention and publicity he dreaded. Now Gibbs had to defend his reputation to Congress. If condemned as a war profiteer, his firm, his work, and his dream to build a superliner after the war were doomed.

By outside appearances, William Francis and Vera Gibbs had prospered during the war years. In addition to an apartment on the Upper East Side, they also owned a fourteen-room mansion in Old Brookville, Long Island, not far from the Cravath "Still Place" compound in Locust Valley.[41] While her husband busied himself with his Navy work, Vera hosted lavish benefits for the Metropolitan Opera, and young son Francis had recently ridden a $2,500 pony in a Madison Square Garden equestrian show.[42]

William Francis Gibbs actually hated horses the same way he hated publicity. "He always suspects they're ready to bite him," a family member, probably son Francis, was once quoted as saying.[43]

Much of the money funding this lifestyle came from an inheritance from Gibbs's father-in-law, Paul Cravath, who had died in the summer of 1940. The most powerful corporate attorney in America left his daughter an estate of over $2 million, the equivalent of about $20 million today.[44]

No matter. It appeared that William Francis Gibbs was living large off the American taxpayer, and Congress was eager to crack open the black box that was Gibbs & Cox Inc.

In May 1944, William Francis Gibbs arrived in a Washington that was no longer a sleepy southern town. The wartime and New Deal bureaucracies had sent the population soaring past half a million residents. The city was gray and bleak. Men in uniform were everywhere, standing guard on the Capitol steps and the Lincoln Memorial. Strict gas rationing had cut down on car traffic. The government urged the thousands of workers at the Washington Navy Yard to carpool. "When you ride ALONE, you ride with Hitler!" exhorted one poster.[45]

A summons to testify before the House Naval Affairs Committee had recently landed on Gibbs's desk in the Glass Menagerie. Shocked, he promised to "iron out any irregularities, if there be any, and provide for a course of action that will be generally satisfactory." He knew that the chairman of the committee, Georgia congressman Carl Vinson, was an interrogator who would have no qualms about destroying the career of anyone put in front of him.

After Franklin Roosevelt was inaugurated in 1933, Vinson had become the leading advocate on Capitol Hill for the modernization of the Navy. Roosevelt, a lover of ships who also needed the support of southern Democrats to stay in the White House, enthusiastically endorsed Vinson's views. A weak and outdated Navy, Vinson said, was "not a political question, but a national problem—namely, an adequate national defense—a first-class Navy for a first-class nation."[46] The $600 million bill (about $9 billion in 2012) became known as the Vinson Naval Plan and funded the destroyers Gibbs & Cox designed to be the most modern in the world. During the 1930s, two more multimillion-dollar "Vinson Acts" were passed, further increasing money available for naval construction.

Now Vinson was charged with investigating one of the chief beneficiaries of his legislation: Gibbs & Cox Inc.

Even before Pearl Harbor, Congressmen Vinson and other Democratic politicians were under intense pressure to investigate possible abuses by government contractors profiting from the "war business." The 1934 Navy modernization plan had turned into a $24 billion money machine, and competition among contractors for government favor grew. Shipyards lavished baubles on the wives of important maritime officials and gifts to politicians who christened ships. *Time* magazine noted that, "after smashing her bottle, each ship sponsor took home a souvenir gift. Some of the trinkets would have lit fire in the eyes of a Follies girl." Female relatives of Maritime Commission vice chairman Admiral Howard L. Vickery "had been in special demand as bottle smashers. Five had received $6,457.65 in shipyard gifts; Daughter Barbara's share included two diamond bracelets."[47]

Investigating wartime abuse produced great press attention for ambitious politicians like Vinson. Leading the pack of Washington crusaders were two southern allies of President Roosevelt: Senator Harry Truman of Missouri and Comptroller General Lindsay Warren, who until 1940 had served as a congressman from North Carolina. Since America entered the war, Truman's special committee investigated suspect defense contracts and rooted out a staggering $11 billion in waste.[48] As for the comptroller general, the press had dubbed the bull-necked, poker-playing Warren the "Watchdog of the Treasury."

The press and the public were now eager to learn if the secretive William Francis Gibbs, the so-called "Technological Revolutionist," was no hero at all, but instead an opportunistic criminal who, after collecting a near monopoly of wartime naval design contracts, was reaping huge profits from inflated fees.

The hearings began on May 8, 1944, in Room 303 of the Cannon House Office Building, a neoclassical marble pile dating from the Theodore Roosevelt administration. The main entrance, with its columns and steps, resembled the portals of the House of Morgan that William Francis Gibbs walked into thirty years earlier, the plans for his dream ship under his arm. This time, he was carrying a stack of notes on his firm's finances prepared by his brother, Frederic.

The hearings opened with a statement by the committee's general counsel, Robert E. Kline Jr. "They have set up in New York, at 21 West Street, a huge organization—without doubt one of the largest engineering organizations in the world," he began ominously, "whose job it is to act, directly or indirectly, as an outside design and procurement agency for the Navy Department."[49]

Kline and the Naval Affairs Committee made it clear that the quality of the firm's work was beyond reproach. But the existing business relationship, they charged, was suspect. Rather than accepting a flat fee for each class of vessel, William Francis and Frederic Gibbs—the firm's sole stockholders—were charging a fee for each ship launched, leading to, as Kline charged, "a multiplication of fees far beyond any original expectation."[50]

The first witness called was Rear Admiral Edward L. Cochrane, the chief of the Navy's Bureau of Ships. It was under his watch that this arrangement had been made with the Gibbs firm. Cochrane made clear that the arrangement with Gibbs & Cox had led to a "great deal of misunderstanding for some time," and that his decision to retain the company as the Navy's lead design firm had the full approval of the secretary of the navy, Frank Knox, who had died the previous month.

The job of the Bureau of Ships, Cochrane argued, was to help the secretary of the Navy, the president, and the Navy's General Board to develop a set of specifications for the ship the Navy wanted to build. The basic specifications, which might run to as many as 115 sheets, would then be sent to prospective bidders. Once the contractor had been selected, Gibbs & Cox would then turn these specifications into as many as eight thousand working plans for shipyard use, as well as to make sure the yard got what it needed from its suppliers on time.[51]

"In our judgment," Cochrane added, "it cost much less to do it the way we have done."

Chairman Vinson cut the admiral off. "I understand that," he said, "but you do designate an outsider to go out and buy, at the expense of the taxpayers of this country, supplies and materials to go into ships that our boys are going to fight in, and pay enormous fees for them to go out and buy, and thereby you admit to the public, as far as I see it, that you are not able or competent to go out and do the job yourself, when you have all the resources of the country; you can call in any man you want and put him in service."[52]

Admiral Cochrane left the stand defeated.

William Francis Gibbs stood at the witness table the following day. As he raised his right hand and swore to tell the truth, the balding, jowl-faced Vinson peered back at him sternly through round reading glasses.

Summoning his dramatic flair and legal training, Gibbs gave a detailed history of his firm's business with the Navy and the Maritime Commission. He then asserted that his company was run in the most conservative, prudent manner possible.

"Gibbs & Cox is in an independent position," he said in his firm, patrician voice, "having no connections with any shipyard or apparatus, equipment, or material manufacturer. The company has never borrowed money from the government. I believe that the company has maintained a sound and conservative financial policy. The salaries of the executives have not been changed since 1937. No dividends have been paid. The company has never made any outside investments of any kind. All of the profits of the company have been added to the surplus. The entire surplus, with no deductions, has been retained for the needs of the business."

The company's profits, Gibbs said, had been reviewed by the Navy Renegotiation Board, which had declared that the previous year's figures were sound and needed no renegotiation.[53]

"You will find there is something unique about what we are doing," Gibbs told the committee. "In other words, here is a project conceived by individuals, backed only with individual initiative that comes forward in a time of great national emergency and performs a signal service for the United States." He then vowed to "present to you this case with clarity and conviction so that when I go back to New York and face those 2,850 people [the Gibbs & Cox employees] who with heart and head and spirit have backed what we have tried to do for the United States, they would not say to me that I had not presented the case skillfully as well."

When challenged about how Gibbs & Cox was performing a service that could have been provided just as well by the government, Gibbs insisted the Navy simply did not have the manpower in-house to perform all the technical work needed. "The Navy does its own ship designing and takes full responsibility for it," he said, "but if you are going to construct ships in a shipyard, it is essential that you have relatively strong technical organization attached to the shipyard and responsible to the shipyard."

Gibbs then explained how his firm performed a role very similar to how General Motors built tanks for the Army. "Today in shipbuilding," he said, "if a shipbuilder gets an order for a ship accompanied by sufficient plans to clearly describe that ship and what the Government wants, it is his job to take that information and translate it into a ship. He translates it to a ship with the approval of every step by the Navy."[54] To prove that the shipyards found his firm's expertise worth paying for, he pulled out a letter written in 1938 by the president of Maine's Bath Iron Works, builder of many of the new destroyers. Even if Gibbs & Cox cost more money than other independent naval architects, it read, the quality of the product spoke for itself.

"The letter can be duplicated by every shipyard that employs us," Gibbs asserted (not saying that many shipyards, like Newport News, hated him and his meddling).[55]

Gibbs then tackled a matter of public perception: the aura of mystery that surrounded him and his firm. A lot of this, he admitted, was by design. "I never speak to the press," he said. "I never hold an interview, I don't advertise, I don't employ any publicity assistants, and the result is that there is no source from which anybody can get information. The result is that they are always thinking that there must be some strange, occult power in the person of some official of Gibbs & Cox which enables them to get work to do."[56]

When grilled by the committee about his profits, Gibbs produced the numbers Frederic had given him. For the past two years, the firm had grossed $15 million, but only made a 4.91 percent profit after taxes.

Kline and the thirty-one members of the House Naval Affairs Committee were stunned. He had to be artificially deflating his profits.

Chairman Vinson jumped in, claiming that Gibbs made a gross profit of 32 percent after taxes.

All the questions about the accuracy of his records infuriated him—he and his brother had maintained meticulous financial records. He hated the insinuation of fraud and must have remembered what his father had been accused of three decades earlier. But Gibbs kept his cool and stuck to his guns, and then offered his own breakdowns, added that the Navy Renegotiation Board had examined these figures and determined that the accounts for 1941 and 1942 "are proper and sound, and that no renegotiation is necessary."[57]

"I have perforce to think that when the Renegotiation Board makes a statement of that kind, they have looked into it and are correct," Gibbs said proudly. He then rattled off a list of distinguished naval officers who sat on that board.

Vinson realized that the last thing he wanted to start was a credibility war between Congress and some of the most powerful men in the military. "Then I would say, also, that we have to be guided by the conclusions they reached in regard to the renegotiation," Vinson said.

William Francis Gibbs knew that he had backed Vinson into a corner. He then extended an invitation to the entire Naval Affairs Committee to visit his office in New York.

"We are coming up there," Vinson replied.[58]

"The reason that I was anxious for you to see that before we discussed this subject was that you will find there a frugal establishment," Gibbs continued; "you will find no fancy offices. You will find me seated on a high stool behind a drawing board. You will see nothing that smacks of panoply and pomp. A simple place, done in the most frugal way. So there are no questions of expenditures made on an extravagant basis. We don't do those things."

After Chairman Vinson agreed that the company had had a clean bill of health since America entered the war, Gibbs laid out his view of how his business participated in the American system of free enterprise. "If you say to me that the profit after taxes cannot keep pace with the gross business," Gibbs said, "you are automatically limiting the size of business in the United States."

"Let's get away from that entirely," Vinson said, interrupting him. "You have the permission of everybody to do as large a business as you can."

"All right," Gibbs agreed. "Then you have to let me run a business that keeps pace in gross earnings with the size of the business."[59]

The naval architect trained as a lawyer then rested his case.

* * *

The next day, Chairman Vinson once again faced William Francis Gibbs and passed judgment on the wartime conduct of the Navy and Gibbs & Cox.

"I will make this statement for the record," Vinson began. "I think, Mr. Gibbs, beyond a shadow of doubt, that you have rendered to the country outstanding service in organizing a staff of technical men as skilled and as learned as they are, and you have been of great aid and value and assistance to the ship production program. . . . I think the Navy would have been bogged down completely had it not been for men of your vision and ability who came in and aided the Navy in its ship building program.

"I want to commend you and I want to compliment the Navy Department on having the vision to employ people of your ability and your firm's outstanding qualifications do this work," Vinson concluded.[60]

He then asked Admiral Cochrane—routed a few days earlier—to take the stand again.

The admiral clarified some additional questions about fee structures and vessel classes for the committee. But before yielding the floor, Cochrane asked if he could make one more statement about the work of Gibbs & Cox.

"One point which perhaps has not been adequately made in the hearings," Cochrane said, "is that Gibbs & Cox is a unique organization and there is nothing equal to it in any field of marine engineers and naval construction. . . .

"We at the Bureau of Ships," Cochrane went on, "feel that Gibbs & Cox has done an outstanding job and has been an essential factor in the over-all war effort of the country. Their contribution has been outstanding. We feel that there has been and will be general agreement in this opinion by those who take the time to read the record of these hearings."

"I agree with your conclusions," Chairman Vinson said.[61]

William Francis Gibbs went to Washington as a suspected war profiteer and he left the city a war hero. On his way back to New York, he may have thought that his father, who called engineers impractical, inarticulate, and short on business smarts, had been proved wrong.

The law was not the only way to earn the respect of people who mattered. Thirty years after he quit the law, William Francis Gibbs had won over Carl Vinson, a powerful Washington figure who was among those overseeing the greatest collective effort in the history of humanity, a war against the Axis powers. He was a man who mattered, not a Philadelphia social arbiter or Harvard rich boy. And forty-two years after the *New York Times* attacked his father for running a fraudulent business, Gibbs showed the country he himself was honest, honorable, and patriotic. Where his father bilked American Alkali out of tens of thousands of dollars, Gibbs demonstrated in a public forum that his firm was squeaky clean.[62]

A day after the Washington hearings, a headline appeared in the *New York Times*: "House Unit Clears, Lauds Gibbs & Cox: Absolved of Excess Fee Charge—Firm Called Big Factor in Record Ship Output."

William Francis returned to his office at 21 West Street and got back to work. "Nothing educates a man like being forced to look up the answer to every possible question that could be asked about his business," he said about the investigation.[63]

Less than a month later, 130,000 American soldiers prepared to storm the beaches of Nazi-occupied France. On the morning of June 6, 1944, William Francis Gibbs asked his staff to "pause for a minute and contemplate the importance of this hour, and make a short prayer for the success of the operation and the minimization of its cost."[64]

The men and women of Gibbs & Cox had designed the landing craft that carried the American soldiers onto the beaches of Normandy on D-Day.

BOOK II

BUILDING THE DREAM

A VERY PLEASING APPEARANCE

As the complete collapse of Germany and Japan neared, William Francis Gibbs was at the pinnacle of an astonishing career that he had created for himself. He was indisputably the nation's most successful naval architect. The accolades cluttering his walls included honorary membership in Phi Beta Kappa at Harvard and the David W. Taylor Award from the Society of Naval Architects and Marine Engineers, named after his mentor.[1]

When the war did end, William Francis Gibbs could say that he had assembled one of the finest groups of naval designers and engineers, perhaps the finest, in maritime history. The men and women of the firm had worked together for five long years under the strict discipline of wartime service and secrecy, and they had good reason to be proud: Gibbs & Cox's ships had brought advanced design and innovative construction to the war at sea, and helped the allies win it. They had designed and supervised the construction of more than 70 percent of all naval ships during World War II.

But Gibbs warned his staff about becoming too comfortable. "It is true of any endeavor where groups of people come together for a common purpose," he said. "Praise is dangerous because of its possible effect of making us over-confident."[2]

Like William Francis Gibbs, the United States of America had achieved the seemingly impossible. After years of economic hardship, the nation was emerging from the war stronger than it had ever been in its history. Its soldiers were victorious, its industrial infrastructure robust, and its workers flush with cash. With most of Europe in ruins, America had no real economic competitors. Following Japan's surrender on August 15, 1945, the great victor nation was now fully poised to be a true two-ocean power.

Finally, the stars seemed to be aligning for William Francis Gibbs's childhood dream, and he now felt ready to build a big passenger liner that he could pitch to the United States Lines and the American military. He knew that despite all his accomplishments, willing his so-called Big Ship into reality would be the greatest uphill battle of his life. After four decades of hard work and perseverance, at the peak of his career, Gibbs was both a visionary and a tough negotiator, unafraid of naysayers among rivals in his field and enemies in Washington.

"I am used to hard fights," he once said. "You never do anything in this world that is different that you do not have nearly every man's hand against you. It does not make any difference whether it is medicine, or religion, or what it is. You will find that as soon as a man comes up with a new scheme of things, he makes everybody mad because they have to think. Then when they go through that process they join all together against this individual to try to knock him out."[3]

Designing the greatest ship in history meant more than a lot of hard work from him and his staff. It had to be driven, as he said, by "an inward urge to crusade."[4] Producing the blueprints of the ship was like composing a score for a great symphony, William Francis said, and a great luxury liner plowing at speed through the ocean was comparable to listening to a great orchestra or church pipe organ; observing its mighty machinery was like looking at the music of Bach under glass.[5] Perhaps William Francis had in mind his own favorite classical compositions as he set to work: heroic pieces from the Romantic period—Beethoven's Fifth Symphony, Tchaikovsky's Sixth Symphony (the "Pathétique"), and Wagner's "Spring Song" from *Die Walküre.* Or Ravel's "Bolero," which, he said, "begins very soft and then rises higher and higher to a tremendous climax. . . . By God, it's good!"[6]

But composing the "score" was the easy part. He would need to be not just the composer, but the conductor as well, supervising and coordinating every aspect of the ship's design and construction. First he had to meet with his old acquaintance John Franklin, the president of the United States Lines and recently returned from war duty.

Even before the war ended, William Francis Gibbs's mind was drifting away from destroyers and landing craft and back to his real love: passenger liners. Though his staff was overwhelmed with war work, he set about designing a prototype for a medium-sized passenger ship, a scenario that would be more palatable to conservative American steamship executives than a superliner. The resulting prototype, titled Design 11811, was no superliner—far smaller than the 1,019-foot *Queen Mary,* and only about 65 feet longer than *America,* the Gibbs-designed liner that had been the last significant American passenger ship built before war broke out. She was still a ship to be reckoned with, in her advanced technologies and speeds that benefited from Gibbs & Cox's learning curve in the war years. The estimated cost to build her was $31.5 million. He submitted Design 11811 to the Maritime Commission for consideration.

But shortly after the war ended in Europe in May 1945, Gibbs learned through back channels that his design was one of four under consideration by the Maritime Commission. Two other naval architects presented slightly smaller and slower (25-knot) vessels, which would be less expensive to produce. It was the fourth design—prepared by James L. Bates, chief of the Maritime Commission's Bureau of Engineering—that posed the biggest threat to Gibbs. Bates's "Design IV" called for a true superliner: 930 feet long, with a top speed of 30 knots and a cost of $39 million. This proposed liner promised to be a *Queen Mary* competitor in size and speed. It also meant that Gibbs's Design 11811 was neither the grandest nor the cheapest of the four ships before the Maritime Commission. And Bates no doubt would have the support of Admiral Emory Land, who still chaired the Maritime Commission and who was no friend of William Francis Gibbs.

Gibbs knew he had to move beyond Design 11811. He needed "to build a much larger ship than 11811, and not to increase the speed well above the *Queen Mary* and *Queen Elizabeth* would be to overlook a most valuable commercial asset in North Atlantic competition."[7] Not only that, but he had to convince John Franklin, who was undoubtedly seeking a federal subsidy for operating a new liner, to accept his design, not one produced by the government.

Sometime in early 1946, William Francis sat down at his drafting table in the Glass Menagerie and drew up some preliminary drawings and specifications for a true superliner that would best Bates's proposal. It was not that difficult for him—it was pretty much the same ship he and his brother had designed three decades ago in their parents' attic, only this time it incorporated all the breakthrough innovations and design refinements gained during years of experience. Above all, Gibbs specified that this ship would have, as he said later, "the power of survival," specifically "staying afloat and not burning up in case of an accident, collision, or fire on board."[8]

When he felt that he had perfected his conception of a thousand-foot-long liner, he strode across the drafting room, sketches in hand, and approached Thomas Buermann, a young designer in the Hull Scientific division. Then, standing next to Buermann's drafting table, William Francis laid out the parameters for the design of a big ship, bigger than anything the firm had ever worked on. But as Buermann listened, he realized that his boss already knew exactly the ship he had in mind.

Gibbs said he wanted to see preliminary designs for a ship 988 feet in overall length and 940 feet at the waterline, with a side-to-side beam of 101.5 feet, and a 31-foot draft from the waterline to the bottom of the keel.[9]

"This was on a Monday morning," Buermann recalled. "He said, 'I will be back Saturday afternoon to get the product.' "[10]

It had to be ready for a meeting that the Gibbs brothers would have with the president of the United States Lines, Brigadier General John

Franklin, who needed to be convinced to build not just another passenger liner, but the finest, fastest, safest, most beautiful ship ever built.

Franklin himself was looking for a new design for an ocean liner, but he still was not sure that building a big one was a good idea.

Unlike William Francis Gibbs, who tried unsuccessfully to maintain joint positions at his private firm and the government-run war effort, John Franklin took a leave of absence from the presidency of United States Lines to take command of the Water Division of the Army's Transportation Corps. Franklin traveled as far afield as Australia to coordinate the movement of Allied troops on all fronts. He knew the importance of the job, from his own service in World War I, when as a young sergeant he helped his commanding officer, Captain Dwight Eisenhower, pull strings to get their heavy tank battalion aboard *Olympic*. Franklin also had a hand in the creation of some five thousand vessels of all kinds. In his memoirs, he calculated that he had helped the Maritime Commission construct "50,000,000 tons of ships" for a total cost of $15 billion. "So as you can imagine," Franklin wrote, "the whole thing got to be pretty much of a rat race at times."[11]

For his efforts, Franklin was appointed a brigadier general.

During Franklin's time in the military, his longtime colleague Basil Harris served as interim president of United States Lines. At war's end, the United States Lines Company would operate only American-flagged passenger liners and cargo vessels, as opposed to ships under British registry.[12] Vincent Astor, the firm's largest stockholder, threw his financial weight into the business. In Franklin's absence, Harris began laying plans for the company's postwar expansion. "Monopoly on the seas is not the aim of American shipping," he said at a speech at Rutgers University in late 1944. "We want our fair share, not more. But with a greatly increased foreign trade after the war, the British need have no fear that since the United States gets its fair share, Britain's slice will be smaller. There will be ample for all."[13] Out of all the shipping companies, United States Lines had borne the heaviest share of the war effort, dedicating to military service the three largest vessels flying the American flag. After several years of wartime battering, *Washington* and *Manhattan* were in no shape to be restored to their prewar luxury. The former vessel was partially rebuilt as an austerity service passenger liner, while the latter was mothballed. Only the relatively new *America* was fit for a complete overhaul and restoration for transatlantic service.

Rebuilding the merchant marine had been one of President Franklin Roosevelt's priorities during the waning months of the conflict. In October 1944, Roosevelt had called for a "bold and daring" strategy to rebuild America's cargo and passenger fleet. The ailing president directed Admiral Land and the Maritime Commission to prepare a plan for rebuilding the American merchant marine with a new fleet of modern, comfortable passenger liners and cargo ships.

William Francis Gibbs put the news story about Roosevelt's directive to the commission in his memorandum book.[14]

After Roosevelt's death by a massive stroke on April 12, 1945, his successor as president, Harry Truman, decided to drop atomic bombs on Hiroshima and Nagasaki. America thus saw the end of World War II and detected the first hints of the Cold War. Unlike Franklin Roosevelt, Harry Truman did not grow up sailing off New Brunswick's Campobello Island; nor was his office, like FDR's, filled with ship models and nautical paintings. In fact, the failed haberdasher and landlubber from Independence, Missouri, had virtually no idea of what it meant to love the sea and ships, though as an enlisted soldier during World War I he had traveled to the trenches aboard a seized Germany liner. In any case, as a matter of national prestige and security, Truman engaged himself in the rebuilding of the American merchant fleet with gusto, supporting the expansion of the passenger vessel fleet as "an important element of national security." In addition, he ordered that the many recent advances in naval technology "must be incorporated" into America's passenger liners, for "a well-balanced modern merchant fleet."[15]

On May 16, 1945, a week after V-E Day, Harris wrote Maritime Commission chairman Vice Admiral Emory Land with a specific request. "We are now convinced that an investigation should be made into the feasibility, practicability and commercial possibilities of constructing two 30-knot vessels to maintain a weekly express service to Channel ports," Harris wrote, and that "the United States Lines has the experience and facilities, and would undertake to operate the service on some mutually satisfactory basis."[16]

Admiral Land responded to Harris two weeks later. He scolded the United States Lines president for not being specific enough in his request for an economic study. "One conclusion that might be drawn from your letter," Land wrote back, "is that the 'burden of proof' is passed from United States Lines Company to the U.S. Maritime Commission. Another conclusion that might be drawn from it is that the United States Lines Company is not prepared to make a definite commitment to the project."[17] Land wanted a firm capital commitment from the United States Lines before he would think about committing government money to a feasibility study.

William Francis knew that Land had a staff of naval architects on the government payroll and that James L. Bates, the head of the Maritime Commission's Bureau of Engineering, was already working on his own designs for large transatlantic liners. If Land remained chairman of the Maritime Commission after the war, Gibbs would probably not be the man to design the new liners. William Francis hoped that Land, who had already served a full career as a Navy officer, would resign from the Maritime Commission when the war ended.

When General Franklin retook the reins of the United States Lines, he sensed that finally the federal government seemed willing to help him build at least one new passenger liner. But the cautious shipping man was still concerned about cost, and he wanted as big a construction subsidy as possible in order to move forward.

As the United States Lines management made plans for postwar expansion, the Gibbs & Cox team refined their designs for the superliner prototype in preparation for the big meeting between William Francis Gibbs, his brother, Frederic, and General John Franklin. They decided that the best place to start was the hull of the *America*. That ship's design drew on *Normandie* and *Bremen*, leading to a medium-sized liner that was sturdy and capacious, but also had a relatively slim hull below the waterline. The crucial design metric was the "prismatic coefficient," a measure of the fineness of the underwater portion of the ship's hull. This crucial metric is defined as the volume of the hull divided by the multiple of the waterline length and the cross-sectional area of the mid ship (widest portion) of the vessel. A vessel with a high prismatic coefficient has a hull similar to a rectangular box, with a blunt bow and stern. Such vessels were built for capacity rather than speed, and were not very efficient as they moved through the water. A vessel with a low prismatic coefficient, like *America,* had a sharp bow and stern and a wide center section, and often sacrificed stability for speed.

For his new liner, William Francis wanted a very low .559 prismatic coefficient, which meant that the new design, Design 12201, had an even narrower underwater bow than *America,* as well as less mass in the stern section.

America's hull also had one severe design flaw: the underwater portion of the prow was almost perpendicular, leading to excessive pounding in heavy weather and a reduction of speed. To fix the problem, Gibbs ordered Buermann to design the underwater portion of the new ship with its prow raked forward, decreasing the drag on the hull.[18] And rather than a large bulbous bow like those of the European vessels, the superliner was given a very small bulb. It was a refinement of the feature pioneered by Gibbs's mentor, Admiral Taylor, nearly forty years before.

A low prismatic coefficient meant that something had to be done to lighten the ship's upper works, namely the superstructure—those several decks rising above the hull that house a passenger ship's most expensive accommodations. To lower the ship's center of gravity and to increase its stability, 12201's designers decided that its superstructure would be built of a lightweight material: aluminum. This would not hurt the ship's structural integrity, since the hull itself would be supported by the solid steel strength-deck below the superstructure. The ship would also have a relatively low profile compared to European liners: a hull depth, from the strength deck to the keel, of 74 feet, 3 inches. By contrast, *Queen Mary*'s hull had a depth of 92 feet, 6 inches.[19] She also had more and heavier superstructure decks than the Gibbs & Cox prototype. *Queen Mary*'s upper decks were made of steel and packed with heavy woodwork and overstuffed furniture.

William Francis would later hint that he used the proportions of two famous German vessels, not the British ships, as models. One was *Leviathan,* a ship that he had refurbished in the 1920s and knew well. The other was *Europa, Bremen*'s surviving sister ship. Gibbs acknowledged "the remarkable agreement in length and beam" between Design 12201 and the German ships, "the variation being largely in the strength deck and the greater height of the superstructures above the strength deck, in the case of the *Leviathan* and *Europa.*"[20] His lower, lighter superstructure would be an advantage. He also gave his ship a draft shallower than either of the two older ships': a mere 31 feet.

The shallow draft was possible because of the aluminum superstructure, which also eliminated the need for upper deck expansion joints that had proved so dangerous on older liners. Gibbs's chief naval architect, Matthew Forrest, said later that if Design 12201 had used conventional construction techniques, five such joints would have been required—"a never-ending source of trouble . . . they leak, they creak, and they groan . . . a nuisance requiring continual upkeep."[21] Nor would the ship have split funnel uptakes, which had so compromised the superstructure decks in *Leviathan* and *Majestic* and made them vulnerable to cracking in heavy seas.

Finally there was the matter of compartmentalization. Most ships of the time could remain afloat with as many as two compartments open to the sea; more meant serious danger. Gibbs told Buermann and Forrest that 12201 should be a "five compartment" ship, meaning that with as many as five of twenty watertight compartments flooded, the ship would remain afloat and stable. This far exceeded the standard he set with the construction of *Malolo* in 1927, which had survived a catastrophic impact comparable to the one that sent *Titanic* to the bottom. Not only that, but the fuel and ballast tanks on Gibbs's superliner created a double skin that extended high up the sides of the ship, protecting her from sinking after grounding or colliding with another ship. As an additional safety measure, the ship would have two separate engine rooms. If one was hit by a torpedo, the ship would still be able to sail under her own steam.

The end result of all the work was a rakish, sleek superliner nearly 1,000 feet long but displacing only 45,400 tons of water. By comparison, *Queen Mary,* just 30 feet longer and 17 feet wider, displaced nearly twice the amount of water, 77,400 tons.[22] This would give Design 12201 a speed advantage over the older British ship, even if her engines did not.

But on the other side of the drafting room floor, the Engine Department was already designing a power plant that would provide all the speed the boss wanted. Presiding over the effort was department chief Walter Bachman, whom Gibbs called "the greatest marine engineer in the world." Bachman knew that Forrest and Buermann's narrow hull design, while offering important advantages, would also pose problems for engine design. Torque caused by the propellers had sometimes led to an almost unbearable shaking in the stern section. *Normandie* had one of many bad cases of seagoing St. Vitus' dance. For Design 12201, Bachman and his engineers planned on four steam turbines, producing as much as 60,000 horsepower per propeller shaft, all of which would be handled by four sets of double-reduction gears. Because the American high-pressure, high-temperature turbines were smaller and more efficient than earlier engines, they could easily be placed inside Design 12201's narrow hull, where they could operate with minimal vibration. After years of seeing the engines perform superbly in the Navy destroyers he had designed, Gibbs was certain the technology was safe. Gibbs and Bachman fervently believed the American turbines could easily outperform *Queen Mary*'s four British Parsons turbines, which when connected to propellers by double-reduction gears, produced about 158,000 horsepower at full throttle. So for Design 12201, the Engine Department confidently mocked up an array of American-built steam turbines in the model shop.

By the end of the week, the Gibbs & Cox marine engineers gave

William Francis what they had produced. The preliminary design showed the ship having a sea speed of 34 knots—unprecedented in liner history. The engines would produce a normal power of 200,000 shaft horsepower (total horsepower delivered to all four propeller shafts) and a maximum shaft horsepower of 240,000. Because of the high-temperature, high-pressure steam technology, the power plant was extremely compact. The four turbines and eight boilers could fit into the space reserved for *Leviathan*'s boilers alone, while producing nearly three times more power than the older ship.[23] Thus, even with a slimmer hull, vast amounts of space would be freed up for cargo, fuel, and passenger amenities. And Gibbs's smaller, more lightweight ship was projected to be able to carry almost as many passengers as the larger *Queens*—about two thousand in three classes.

The Gibbs brothers knew that General Franklin was on record as a supporter of moderate-sized liners only, and was also very conservative with his company's money. Their initial projections were that their Design 12201 would cost $50 million to build, $11 million more than Bates's Design IV. The brothers needed to sell this bigger, more expensive ship, so they decided to first present their company's moderate-sized Design 11811. Its advances over United States Lines' *America* would underline Gibbs & Cox's leadership in naval architecture. Then, if Franklin gave them an opening, they would bring forward Design 12201—the thousand-foot-long ship.

With the initial proposal complete, William Francis prepared his sales pitch for General John Franklin, and in February 1946, he decided to make a lunch date with the United States Lines president at the Broad Street Club in lower Manhattan. The timing was critical—his enemy Admiral Land had resigned as chair of the Maritime Commission only a few weeks earlier, and William Francis Gibbs, ever the opportunist, saw a rare chance to make his case.

PRIMACY ON THE SEAS

On February 6, 1946, General Franklin and the Gibbs brothers met for lunch at the Broad Street Club. Below the windows of the swank interior, tugboats, freighters, and ferries glided across the shimmering expanse of New York harbor. Nine months after V-E Day, troopships packed with soldiers were also on the water, as millions of American troops continued to trickle home. "Welcome Home!" banners on the pier heads greeted the ships as they pulled into berths on the Hudson River, and the bellow of ships' whistles met the cheers of waiting families.

The most famous of the troopships was due into New York once more in just a few days: the great British ocean liner *Queen Mary*. "At a speed never before realized in war," Winston Churchill wrote, "they carried over a million men to defend the liberties of civilization." Now ships as well as people were returning to peace. The troops on board no longer lay awake at night worrying about attacks from submarines and Luftwaffe bombers. Nor did they dread the upcoming combat they knew they would face on the beaches of Normandy. As of early 1946, *Queen Elizabeth* had just been released from military duty and returned to the Cunard Line to finally be completed as a luxury passenger ship. *Queen Mary* and other liners would soon follow her.

William Francis knew that in the coming months, workers would swarm aboard the European ships that survived the war, reinstalling furnishings and artwork long languishing in storage, tuning up hard-worked turbines, and sanding soldier graffiti from decks and railings. Hulls would no longer be rust-streaked wartime gray, but gleaming peacetime black and white. Funnels would again bear colorful company livery: the red and black of the French Line, the orange and black of the Cunard Line, and the red, white, and green of the Italian Line. Salons and corridors would soon again echo with music, laughter, and clinking champagne glasses, decks lined with thousands of cheering passengers.

As he sat down to lunch, William Francis Gibbs believed his dream ship would become the leading lady of the great pageant to come. The victorious United States was now the wealthiest, most powerful nation on earth, and deserved a ship worthy of bearing her country's name and colors.

Clearing the lunch table, William Francis Gibbs—whose paper-white face and patched jacket stood out in a dining room full of ruddy complexions and blue suits—solemnly presented a set of drawings and specifications. They were for Design 11811—a ship beautifully conceived for United States Lines passenger service, but not in the same league as the European giants such as *Queen Mary.* After taking in Frederic's financial presentation, General Franklin seemed interested, but remained noncommittal.[1]

Sensing Franklin's lack of enthusiasm, Frederic shifted tack and said that "a far faster and more outstanding ship could readily be designed." And one that also made sense financially, Frederic added, speaking as one risk-averse man to another. The topper: "It might be that one such ship which would place the U.S. Lines in a preeminent position."

William Francis Gibbs jumped in and told Franklin that his team had already produced plans for that ship: Design 12201. This was to be an ocean liner, he said, that would not only be the fastest ship in the world, but also the safest and the most beautiful ever built. It would be a ship big and luxurious enough to steal travelers from the *Queens*. And it would have the design and power needed to snatch the fabled Blue

Riband from *Queen Mary,* clinching an honor no American-built ship had held for almost a century.

Franklin left the Broad Street Club that afternoon far from convinced about going ahead. But the Gibbs brothers had given him something to think about. The general understood the financial commitment involved in the superliner, but he also knew his company had the backing of Vincent Astor, still the United States Lines' largest single shareholder. Astor, recognized as one of the nation's leading venture capitalists, gave his friend Franklin permission to look into things further.

One month after the Broad Street Club lunch, on March 4, Franklin met with the Gibbs brothers again. Franklin said that he had given their proposal "careful consideration" and had decided he would be willing to move ahead, provided the Gibbs brothers could design a liner "superior to the *Queen Mary* and *Queen Elizabeth* in speed" and deliver it for $50 million, equivalent to about ten times that today.[2]

"We will have to rejuvenate our passenger service," William Francis recalled Franklin telling them. The superliner must be one "the public can get behind, like a Cup defender—a sort of mythical flagship of our fleet."

But as Franklin also insisted, and as he told his board, the project could only get the blessing of government if she was "quickly convertible to a transport."[3] The liner would have to be able to switch from a luxury liner into the world's fastest troop transport within forty-eight hours, ready to carry 14,000 soldiers anywhere on earth without refueling. The conversion capability, the Gibbs brothers insisted, was not a problem. Unlike previous liners, peacetime furniture and fittings could be easily removed. As for speed and cost, they said, a ship with a "sustained sea speed of 33 knots, corresponding to the scheduled 28.5 knots of *Queen Mary,* could be designed and built at a figure of about $50 million."[4] The ship would match the dimensions Gibbs had laid out to his chief hull designer: 990 feet long (over three football fields in length) and 101.5 feet wide, with a draft of 31 feet. She would weigh in at about 50,000 tons, significantly lighter and hence more fuel efficient than the 80,000-ton *Queen Elizabeth* and *Queen Mary.*

The brothers also promised she would incorporate technologies perfected during World War II, features that would give her the greatest power-to-weight ratio in the history of commercial shipping, fast enough to outrun any submarine. Although the new ship would be luxurious and boast superlative service, unlike her European predecessors she would be designed for wartime use first and foremost. And her sleek and sensuous lines would take the American public's breath away.

Franklin was sold. Now he had to find a way to pay for her. On March 26, he sent a proposal from his office at One Broadway to the United States Maritime Commission, the government agency in Washington that would determine how much money, if any, his company would receive for the project. Franklin told the commission that his company sought "to develop and construct, with the aid of a 50% subsidy, a new fast trans-Atlantic passenger liner to be operated on the North Atlantic . . . on a two-week turnaround from New York to Plymouth, England, Havre, France, and return by way of Southampton, England."

The new ship, Franklin promised, would incorporate "the advances made during the war in superheat, high pressure, strong light metals, ventilation and air conditioning," as well as fireproof construction. She would be so far ahead of her time, Franklin believed, that she would ensure American dominance of the prestigious North Atlantic sea-lanes for the next twenty years.[5]

William Francis and Frederic Gibbs returned to the Gibbs & Cox offices at 21 West Street. They were elated but knew that during the century before, ambitious men had also staked fortunes, reputations, and national honor on bigger, faster, and grander ships designed to dominate the North Atlantic. A fine ship was a moneymaker, a wartime asset, and a point of immense national pride. A bad one was a money loser, a mechanical nightmare, and a national embarrassment.

The ship the Gibbs brothers had sold to the United States Lines in the spring of 1946 would be more than just a proud bearer of the American flag on the high seas, more than just a massive commercial endeavor. The superliner on his drafting board, William Francis Gibbs vowed, would be the greatest ship in the world, maybe even the greatest ship ever to sail.

Within a month, Gibbs & Cox and the United States Lines came together in agreement. They would go forward with plans for this radical new design: the world's most advanced and fastest superliner.

When the Maritime Commission's naval architect, James L. Bates, heard the United States Lines had accepted the Gibbs & Cox design, he refused to take the decision as final. Putting aside his "Design IV"—the big ship that had goaded William Francis Gibbs into Design 12201 in the first place—Bates prepared to leapfrog the Gibbs & Cox design with plans for a liner even bigger than 12201 and almost as big as *Queen Mary.* On April 6, he brought his proposal to General Franklin.

But Bates no longer had his pugnacious boss Admiral Land to pull strings for him—he had retired from the Maritime Commission only a month before the Gibbs brothers and Franklin had met for lunch. Moreover, the president of the United States Lines had no plans to build another *Queen Mary* or *Queen Elizabeth.* Franklin didn't want a bigger ship, but an exciting, fast, and efficient one. He also wanted economy, especially in fuel consumption: government subsidies supplemented crew wages and other labor costs, but not fuel. Franklin had told the Gibbs brothers that construction costs must come in at $50 million, which was considerably below the estimated $75 million for Bates's new design. All in all, Gibbs & Cox's Design 12201 fit his requirements, and Franklin saw no reason to change his mind.

Meanwhile, Gibbs and his staff had been leafing through the papers Bates had just published. This included a recent paper to the Society of Naval Architects and Marine Engineers titled "Aspects of Large Passenger Liner Design." William Francis always made sure he understood his competition, and this was no exception. The Gibbs & Cox team carefully studied Bates's design specifications, and then had the firm's model shop prepare a series of hull models. The first would be Bates's Design IV, the 930-foot liner; the second would be Bates's *Queen Mary*–like,

1,006-foot liner; and the third would be their own, 990-foot Design 12201.

Gibbs then asked his firm's chief engine designer, Walter Bachman, to calculate how much horsepower would be needed to drive each ship at various speeds. On April 9, 1946, only three days after General Franklin had met with James Bates, Bachman presented his two key findings:

- At 30 knots, Bates's 1,006-foot design would require 124,000 shaft horsepower, 4.5 percent more horsepower than Design 12201, and consume 15.5 percent more fuel.
- At 35 knots, Bates's 1,006-foot design would require 204,000 shaft horsepower, 2.9 percent more horsepower than Design 12201, and consume 10.9 percent more fuel.

Bachman's calculations and model tests confirmed to Gibbs that Design 12201 was by far the more efficient ship.

It was another year until President Truman gave the Maritime Commission approval to proceed with plans to finance this new passenger liner. In the meantime, another interesting document landed on Gibbs's desk, this time from his friends in the Navy Department. "Through Naval Intelligence," William Francis wrote to his brother, "information has been obtained with respect to a German express liner under design by Dr. Gustav Bauer of the Weser Yard at Bremen in 1938, the purpose being to win the blue ribbon back for the Germans in competition with the *Queen Mary* and *Queen Elizabeth.*"

The Gibbs & Cox design team found themselves facing a fearsome-looking beast of a ship, each line and curve oozing with 1938 Munich-era Nazi swagger. It would have stretched 1,024 feet in length and had a beam of 111 feet. Judging by its size and engine power, cost had been no object. "This vessel," Gibbs observed, "was provided with five propellers driven by a turbo-electric plant. The normal power was slated to be 240,000, for a speed of 34 knots and a maximum power of 300,000."[6]

The proposed ship would have been a super-*Bremen,* and Gibbs had been an ardent admirer of German engineering and sleek styling. The war had stopped its construction, but Gibbs still wanted to know how the newly discovered Aryan *übermenschen* vessel might have stacked up against Design 12201. He asked Matthew Forrest and Walter Bachman to find out. The two ran the model tests and calculations at the David W. Taylor Model Basin in Maryland. On May 3, 1947, Bachman phoned his boss and reported that Design 12201 needed 12 percent less horsepower than the German design to achieve a cruising speed of 33 knots. The following month, another member of the team provided a more detailed report. To achieve 35 knots, Design 12201 required 231,000 horsepower, while the German ship needed 235,500.[7]

After looking over the figures, Gibbs believed he could confidently tell the United States Lines and the Maritime Commission that his design 12201 was a more efficient design than both the German vessel and the two proposed by James L. Bates.

One lesson from Gibbs's design philosophy was clear: bigger did not necessarily mean better, let alone faster. He drew an analogy to the progress made in locomotive design during the past forty years. The new lightweight diesel locomotives, he said, had a 62 percent power-to-weight ratio increase over the old steam locomotives of forty years before. Likewise, Design 12201's tonnage would be a fraction of her European rivals' of the late 1920s and the 1930s. *Queen Mary*'s ratio of horsepower per ton of ship was 2.04, *Normandie*'s was 2.34, and *Europa*'s 1.68.[8]

"It is submitted," Gibbs concluded, "that these larger displacement ships are consistent only with the past and the limitations of obsolescent engineering, and not with the general principle that higher speed with economy of fuel necessitates light weight."

To General Franklin, who had been in charge of transporting American soldiers by sea during the war, a critical advantage of Design 12201 was its potential as a troop transport. To back that up, Gibbs sent Design 12201 plans to Admiral Frederick E. Haeberle, a top U.S. Navy ship constructor with extensive experience in designing large naval vessels. The admiral gave a nod to the superliner in an April 27, 1947, report. He echoed Churchill's sentiments about the military importance of the *Queens*: a superliner that could be converted into a troopship was an essential addition to the United States armed forces. The construction of moderate-sized liners might be all well and good for commerce—that wasn't up to Haeberle to say—"but if questions of national defense and security are taken into account and given proper weight, I believe that most of the arguments in favor of the larger type of vessel are valid."

About Design 12201, the Navy man had one criticism: he thought the two stacks were too big. But Haeberle wrote that "the vessel on the whole has a very pleasing appearance both above and below the water. There is nothing radical about it and it is clean-cut and neat." As for the ship's designer, the admiral concluded that Gibbs was the man for the job because of his knowledge of both commercial and naval construction. He "knows the problems and personnel of both fields very well," the admiral commented. "I think he has combined the two in a splendid ship."[9]

Finally, on March 11, 1947—nearly a year after General Franklin accepted the Gibbs superliner design—President Harry S. Truman formed a five-person advisory committee to develop postwar plans for the expansion of the American merchant marine. Any expansion, Truman insisted, was to serve the nation's domestic economy and its national security. Heading the President's Advisory Committee of the Merchant Marine was K. T. Keller, chairman of the Chrysler Corporation. Among the others was Vice Admiral Edward L. Cochrane, former chief of the Navy's Bureau of Ships. Still one of William Francis Gibbs's most ardent admirers, Cochrane had defended Gibbs & Cox during the House war profiteering investigation in 1944. And like his predecessor, Admiral David W. Taylor, Cochrane had an intense interest in Gibbs & Cox's Design 12201.

"We used all our pre-war passenger vessels, one-third of which were more than 20 years old, as transports or fleet auxiliaries," Truman wrote in his mandate to Advisory Committee members. "Some of these were sunk or badly damaged, and many others were so drastically altered for war use that their complete reconversion to peacetime needs is not economically justified."[10]

That the merchant fleet was not "well balanced" had never been more evident. Because of the enormous output of cargo vessels during the war, freight divisions of the major shipping lines now had hundreds of high-speed Victory ships and tankers perfectly suited to be operated on peacetime trade routes. Passenger liners were a different story. Attacks and accidents had reduced the passenger fleet from 127 ships with berths for 40,000 before the war to 36 ships with berths for less than 9,000 passengers.[11]

By then the United States Lines had put its remaining passenger and cargo fleet back into commercial service, with *America* making her maiden transatlantic voyage in late 1946. Those boarding her on that trip found her Art Deco interiors restored to their prewar luxury. The United States Lines had also poached kitchen staff from European lines by offering higher American wages, and boasted that chefs who "once cooked for the King of the Belgians and the *Ile de France* now work on the *America*."[12] Her new captain was also the commodore of the United States Lines: Harry Manning. But apart from the stripped-down *Washington,* the popular liner ran alone. She desperately needed a suitable, much larger running mate.

As *America* settled in to her schedule of crossings, Commodore Manning would find himself fuming about the country's failure to support shipbuilding. He blamed midwestern and southern politicians who did not grasp how important an American shipping industry was to the nation's commercial strength and international standing. "Our merchant marine policy history of the past has been like a hospital patient's fever chart," he would one day say. "It leaps up frantically when something serious goes wrong. Then it falls. Interest dies. It costs an awful lot of money to run a shipping enterprise that way."[13]

Meanwhile, the Cunard Line looked forward to a rosy postwar future. *Queen Mary,* her troopship duties over, joined her sister, *Queen*

Elizabeth, on the North Atlantic run on July 31, 1947. In the months that followed, the *Queens* were both booked to capacity, ferrying the likes of Winston Churchill and Duke Ellington across the Atlantic, as well as thousands of immigrants fleeing war-devastated Europe for a new start in America. Cunard had decided that the Blue Riband of the Atlantic would remain with *Queen Mary,* and the bigger *Queen Elizabeth* was never opened up to full speed. Still, Cunard chairman Sir Percy Bates's dream of a weekly, two-ship express service between Southampton and New York had become reality. The hard-nosed businessman did not live to see it, however; he died just before *Queen Elizabeth*'s peacetime maiden voyage.

William Francis Gibbs followed the *Queens*' weekly runs, methodically keeping track of their travel times and average speeds. He would cut out stories from the shipping news for his memorandum book—bad weather delays, broken bones caused by their notorious rolling, and any mechanical breakdown. But with his own superliner finally promising to leave the drawing boards, he was a happy man. The radio play-by-play of a Yankees game frequently blared in the background as he and his team worked and planned.

More liners arrived in New York harbor. Some were old favorites. Overcoming bomb-ravaged shipyards and factories, the French Line reintroduced the legendary *Ile de France* for passenger service in 1949. The Dutch revived their flagship, *Nieuw Amsterdam,* which had debuted only months before the Nazis invaded Poland. Others were new arrivals. The Italian Line, who had lost most of their vast fleet to torpedoes and bombs, launched an ambitious new shipbuilding campaign. Within a few years, the Italians would launch two elegant new sister ships: *Cristoforo Colombo* and *Andrea Doria.*

For a brief period, United States Lines appeared to have had the option of running another refurbished German vessel with *America:* the Norddeutscher Lloyd liner *Europa,* which the U.S. Army had captured in 1945 at her Bremerhaven dock (aside from *Queen Mary, Queen Elizabeth,* and *Ile de France,* the only other European superliner to survive the war). But the once-proud German Blue Riband holder was found to be badly neglected and obsolete—not to mention a firetrap. Her old electrical system consisted of solid copper electrical cables wrapped in rubber and cotton and nestled in wooden raceways. The setup caused several small electrical fires during several trooping voyages bringing American GIs home from the front.[14] Cracks also began forming in her hull, caused in part by her split funnel uptakes.[15] The United States Lines rejected her, and the United States government gave *Europa* to France in 1946 as reparations for the lost *Normandie*. The French, unable to build a new ship themselves, were willing to spend the money to completely transform the old German liner into the Gallicized *Liberté*. Her Germanic Art Deco interiors were ripped out and replaced by furnishings and art salvaged from *Normandie*.

General Franklin easily accepted the loss of the money-draining former *Europa*. It was best to use the company's resources for new construction, not a twenty-year-old German "also-ran."

William Francis Gibbs could not agree more.

Yet even as new and refitted ocean liners were prospering, there were signs that Atlantic travel might be on the cusp of a new era. From his office, Gibbs could see propeller-driven airplanes taking off and landing from the newly opened LaGuardia Airport. Aircraft design had advanced by leaps and bounds in the crucible of World War II, and *Marine Age* magazine warned that out of 640,000 people who crossed the Atlantic in 1947, 464,000 traveled by ship and 160,000 by plane. Experts began predicting that planes would continue to eat into the liners' share of the transatlantic business.[16]

William Francis read the article but was not worried about his yet-unborn ship. Of course the public wanted speed, he figured, but wealthy travelers also craved comfort, and few tourist-class passengers could afford to fly. Also, many people still thought that flying was unsafe. Finally, the new propeller planes were cramped, loud, and bumpy, especially during wintertime, over-ocean flights. Aircraft might have been fast compared to ships, but for travelers, being trapped in a flying metal tube for twelve hours or more—without any of the creature comforts available at sea—could seem very long indeed. All of this held down demand for air travel. "It is believed that air traffic will supplement this ship and not harm it," the naval architect wrote in a memo to his brother, "since the ability to leave New York on Wednesday and do business in London on Monday will mean that the average person will welcome the sea voyage either one way or the other, and for some time to come it is obvious that safety will be on the side of the ship."[17]

At United States Lines headquarters on Broadway, General Franklin thought that expanded passenger service was vital to the company's future, and that this new ship would do it.

With the basic superliner design pretty much complete, Franklin was ready to sell it to his company and to the U.S. government. On March 13, 1947, General Franklin presented the Gibbs & Cox design to the United States Lines board of directors, headed by Vincent Astor. In the thirteen months since the Broad Street Club lunch, his superliner Design 12201 had been measured against the competition and come out ahead. Harry Truman had only two days earlier shown a new government commitment to building technologically advanced passenger ships, and Admiral Haeberle at the U.S. Navy was about to come out with his positive report on superliners and, in particular, Design 12201. The response by the United States Lines board was also positive. But the company directors insisted on one important caveat: the company's contribution could not exceed half of the estimated $50 million price tag. The rest, the board said, must be borne by the federal government and the subsidies provided by the 1936 Merchant Marine Act.

The day after the board approved the superliner concept, John Franklin contacted Gibbs & Cox and told them to get a complete set of plans ready so that he could make the case to the Maritime Commission for a 50 percent construction subsidy. This would be a reach because the standard subsidy under the Merchant Marine Act was only 33⅓ percent. A 50 percent subsidy would be awarded only in extraordinary cases, but Franklin was convinced that Design 12201 represented an extraordinary passenger liner.

As his older brother supervised the design of the vessel, Frederic Gibbs steadily worked his adding machine in his solitary, austere office at One Broadway. Frederic's first goal was to demonstrate that operating a large passenger liner could turn a reasonable profit for its owners. His second was to convince the Maritime Commission why the maximum construction subsidy was justified—namely that high American shipbuilding costs made it prohibitively expensive to produce a ship with private funds alone. His third was to show why the government should provide an operational subsidy—to compensate the vessel's owners for high American labor and maintenance costs in order to make such a big ship competitive with European rivals. It was tedious but essential work.

United States Lines vice president Colonel Raymond Hicks tweaked Frederic's calculations and projected that an average, two-week round-trip transatlantic voyage would bring gross receipts of $464,720. Next came a long list of expenses, including unionized crew wages and overtime. The annual operating subsidy would defray these labor costs by slightly more than 50 percent; the figure was the same for marine insurance and other miscellaneous costs. Repairs and maintenance, such as annual dry-docking and overhaul, would receive a nearly 60 percent subsidy. But fuel, the biggest single expense, was the sole responsibility of the company. Frederic estimated that the ship would burn 8,000 tons of fuel oil each round trip, which meant $83,000 every two weeks to keep her bunkers filled.

All told, the annual, unsubsidized cost of running the ship came to about $9.5 million. Subtracting operating expenses from revenues, United States Lines figured on annual earnings of $4.3 million, with a subsidy. The company hoped for a 13 percent annual return on the company's total investment of $25 million.

Frederic Gibbs did not stop with the one-ship scenario. Like his brother, he strongly believed that having a two-ship schedule of weekly sailings was essential to compete effectively with the two British *Queens*. There was every reason to believe that a new weekly service by two fast, modern ships could capture the lion's share of the transatlantic traffic from the prewar *Queens*, increasing occupancy, revenues, and the bottom line.[18]

This seemed excellent news for a government looking to support a healthy shipbuilding industry and merchant marine. But there was another way to look at the project. Antispending pit bulls like Lindsay Warren, Truman's comptroller general, strongly believed that hefty government payments would contribute too much to the company's bottom line.

To shipbuilding advocates, the subsidies were essential to creating high-paying union jobs and business for shipyards and marine suppliers starved for work after the Truman defense cutbacks. But Warren raised a fair point, and perhaps a compelling political one. Was taxpayer money going to subsidize wages and the national economy, or was it lining the pockets of rich company shareholders like Vincent Astor?

During the next year, Frederic carefully reworked his projections, this time on the basis of one ship of the 12201 design. He came up with slightly higher ship earnings, but he had bad news on the construction front. Thanks to rising postwar labor and material costs, building one ship would cost not $50 million, but $70 million. Even assuming the higher-than-average, 50 percent government subsidy, the United States Lines' contribution would climb to $35 million—$10 million more than General Franklin had said the United States Lines could make.

Frederic presented his revised projections to John Franklin on July 17, 1947, and the general balked. A $35 million contribution was simply unacceptable. The interest payments on that amount would be enough to kill the deal for him. Franklin had to stay on Truman's good side. He knew the president was a tough customer.

After World War II, *Queen Mary, Queen Elizabeth,* and other great liners had brought thousands of soldiers back to American shores, but Truman realized that many other American troops would not be coming home any time soon. To maintain political stability in an economically and physically shattered Europe and Asia, America would be maintaining permanent military bases all over the world. To do otherwise might easily lead to the rise of new demagogues, Truman concluded.

And then there was the growing threat of the Soviet Union. Truman was present when a British politician delivered a speech at Westminster College in the president's home state of Missouri on March 5, 1946.

"From Stettin in the Baltic, to Trieste in the Adriatic," former prime minister Winston Churchill asserted, "an iron curtain has descended across the Continent. Behind that line lie all the capitals of the ancient states of Central and Eastern Europe."[19]

Earlier, George Kennan, chargé d'affaires at the U.S. Embassy in Moscow, had written his famous "Long Telegram" of February 22, 1946. "We have here a political force committed fanatically to the belief that with U.S., there can be no permanent modus vivendi," Kennan wrote; a political system believing "that it is desirable and necessary that the internal harmony of our society be disrupted, our traditional way of life be destroyed, the international authority of our state be broken, if Soviet power is to be secure."[20]

On November 1, 1947, the Merchant Marine Advisory Committee sent Truman its report. It found that tensions abroad meant prudence at home; namely, the country should not again find itself in a massive emergency buildup of its merchant marine. Its decline during the 1930s, caused by the Great Depression, scandal, and labor problems, left the country woefully unprepared to transport cargo and troops to the battlefields of Europe and the Pacific. A strong, postwar merchant marine, anchored by refitted wartime vessels and bolstered by new construction, would be integral to keep the peace and to wage war if necessary.[21]

The Advisory Committee then argued for not one, but two superliners. "The building of two express-type passenger carrying ships for the New York to Channel-port service," the report noted, "will help to restore the heavy war losses in transatlantic express passenger tonnage and will provide a type of vessel particularly useful in time of emergency."[22]

This was exactly the rationale, based on military need, that General Franklin and the Gibbs brothers wanted the president to read.

The Advisory Committee also supported continued subsidies, to encourage companies to invest in passenger ship construction. If the transatlantic liner business collapsed as it did during the Depression, the Advisory Committee wrote, "the commercial market value of such a vessel except for scrap because of complete loss of earning power . . . would become zero." However, as a military asset, the ship "might exceed several times the unpaid balance of the purchase price."[23] *Queen Mary* and *Queen Elizabeth,* the Advisory Committee was saying between the lines, were floating testaments to the idea that a big ship was more valuable than her commercial earning power in peacetime.

The message of the Advisory Committee was clear on ships. But it presented a problem for the president. Truman was committed to a strong defense, but two years after war's end, he was adamant that it was time to rein in the free-spending ways of the military. In the last years of war, the annual defense budget had risen to about $90 billion a year; Truman wanted to ratchet it down to around $15 billion.[24] Moreover, he believed that corruption easily resulted when the government and private companies worked in tandem. And as a former U.S. senator, he knew that the shipping lines had a particularly bad track record when it came to gorging on subsidies provided by the taxpayers.

Partly to get control of the military budget, Truman had advocated a unified Department of Defense, bringing the three military services (Army, Navy, and the newly independent Air Force) together in one National Military Establishment. The man Truman chose as his first defense secretary was James V. Forrestal, a son of Irish immigrants who had driven himself mercilessly to join the nation's elite. Fortunately for Gibbs, Forrestal was one of William Francis Gibbs's greatest champions during his time as secretary of the Navy, writing to the naval architect that his wartime designs were a "great national achievement."[25]

After the war, as defense secretary, Forrestal was the principal advocate for spending on defense, including the construction of a large troop carrier that could ferry 14,000 soldiers anywhere in the world. This, he thought, would constitute a powerful deterrent to the spread of communism in both Europe and Asia. For his part, Truman wanted Forrestal to show a tight civilian grip over the military and to support his defense cuts. The two men quickly began bickering over how much to cut and where. Forrestal's days as defense secretary were numbered, and his health declined rapidly as well.

As discussions continued on how, and how much, the government would pay for the new ship, United States Lines president John Franklin was meeting with the Maritime Commission and Gibbs & Cox to review the two construction bids received on the project. Both were far higher than the original $50 million Franklin had insisted upon when he had told the Gibbs brothers he would support Design 12201.

Newport News Shipbuilding and Dry Dock Company came with the low bid: $67,350,000. When hotel equipment (dishware, linens, galley equipment), Gibbs & Cox's design fee, and a small change order allowance were added in, however, the total estimated cost of construction came to $71,081,100.

The big number would kill the deal, something William Francis Gibbs could not accept. Now was the time to call on his brother Frederic's accounting skills. On December 14, 1948, the brothers met with General Franklin and other United States Lines executives at Frederic's office at One Broadway.

To prepare for the meeting, William Francis had asked Newport News for the cost if Design 12201 were a run-of-the-mill, twin-screw vessel—a "usual ship"—*without* any national defense features such as reinforced decking, Navy-grade wiring, and additional speed. The yard had come back with a figure of $44,470,700.[26] It was clear that the following of stringent military design standards—fireproofing, extra compartmentalization, and Navy-grade steel plating—caused the ship's cost to spiral frighteningly upward, and that it was only fair that the government should make up the difference.

Frederic then presented a financing scenario that allowed the United State Lines to keep its contribution at $25 million. First, the government would provide a 45 percent construction subsidy for building that $44.47 million "usual ship." This amounted to $20,011,815. The United States Lines would then contribute $24,458,885. Then, under the Merchant Marine Act of 1936, the government would pay for all extra "national defense" features—all the improvements that differentiated Design 12201 from that "usual ship"—which would come to $26.61 million. The total: $71.08 million, the Newport News bid.

On April 7, 1949, the four members of the United States Maritime

Commission began a three-day marathon meeting with General Franklin and William Francis Gibbs. On the agenda was a verdict on the construction of an American superliner. Also around a big table at the United States Lines' offices at One Broadway were the shipping company's directors, including Vincent Astor, United States Lines' vice president Raymond Hicks, company counsel Cletus Keating, and William Blewett, executive vice president of Newport News Shipbuilding. A large model of the ship (but one showing nothing below the waterline) sat perched on a shelf at one end of the conference room.[27]

After some pressure from the commission, General Franklin agreed to up his company's contribution to $28,087,216, with $1.3 million going toward the "national defense features."

One stumbling block was not yet addressed. The contract included an escalator clause that allowed for additional costs of about $5 million. The commission suggested the United States Lines should bear the brunt of those costs. This could push the shipping line's share above $33 million.

United States Lines balked. Company lawyer Keating said, "Mr. Commissioner, you have named a price which, in my personal opinion, is astronomical."

"I've had the devil's own time processing these figures to reach that low price," Maritime Commissioner Grenville Mellon replied in frustration. "If you really mean that, I am willing right now to reconsider the whole thing."

General Franklin weighed in with a little diplomacy. He argued that any increase in cost due to escalation would cause his company directors to balk, as "our company has a limited net worth." What Franklin wanted was a commitment to a subsidy of at least 50 percent, and that the United States Lines' financial commitment would remain fixed.

Commissioner Mellon threw up his hands and caved. "I think Mr. Keating's request is very extreme, General," he said. "I don't want to delay this thing further," he added, offering to provide the fourth vote needed for the project to go forward.[28]

One commissioner still objected. Raymond McKeough, disgusted with what he saw as sloppy analysis, maintained that Franklin had simply fixed $28 million as the maximum amount his company would contribute, period. It was, as McKeough vented later, "a reckless abandonment of orderly oversight to the required determinations," as well as a "neglect of an objective, unprejudiced, and uninfluenced professional search for the correct answer."[29]

Still, McKeough grudgingly voted in favor of the project. He would later say that he changed his vote under pressure.

The actual cost of the new superliner would eventually balloon to $79,422,469.[30]

A few hours after the meeting concluded, the Maritime Commission fired off a telegram to Newport News Shipbuilding. The contract was closed and construction was to start as soon as possible.

In the end, the United States Lines, the private corporation that would profit from the vessel, would put up only 35 percent of the total construction cost. The U.S. government would pay the rest, as well as an operational subsidy to offset the high costs of American labor. The government agreed to fund the national defense features that would make the project financially feasible and, it appeared, legally possible.

William Francis Gibbs's lifelong dream was now one major step closer to fruition. His shrewd negotiation skills had shepherded the project through more than three years of complex negotiations. But he did not live to negotiate contracts—he lived to build ships.

The contract for the construction of the superliner was signed on May 8, 1949, at the headquarters of the Maritime Commission. Those signing were Grenville Mellon of the Maritime Commission; J. B. Woodward Jr., president of Newport News Shipbuilding; and General John M. Franklin, president of the United States Lines.

A model of Design 12201 was then brought out for the members of the press, who immediately began taking notes and snapping pictures.

After forty years of planning, setbacks, and persistence, William Francis Gibbs was on the cusp of building the superliner that he first conceived in his parents' attic. Everything was falling into place: he had won not only the support of the nation's biggest shipping line, but the signed financial commitment of the United States government. He was not going to let the project die again, not now.

Although he was exultant, Gibbs stood stone-faced and silent by the model's side. A *New York Times* reporter asked Gibbs how long had he been designing the vessel. "I began jotting down notes for a superliner in 1916," he responded flatly, "but two wars got in the way."

Rumors were circulating that the Maritime Commission and the United States Lines were both reeling from sticker shock but had decided to go through with the deal anyway. "Sources close to the transaction," the *New York Times* reported shortly before the contract was signed, "expressed the opinion yesterday that the difference in figures between the company and the commission would be too small to militate against it."[31]

There was also a lot of speculation in the press about the name of the new ship. Suggestions appearing in the newspapers included *Mayflower, Columbia,* even *American Engineer.*

It was at the May 8 contract signing that the official name was announced: *United States.*[32]

With the signed contract in hand, Gibbs finally agreed to an interview. The American public, especially the well-to-do traveling set, was eager to hear from the ship's designer. Gibbs sat down for an interview with George Horne, shipping editor for the *New York Times,* who was eager to know everything about the new ship.

But Gibbs was extremely cagey. Asked about the published dimensions of the proposed vessel, he said they were "inaccurate." Asked if the ship would be powered by a nuclear reactor, he stonewalled.

"I'm not going to tell anybody anything," Gibbs finally said. "We have spilled too many secrets in this country. We have something here, and why tell them what we have found. Why should we save them the ten years of research we have gone through?"

The naval architect freely admitted, the reporter wrote, that he "intends to use every stratagem to heighten the mystery" surrounding the project. The rationale for the secrecy, Gibbs claimed, was two-pronged: "possible national enemies who do not yet dream of the revolutionary advances represented in the design, and friendly maritime rivals in the same un-blissful state of not knowing."

Gibbs also knew silence would build public interest in his ship. The less he said, the more America wanted to know. He would play the cat-and-mouse game with the press as long as possible.

But what he would tell Horne and all who asked was whom the ship served. The new superliner, he argued, would be "a break for her owners, the people of the United States of America" and "when they see what she can do, the people won't be sorry." The men and women of his firm, he would say later, wanted something simple: "They knew they were trying for the greatest ship in the world, and they knew they were doing it as trustees for the people of the United States."[33]

Truman's comptroller general, Lindsay Warren, did not share Gibbs's views. He had gotten his hands on the contract. After reading it, Warren and his staff got to work on a report to Congress.

In the meantime, William Francis started to assemble a scrapbook of newspaper clippings on the Big Ship. He basked in the public support, but also delighted at how the reporters were kept in the dark about the ship's specifics.

The same day that the contract was signed, hundreds of workers bent over the drafting tables at 21 West Street heard a firm, patrician voice over the public address system. Engineers dropped their pencils and slide rules. Telephone receivers were put down, and secretaries stopped working except for the one taking down Mr. Gibbs's remarks.

"We can all take pride and we can all be thankful for the events of April 7, 1949," he said. "It marks, as I told you, a great event in the maritime history of the United States because on April 7, the United States, a giant among nations, shook off its lethargy in connection with things maritime and determined to take its rightful place in primacy on the seas."

The negotiations, he said, were "complicated and difficult," but ultimately "satisfactory." The ship, he estimated, would be delivered in 1952 and "may be the possessor of the famed Blue Ribbon as the fastest passenger ship afloat." After reminding his staff that it had been a hundred years since an American ship had captured the Atlantic speed record, he continued, "You have all had a part in the performance and the effort that brought this result and advantage to the people of the United States. Each of you who by heart or hand or enthusiasm have stood beside and behind those of us who were on the firing line has contributed a real and lasting part in the result.

"On behalf of myself and the people of the United States," he said in closing, "I thank you from the bottom of my heart."[34]

But trouble was brewing. As Gibbs, Franklin, and the Maritime Commission were finalizing the superliner contract, President Truman decided to replace the intransigent and increasingly troubled Secretary of Defense Forrestal, who had so admired Gibbs's work. The sallow-faced Forrestal complied on March 28, 1949. Shattered and dazed, one of the Navy's greatest champions suffered a complete emotional breakdown. He checked into the Bethesda Naval Hospital just outside of Washington. On May 22, Forrestal jumped out of a sixteenth-floor window to his death.

There was also more turmoil in the military establishment that threatened to impact the Big Ship. Two weeks before the contract was signed for the construction of *United States,* the Truman administration had canceled a contract for a warship even more ambitious than a transatlantic superliner.

World War II might have proven the military value of the superliner, but a much bigger revolution had taken place in the skies. Long-range fighter planes, based on an aircraft carrier, could devastate unprotected fleets. Even the mightiest battleship, bristling with antiaircraft guns, was vulnerable to a well-coordinated torpedo attack from the air. Just after Pearl Harbor, for example, Japanese bombers sank the British battleships *Prince of Wales* and *Repulse* off Singapore, with a huge loss of life. The future of the Navy, American military brass reasoned, lay in a well-coordinated use of air and sea power, which meant building bigger, faster, and technologically advanced aircraft carriers. In the age of the atomic bomb, a carrier also eliminated the need to plant air bases overseas should America choose to launch a nuclear attack on the Soviet Union.

Before his death, Secretary of Defense Forrestal and his allies in the Navy were pushing for the construction of a new generation of super aircraft carriers. His successor, Louis A. Johnson, found himself looking at a contract with Newport News Shipbuilding to build the largest carrier in the world. Costing a staggering $189 million, nearly three times as much as Gibbs's superliner, the first of this new class of supercarrier, coincidentally named USS *United States,* would be able to carry a staggeringly lethal number of planes: 54 fighter jets and 12 nuclear-capable bombers.

William Francis Gibbs had never designed an aircraft carrier, and never wanted to. Although he sent his chief engineer, Walter Bachman, to investigate the proposed ship, Gibbs remained totally uninterested in combining air and sea power. "His attitude? A wet blanket," one colleague said.[35] But the aircraft carrier was slotted to take up the biggest slipway at Newport News, a real problem for Gibbs's timetable. On April 18, 1949, a few weeks before the contract for the passenger liner SS *United States* was signed, the keel of the supercarrier USS *United States* was laid on the floor of Slipway #10. Five days later, Secretary Johnson canceled the contract. The project, he said, was just too expensive. Newport News had no choice but to break up the carrier's keel plates.

The slipway was now clear for Gibbs's superliner. But the fallout was a huge public relations disaster for the Truman administration. Secretary of the Navy John L. Sullivan declared that what Johnson did to the supercarrier USS *United States* was outrageous—"drastically and arbitrarily changing and restricting the operations plans of an armed service without consultation of that service."[36] In the "Revolt of the Admirals" that followed, Secretary Sullivan resigned. So did Chief of Naval Operations Admiral Louis Denfield, while many others in the Navy spoke out publicly—a shocking action against a decision backed by the president and his top civilian defense official. Truman only dug in harder.

Johnson's decision to kill the supercarrier opened the floodgates for further review of big government spending projects, specifically the construction of passenger ships. Leading the charge was Truman's comptroller general, Lindsay Warren, who had made a name for himself nailing war profiteers during World War II. Now his really big target was United States Lines, which had just received more than $42 million in construction and national defense subsidies to build a superliner. On July 11, 1949, three months after the United States Lines contract was signed, and as preparations for its keel-laying went forward, Lindsay Warren and the General Accounting Office reported to the Senate that the contracts made with the United States Lines was riddled with "various irregular procedures, inaccurate calculations, and unjustifiably liberal interpretations of statutory language."[37] Warren concluded that the proper response of the United States Lines and the Maritime Commission after learning of the $73 million price tag was to build a smaller ship, or to drop the project entirely. "It is apparent," he said, "that the evidence supporting the Commission's calculations falls far short of the 'convincing evidence' required by the act for an approval for a subsidy in excess of 33⅓ percent."[38]

After Warren asked for a congressional committee to investigate the Maritime Commission, the Committee on Expenditures, chaired by Congressman William Dawson of Illinois, set to work that fall. They hauled in each of the four maritime commissioners for intense questioning. The answers were not good ones. Commissioner Grenville Mellon, the man who signed the superliner construction contract, admitted that he understood "only vaguely the procedures involved in the staff's calculations, [and] he claimed to have detected in them mathematic errors which he failed to correct because they 'were all recited, they were in the record.'"[39]

Congressman Dawson could barely contain his rage. "You are not given a blank check," he said, "to say 'I believe this is convincing, and upon this I am going to do as I please with the Government money.'"[40]

The commissioner who completely broke ranks was Raymond McKeough, who now rehashed all the objections he had made before finally voting to approve the contract. McKeough told the committee that the commission's goal of a fixed $28 million contribution from the United States Lines resulted in a "reckless abandonment of an orderly, systematic approach to the required determinations."[41]

Truman, meanwhile, thought about abolishing the Maritime Commission as an independent body and putting it under the direct control of one of his cabinet secretaries. He asked his attorney general to look into the House's report and make a legal recommendation.

After reading Warren's report, Frederic Gibbs fired off a rebuttal to Commissioner Mellon. Frederic asserted that the large national defense allowance in the language of the contract was not some kind of ruse, but the result of demands placed on the designers by the Navy, which made it "inevitable that national defense features were interwoven into practically every part of the ship." He also insisted that more stringent American naval and commercial regulations regarding safety, especially fireproofing, did not benefit the owners. "The owner does not make a penny more because the vessel is fireproofed," Frederic wrote.[42]

In short, the contract price was legitimate in that it allowed United States Lines to build a superior dual-purpose ship that was otherwise not possible.

In spite of the controversy, on February 8, 1950, a giant shipyard crane gently dropped a fifty-five-ton steel plate onto the same huge dry dock floor where the keel of the carrier USS *United States* had lain ten months before. It was the first keel plate for "Hull Number 488," the official yard name for new United States Lines flagship.

The massive plate was lowered into a canyon of wooden scaffolding that lined the entire dry dock. One of the men there was shipwright Tom Paris, whose job was to help set the huge piece of steel precisely in place. "When the first piece of keel came in, it was nothing but a flat plate," he recalled. "I set the plate down on the dock floor, I put the centerline on the frameline, and then tied it down to the dock floor."[43]

There was no keel-laying ceremony to speak of. A group of shipyard executives, naval officers, and reporters, all dressed in winter coats, huddled together on a crude wooden platform overlooking the dry dock floor. William Francis stood at the center of the group, his black lapels turned up against the wind and hands clasped together.

With the keel down, William Francis Gibbs's ship began to rise.

But once construction started, newspaper reporters and photographers were barred from the construction site. The way the ship was built was a military secret. Gibbs ordered that all images and plans be kept under lock and key, and the three thousand people who worked at the site were not to say anything to anybody. If a worker was caught discussing construction details in public, he faced immediate dismissal, a prospect that sealed lips in a company town like Newport News, Virginia.

A few months later, President Truman decided that the best way to resolve the issues raised by the new superliner was to abolish the Maritime Commission. Its powers would be assumed by the newly created Federal Maritime Board and the Maritime Administration, which would be under the control of the Department of Commerce. The Maritime Board would supervise rates, while the Maritime Administration would carefully handle the granting of construction and operating subsidies. The president then tapped Vice Admiral Edward L. Cochrane, the former chief of the U.S. Navy's Bureau of Ships and the man who had defended Gibbs against war profiteering charges. He accepted the position of head of the Maritime Administration.

Gibbs now assumed that his ship, now well under construction, was safe from the fate that had befallen the supercarrier USS *United States*. He was wrong.

A MIGHTY SWEET BABY RISES

All during the spring and summer of 1950, work on the ship proceeded steadily. William Francis made frequent trips to the shipyard to keep an eye on developments, making sure that everything was being built just as he had envisioned it. Shipyard workers knew they could never slack off, because Gibbs had a habit of sneaking up on them. The sight of the darkly dressed man in the floppy hat put the fear of God in everyone from foreman to welder. "Everybody knew Gibbs was there," shipwright Tom Paris said. "They knew not to speak to him, but they knew when he came by, he was making sure that everything was put just as he put it."[1]

As he wandered around the site, blueprints under his arm, Gibbs must have reveled in the complete control he finally had over the project. After forty years of dreaming, the Big Ship was becoming a physical reality.

Once the keel was laid, workers attached huge steel ribs, which spread out from the keel as if they grew from an enormous backbone. The ribs were then slowly sheathed by 152,000 high-tensile steel plates that were welded together, not riveted. Welding—a faster and superior method of construction that created a flat, hydrodynamic hull surface—was used sometimes on big liners before the war, but never on this scale.

Gibbs & Cox had perfected a way to weld the mass-produced Liberty ships together, with most components made off-site by subcontractors. To build *United States,* Gibbs had to coordinate production and delivery with eight hundred suppliers in sixty-eight different locales.[2]

Special attention was given to the stern, which if not built correctly would vibrate badly from the force of the propellers. To make sure that the stern was rigid and strong, workers tied the structure together with an interlocking network of bulkheads, brackets, and girders.[3]

As the steel hull rose, machinery components were installed. The eight Babcock & Wilcox boilers were built off-site in Ohio, but Gibbs insisted that a prototype be tested at the Philadelphia Navy Yard before they were installed aboard the ship. A barge transported the boiler up to Philadelphia, where it passed muster, and back down to Newport News. The Westinghouse plant in Pittsburgh built the turbines, which were also carefully tested before installation.

Even as construction proceeded, a few key design elements continued to be refined. The ship's propellers required special attention and testing. Along with engines and underwater hull design, they were crucial to the ship's speed.[4]

And the propeller design lay in the province of a woman.

Despite his reputation for being uncomfortable around women, William Francis found one who was essential to the revolutionary design of his ship's propulsion system. Elaine Scholley Kaplan started working at Gibbs & Cox during the war, when she was an undergraduate mathematics major at Hunter College. In 1945 she had married a fellow engineer, Howard Kaplan, who also worked at Gibbs & Cox and was exempted from the draft on the recommendation of William Francis Gibbs.

The Kaplans' business and social lives were completely intertwined at Gibbs & Cox's. Men and women went out to dances together, bowled together, and played music together. The company's informal social goings-on, as well as announcements of engagements, weddings, and the birth of children, were published in a bound monthly magazine called *Compass Points.*

Not everyone enjoyed the firm's social side. One young engineer fresh out of the University of Michigan described the company Christmas party at the Downtown Athletic Association as "about as fun as a middle school dance."[5] But Elaine and Howard Kaplan thrived. Bill played the saxophone in the company swing band, the aptly named Destroyers. Both developed friendships with people who understood the demands and satisfactions of Gibbs & Cox's high-pressure, long-hour days.

As Elaine Kaplan rose to be a top propulsion engineer, she enjoyed success that testified to Gibbs & Cox's openness not just to women professionals, but also to those who were Jewish. At the same time, as one of only two women on the fifty-person design team (the other, Rebekah Dallas, was chief of the firm's female personnel), Kaplan understood the conventions of the 1950s that kept many women in support roles. She broke these barriers with intelligence and grace.

"She had an understated elegance," Kaplan's daughter Susan Caccavale recalled. "You could tell she was bright but she was so humble and unassuming that you would never know how accomplished and brilliant she was, which was part of her charm." As a young engineer, Kaplan earned respect not only for her intelligence but for her meticulous work. She quickly came to the attention of Walter Bachman, the company's chief marine engineer. "He was very kind and very patient, but also very brilliant," Caccavale recalled. "He became kind of a father figure to her."[6]

Kaplan's talent eventually gained her William Francis's high regard, although he could not entirely grasp the phenomenon of an attractive, hardheaded woman engineer. "Mrs. Kaplan to me is a complete and perfect mystery," Gibbs said of her. "How anybody can look the way Mrs. Kaplan looks and come up and talk to you on a technical subject is beyond me—I am not over it yet."[7]

The only workplace disagreement known between Gibbs and the young designer was the time he docked her a week of vacation for being late for work. They were very much alike; like Gibbs, Kaplan could explain complex engineering concepts in a way that any layman could understand.[8]

Gibbs thought so much of Elaine Kaplan that he gave her one of the most important assignments in his firm: designing the propulsion system for *United States.* And a simple idea perfected by Kaplan and Newport News Shipbuilding became one of the unique designs that Gibbs guarded as the ship's deepest secrets.

Vibration from engines and propellers remained one of the great problems faced by marine engineers when designing large, fast ships. As Gibbs knew from personal experience, vibration had rattled the sleep and shaken the dinner tables of passengers on *Lusitania, Normandie,* and even the *Queens*. The culprit was cavitation, in which fast-spinning propellers create tiny air bubbles that disturb the blades' grip in the water. Cavitation sends shock waves throughout the hull and wears down the huge, expensive propeller blades. And no ship ever had 60,000 horsepower forced onto each propeller shaft as *United States* would.

The challenge brought Kaplan deeply into the work at the shipyard. She began to make frequent trips down to Newport News by train. What were supposed to be day trips turned into weeklong marathon sessions, where she supervised testing of the propellers in the model tanks and kept close tabs on her engineering colleagues from the shipyard. At the hotel, the fashionable but frugal Kaplan washed her one outfit in the bathtub every night.

Kaplan's original designs for *United States*' propulsion system called for four 18-foot-diameter manganese bronze propellers, each with four blades. The four propeller shafts were enclosed in winglike structures known as bossings, which projected out from the stern and added to the ship's stability.[9] But tests showed that this traditional arrangement, despite all of Kaplan's attention to the size and pitch of the propeller blades, still created bad vibration at high speeds.

The solution that emerged from work by Kaplan and colleagues at Newport News was to configure the two sets of propellers in a new way: the two outboard props would have four blades; the two inboard props would have five. With more surface area, the inside, five-bladed propellers could better grip the water churned up by the two outboard props.

In April 1950, the yard ran a series of self-propulsion model tests using the new five-blade design. The results showed the ship would have an astounding, estimated top speed of about 37 knots at full power, with cavitation greatly reduced.[10]

Characteristically, William Francis did not lavish praise on his engineer. "Let me say that for my money, Mrs. Kaplan is quite the equal of any technical person that we have in this place," he said at a company dinner. He then admonished her: "Now, for God's sake don't get conceited and for God's sake come in on time."[11]

By May 5, as the propeller tests ended, the bottom of the ship was largely complete, and the sides and bulkheads were rising up from the dry dock floor. Tiers of wooden scaffolding surrounded the nascent hull. To an outside observer, she still looked more like a mammoth barge than a sleek ocean liner. The first hint of the ship's beautiful exterior lines was at the front of the ship, where the graceful bow was taking shape.

But eight months after construction began, it suddenly seemed that the Truman administration would yank the ship out from under William Francis Gibbs. The reason: military necessity.

After making a series of frantic phone calls, William Francis and Frederic Gibbs grimly called a meeting with their decorators in September 1950. Dorothy Marckwald and Anne Urquhart, who had been working for months on the ship's interior, came to Frederic's austere office at One Broadway. The two decorators had just been given word: *United States,* a third of the way through construction, was being seized by the Pentagon and would be completed as a troop transport. Military brass called General John Franklin of United States Lines to break the news, and Franklin then told the decorators that their contract was canceled and their services were no longer be needed. The two women were devastated—after two decades of designing interiors for Gibbs & Cox liners, this would be their crowning achievement.

But at Gibbs & Cox, neither brother was fazed. William Francis had suspected for some time that Secretary of Defense Louis Johnson was planning to make the move. The outbreak of war on the Korean peninsula in June 1950 had caught America unprepared. Not until September did General Douglas MacArthur's daring Inchon landing put U.S. forces on the offensive. By then political and military leaders in Washington were frantically working to rebuild a military that had been stripped of men and equipment by the demobilization and budget cuts after World War II. The Pentagon wanted ships, and they knew where to find a big one. In fact, during the summer, the *Newport News Daily Press* reported that a plan was being "kicked around by Washington officialdom . . . to abandon the construction of the superliner" in favor of some kind of military vessel.[12] Still, there was no public announcement about what exactly would be done with the Big Ship until September 16, 1950—when American GIs were still on the beachhead at Inchon.

During their meeting, the brothers assured the two anxious women that they would do everything in their power to get the ship back.

"This will not happen," Frederic Gibbs calmly told Dorothy Marckwald and Anne Urquhart.[13] William Francis promised to make good on the interior decorating contract and quietly asked his staff to keep working.

The seizure electrified the press, and reporters immediately started digging. News accounts appeared that claimed modifications to the ship's design were permanent, not cosmetic, making it very difficult for her to be reconverted into a passenger liner. On October 29, 1950, Walter Hamshar of the *New York Herald Tribune* reported that *United States*' big ballrooms and lounges would be cut up into smaller messes and wardrooms, and that almost all private toilets and showers, necessities for luxury passenger service, would be eliminated from the deck plans. The completion of the ship as a troop carrier, Hamshar reasoned, "will probably cut several million dollars from her estimated construction cost of $70,300,000, but the savings will not benefit taxpayers." This was because while construction costs would be reduced, the savings would be eaten up by the new installations needed to support the estimated 14,000 troops it would carry: additional freshwater tanks, condensers, lifeboats, windbreaks, and gun mounts.[14]

Savings or no savings, Defense Secretary Johnson and his Air Force and Army allies on the Joint Chiefs of Staff were determined to follow through on the seizure. But within days of the decision, Secretary Johnson was gone from office. The president had shared Johnson's cost-cutting approach but was fed up with his bombastic management style. On September 19, he called Johnson into the Oval Office and told him, "Lou, I've got to ask you to quit."[15]

Johnson was dumbfounded, but quickly complied. Within days, a new secretary of defense sat in the office—retired five-star general George C. Marshall, the respected Army chief of staff in World War II whom Churchill had called "the organizer of victory," and more recently, secretary of state and author of the Marshall Plan to rebuild Europe.

Sensing an opportunity, Admiral Edward Cochrane, chairman of the new Maritime Administration (successor to the old Maritime Commission), decided to have a word with the new secretary of defense, who the admiral knew valued efficiency. Cochrane was convinced that if *United States* were rebuilt as a troopship, much of the work could not be undone, at least not cheaply, and that the ship could not be completed in time to meet the urgent troop-carrying capacity of the moment. On November 1, just a month and a half after the seizure, Secretary of Defense Marshall announced that the Joint Chiefs of Staff had "reconsidered their previous recommendation and now recommend that the ship be completed as a commercial passenger liner." Marshall wrote Cochrane the same day, thanking him for his "whole hearted cooperation in this matter and wish[ing] to assure you that we shall endeavor to keep our requests as modest and reasonable as the international situation may permit."[16]

What happened at the meeting between Marshall and Cochrane is not known. What is known was Cochrane's unswerving admiration for William Francis Gibbs, whom he had helped clear from war profiteering charges in 1944. Like Admiral David W. Taylor before him, Cochrane was a staunch supporter of a dual-purpose luxury liner. Cochrane was the *deus* in a deus ex machina rescue of the Big Ship.

Hearing the news, General Franklin was jubilant. "We are delighted to have our ship back!" he told one reporter.[17] But the tug-of-war between the United States Lines and the U.S. government was far from over.

* * *

In the month and a half that the ship's fate had hung in the balance, it was never actually "tools down" at the yard. Once the word came down that the ship would be completed as originally planned, as a passenger liner, work continued with renewed vigor. The riveting and welding continued apace. As the hull rose, crews carefully coated it with red antifouling paint below the waterline, and jet black above. No longer a dull steel mass, she now shone in the morning sunlight as a stream of workers arrived each day, tools and lunch boxes in hand.

As the construction moved from the hull to the superstructure, the unique ship design presented the shipyard with other challenges. *United States* was the largest user of aluminum in any construction project up to that time. Almost the entire superstructure would be made of the light metal, as well as the two funnels, deck ventilators, lifeboats, railings, and davits. There were good reasons. Using aluminum above the steel strength-deck greatly reduced the ship's overall weight, and allowed for a finer, narrower hull than had ever before been possible on a transatlantic liner. The ship would also have a greater margin of stability and lower center of gravity than any other ship in her class, without sacrificing spaciousness in the premium, upper-deck passenger areas.

But the Newport News shipyard workers were frustrated to no end once they completed the hull and began work on the superstructure. "That was the biggest mass of aluminum we'd ever worked with in a shipyard," Tom Paris recalled, "and therefore we had a lot of problems." [18]

Aluminum presented big construction dilemmas that steel did not. Unlike steel, aluminum lost strength when heated—all plates had to be carefully hammered and shaped by hand. Also, aluminum plates could not be welded together, and conventional riveting would weaken the metal and accelerate aging. This meant that in building the superstructure, over a million aluminum rivets had to be cast and then annealed (superheated to over 1,000 degrees Fahrenheit), making them easier to shape while cold. They were then deep-frozen at 40 degrees below zero, which slowed the hardening process as they aged. When needed by workmen, they were dipped in alcohol, put in insulated boxes, and then pounded into the plating. Once put in place, the cold rivets hardened in warm temperatures and locked the aluminum plates firmly into place.[19] It was a tedious, slow process, and frustrating to those used to working with steel. According to Tom Paris, the crews would set a section of aluminum in place one day, set rivets in it that night, and then wait till the following morning before the rivets had set, at which point they could move on.[20] One consolation for the workers was that the aluminum superstructure deflected much of the intense Virginia sun that beat down on their pith helmets.

Still another problem was that when aluminum comes into contact with steel, the result is an electrochemical reaction known as galvanic corrosion. To stop it, an insulator has to be inserted wherever the metals might come in contact. On *United States,* the base of the aluminum superstructure joined the steel hull at the promenade deck. If the two metals are put into direct contact and then exposed to an electrolyte such as seawater, an electrical current forms, and ions from the aluminum superstructure migrate over to the steel hull. The aluminum corrodes and disintegrates, and the ship comes apart.

One way of mitigating galvanic corrosion was to coat the aluminum with a layer of zinc chromate. The other was putting an insulating barrier between aluminum and steel joints. Newport News first chose an insulation known as "aluminastic." But Gibbs did not trust the shipyard's initial judgment and decided to see if the insulation could hold up against the heat generated by the welding of nearby steel components. On December 14, 1950, he ordered a test of the joinery. Workers welded two pieces of steel together, one of which was riveted to a section of aluminum insulated by aluminastic. When the rivets were removed and the aluminum plate examined, the aluminastic had been badly melted and frayed by the heat from the welder's torch. This was completely unacceptable.

The following day, welders performed the same test on a section insulated by a new material: DuPont PAW tape. As the welder reached the top of one of the pieces of steel, the PAW tape spouted out flames nearly two inches long. But the sparks settled, the tape held up beautifully, and the aluminum below it was undamaged.[21]

Above all, William Francis took the most intense interest in the design of the funnels. A ship's funnels, of course, are more than big tubes to exhaust smoke from the boiler rooms. They also bear the company colors, and give an ocean liner an unmistakable silhouette, recognizable from miles away. For *United States,* Gibbs designed the two largest smokestacks ever put on a passenger ship: 55 feet tall and 60 feet across at the base, each crowned by the signature, soot-deflecting fins that he had pioneered on the Grace Line's *Santa* ships and used again on *America.* He wanted to make sure his ship would be instantly recognizable, and he thought the finned, teardrop-shaped funnels would do that.

Like every key design element, the stacks were put through grueling tests. The shipyard placed a scale model of the ship in a wind tunnel, with smoke from burning woodchips vented through the stacks.

But Gibbs's initial stack design proved deficient—the stacks were graceful and striking, but the inclined fins at the top did not deflect enough smoke away from the upper decks. If the stacks were built as Gibbs wanted, soot and cinders would coat the clothes and irritate the nostrils of strolling passengers.

A Newport News shipyard apprentice named Howard E. Lee Jr. approached Gibbs with a simple idea: rotate the fins so that they were horizontal, parallel to the deck.

Lee wanted to test his idea, and Gibbs grudgingly agreed.

As the test ran, Gibbs stared at the smoke blowing from the model's modified stacks, and then according to Howard Lee, "[he] only grunted, and walked away without comment."

Lee claimed he was never thanked for his idea, but the horizontal stack fins were integrated into the final design. William Francis was never happy when someone bested him, especially when it was a young shipyard apprentice, not from Gibbs & Cox.[22]

Once the patterns were sent to the Newport News metal shop, the funnels proved to be the most challenging part of the ship's construction. The funnels were not the simple steel stovepipes of liners past; they were curved, finned sculptures shaped from unforgiving aluminum. According to the *Shipyard Bulletin,* published by the company, "Had they been built of steel, the plates would have been shaped with relatively little trouble by heating and bending. This was not possible with aluminum. All of the plates were shaped cold."[23] The mode of construction meant that workers had to hammer individual aluminum plates over molds, in much the way French sculptor Frédéric-Auguste Bartholdi shaped the copper plates of the Statue of Liberty eight decades earlier.

There was one thing Gibbs handled completely by himself. He spent countless hours testing the ship's three whistles in the New Jersey Meadowlands, pitching them with a tuning fork.[24] Not only did he want their deep tones to sound harmonious, but also to be distinct from any other ship on the ocean: a raucous, husky, masculine bellow.

In early 1951, cranes hoisted the two enormous and carefully shaped funnels into the air and then gently dropped them into place on the waiting ship's deck. A team of painters, clad in bodysuits and respirator masks, then mounted scaffolding and ladders to spray paint them in the vibrant red, white, and blue of the United States Lines.

With the funnels in place, *United States* finally looked like a great ocean liner, towering high above the shipyard that created her. Her graceful lines and elegant proportions stunned those who visited the site. "There's a truly American cut to it," a reporter from the *New York Times Magazine* observed, "in the springing jut of the prow; the great mass in its lines, but no bulkiness."[25]

The workers on-site sensed the transformation once the stacks were riveted in place and her striking profile complete. By then *United States* was more than just another construction project; the great ship was taking on a life and soul of her own.

By the spring of 1951, the hull and superstructure of the ship were largely finished and the construction crews were rushing to install mechanical systems and interior partitions. Cranes danced and bowed around the huge vessel, each carefully dropping whole sections—hull plates, deckhouses, lifeboat davits, winches—into place. With a cascade of white and red sparks, a team of welders would attach them to the main body of the liner.

"The noise," reported one visitor, "is ear shattering."[26]

The pride the three thousand workers felt in *United States* grew with each passing day; the ship was now an extension of them. "She's a mighty sweet baby," said Preston Hicks, a rigger supervisor at the Newport News shipyard.[27]

The "sweet baby's" christening ceremony was scheduled for June 23. For such a magnificent vessel—America's grandest entry into the transatlantic stakes—the highest lady in the land might be expected to christen her, just as Queen Mary had christened *Queen Mary*—or, as William Francis Gibbs knew, just as First Lady Frances Cleveland had christened *St. Louis* back in 1894.

But First Lady Bess Truman refused the honor. Her husband felt the event would not be something she should attend, given the contentious relationship between his administration and the ship.

The refusal did nothing to dampen Gibbs's excitement. To him, *United States* was the country's real first lady, and she was ready to take center stage.

AMERICAN MODERNE

Unlike the dramatic launchings of the great liners of the past, there would be no great slide into the river, no smoke billowing from the tallow-greased ways, no enormous splash of water soaking the spectators.

In accordance with instructions from William Francis Gibbs, until the dock was fully flooded and the hull was hidden from view, the press could photograph only the forward end of the ship. At no time would pictures of the rudder or propellers be permitted. "This is consistent with arrangements usually made with naval vessels," Gibbs's April 30 memo noted.[1] The secrecy that applied to Navy warships would be strictly followed for the new passenger liner.

At 4:30 P.M. on June 22, 1951, Newport News workers opened the sluice gates, allowing water to flood the dry dock.

Gibbs's wife had joined him for the historic day. "I had never seen Gibbs look happier," Vera Gibbs wrote in her diary. "He was attired in his khaki overalls . . . [he] sat down on the dock and just looked. It was rather an eerie feeling to realize we were nearly the last people to gaze on the bottom of this big ship."[2]

Vera had followed the ship's progress closely from New York. She coped with an always high-strung husband, obsessive even during weekends when he came back from Newport News to 945 Fifth Avenue. He once stepped into their apartment, Vera recalled, announcing he was completely worn-out. "God, I'm tired," he said. "I've been deciding on the heights of toilets all day long."[3]

As water from the James River roared into the dry dock, a *Life* photographer caught a scene harking back to Gibbs's childhood: a view from behind of William Francis Gibbs and his older sister, Bertha. The two sat on a hatch cover beneath the ship's overhanging stern on the dry dock's edge. Gibbs was dressed in the characteristic khaki coveralls and brown fedora hat, Bertha in a long dress, an old-fashioned hat tilted on her iron-gray hair. Bertha leaned in closely to her brother, as if listening intently; both stared down at the water swirling around the propellers and rudder. The naval architect was no longer "Sir Francis" of Gibbs & Cox, but the simple "Willy" of his childhood.

By five the next morning, June 23, the ship was fully afloat and ready for christening. By the start of ceremonies that afternoon, a crowd of twenty thousand people had poured through the shipyard gates and lined up along both sides of the dock, all the way down the ship's 990-foot length. The launch had also created the biggest traffic jam in the history of Newport News, and those who had showed up after late morning could not see or hear a thing, so packed were the stands. The temperature was a hundred degrees, and a fierce noon sun beat down on everyone.

Among those in the guest area was mechanical engineer Elaine Kaplan. Just before Kaplan boarded the train in New York for the launching ceremony, her husband gave her an orchid corsage to wear. As the train pulled out of Penn Station, she took it off her lapel and put it away. She wanted to be seen at the ceremony as an engineer, not as somebody's wife.[4]

William Francis Gibbs, his thinning gray hair protected from the sun by his fedora, watched the ceremony, not on the christening stand, but from a seat on the sidelines.

Those who read the launching brochure saw a brief message from the ship's creator. "The S.S. United States represents the combined technical effort of many interests, and the best materials, machinery, outfit and equipment produced by the American people," William Francis wrote. "I salute, with grateful appreciation, all those who, with heart, head and hand, have helped to make our dream of this great ship come true."[5]

As General Franklin mounted the podium, sweat was dripping down his shirt. The scooped-out hull, framed by two latticed shipyard cranes, loomed behind his broad frame. Above him a great red, white, and blue banner billowed from the ship's prow.

Franklin addressed the multitude in front of him. For the first time since the 1890s, he declared, "when the *St. Louis, New York, Philadelphia,* and *St. Paul* were among the crack ships on the North Atlantic, a ship is being launched that will compete with the finest of the foreign-flag ships. . . . [T]he same house flag will fly on the *United States* as flew on the mainmast of these splendid ships."[6]

Franklin also said that *United States* was essentially the same ship that his own father had hoped to build back in 1916, when he met with J. P. Morgan Jr. and two young men from Philadelphia.

The last speaker was Senator Tom Connally, a New Deal Democrat from Texas, whose wife, Lucile, was to perform the christening. Connally matched General Franklin's size and was described as "the only man in the United States Senate who could wear a Roman Toga and not look like a fat man in a nightgown."[7]

With the Korean War still locked in bloody stalemate and Cold War tensions high, Connally painted *United States* as neither a "mighty battle wagon" nor a "death-dealing aircraft carrier," but as a beautiful passenger liner built for peace, repudiating "the falsehood of Communist charges that the United States of America is a war-mongering nation." Yet should communist forces strike, he concluded, *United States* "would be the difference between victory and defeat for the free nations of the world."[8]

After Senator Connally sat down, the 50th Army Band broke out in the snappy "Salutation March." As the final chords echoed through the shipyard, Reverend Paul K. Buckles of the First Presbyterian Church of

Newport News delivered a benediction. The final "Amen," murmured by thousands, was cut off by a thunderous bellow from the shipyard whistle.

On the stand, Mrs. Connally then stepped forward, clad in a pink dress and hat and holding an enormous bottle of champagne in her right hand.

"I christen thee *United States*!" she shouted, and smashed the bottle against the bow.

It was a wallop worthy of a Texan. The bottle exploded in her face, spraying her with foam. The cascade also hit Admiral Cochrane's wife.

A dripping Mrs. Connally faced the crowd with a startled look on her face, and then burst out laughing.

"Sho 'nuff," decorator Dorothy Marckwald said, mimicking Mrs. Connally's Texas drawl.

At that moment, all shipyard whistles blasted in unison, echoed by the horns and sirens of ships up and down the James River. The band struck up John Philip Sousa's "The Stars and Stripes Forever," and the crowd let loose a mighty cheer, waving thousands of American flags.

Almost immediately, the hull of *United States* inched away from the christening stand as tugs began pulling her out of the dry dock. The crowd gawked in awe as the graceful ship backed away and swung into the James River, her two finned funnels gleaming in brilliant red, white, and blue.

Unlike ships before her, *United States* was almost complete on the exterior. Her boilers and engines already lay nested in her hull. Windows and portholes had yet to be cut into her topmost decks, parts of which were still unpainted. But even in her incomplete state, she presented a magnificent face to the world.

The Gibbs & Cox launching party moved on to nearby Chamberlain Hotel for a gala dinner. William Francis shared a table with his wife, Vera, and sister, Bertha. Elaine Kaplan took a seat with Gibbs's son Christopher, now a private first class in the Army, adopted son Adrian Larkin, and the Zipplers.[9]

"I could not help thinking she must be a peace ship," Vera Gibbs wrote in her diary after the dinner party. "She represents America at peace."[10]

United States would remain at the outfitting pier for another ten months. Here another small army of craftsmen, artists, electricians, and plumbers would transform the bare mechanical shell into a luxury liner fit to carry the world's most discriminating travelers. Or, as some would later joke, into the world's most luxurious troopship.

For the past two years, Dorothy Marckwald and her team had been carefully scrutinizing the reams of blueprints churned out by Gibbs & Cox and interior architect Eggers & Higgins. "We would get preliminary sketches from Eggers and Higgins," her partner, Anne Urquhart, recalled, "and then we would make our preliminary drawings, suggesting colors and styles."[11] The New Yorker could not wait to turn her team loose on the bare steel and aluminum bulkheads. She had designs ready for 23 public rooms, 395 staterooms, and 14 special first-class suites.

They had their work cut out for them. Gibbs told Marckwald that he had to okay every piece of furniture, bolt of drapery, and square foot of carpet. He also said that the ship's public rooms and cabins had to be spacious, cheerful, modern, and uncluttered, furnished with lightweight and durable furniture. And there was to be absolutely no wood used in the ship's interior. "The greatest danger in war or peace in big ships," William Francis said, "is the danger of fire."[12]

As a result, luxury would take a backseat to fireproofing and weight saving. The ship was to be comfortable, to be sure, especially in first class. But Gibbs had learned the lessons of the *Leviathan* project: since this ship was to be paid for by taxpayer dollars, nothing about the ship could appear frivolous, opulent, or overdone. Accordingly, all artwork and decorations had to be based on an American or nautical theme, preferably both. There were also to be no sculptures of marble or bronze, as they would add unnecessary weight to the ship.

Unlike land-based decorators, who worked with flat floors and vertical walls, a naval interior designer had to deal with a ship's "sheer" and "camber." Sheer was the upward slant of the decks toward the bow and stern. Camber was the downward slant of the decks away from the ship's centerline. Because of this, door frames, paneling, and even furniture had to be custom-made for each ship. In her studio, Marckwald jokingly called a ship's sheer and camber its "umm" and "umph," respectively.[13]

Marckwald was not fazed by the demands of the *United States* commission. She had prospered by giving her affluent clients a sophisticated, modernist aesthetic. Business at Smyth, Urquhart & Marckwald was booming, flush with interior design work on Park Avenue apartments, Long Island mansions, and private yachts. Among recent commissions was the redecoration of New York's lavish St. Regis hotel, built by Vincent Astor's father, *Titanic* victim Colonel John Jacob Astor IV. Her interiors for *America,* with their polished brass, murals, and bright colors, met with near-universal praise from the traveling public. Excited about the biggest commission of her career, she and her partner whipped up color schemes even before the keel of *United States* was laid. "Quick and snappy," Marckwald told a reporter when asked to sum up her ethos.[14]

What followed was a game of give-and-take with Gibbs, who spent countless hours making sure the ship's exterior lines were sensuous and beautiful, but had a surprisingly utilitarian view about how she would look inside.

Marckwald and Urquhart got to work selecting furniture and color schemes. They quickly ran into trouble when it came to upholstery and draperies. Gibbs kept vetoing their selections. No sponge rubber in the seat cushions, for example; a stray cigarette could ignite it. After much haggling, the team decided to use a synthetic fabric called Dynel. To brighten the material, Marckwald decided to have metallic threads, such as real silver and gold, woven between the strands. It was nonabsorbent, and above all, fireproof. In addition, paint samples on thin metal strips were heated to 2,300 degrees to make sure they were not flammable.[15]

On February 17, 1950, he had Marckwald's designs subjected to what he thought was the most important test. A full-size stateroom model was constructed at the National Bureau of Standards' test facility.

The model was furnished with all of the interior elements planned for the ship: Dynel-clad chairs, metal chests of drawers and bedsteads, flame-retardant sheets and linens, aluminum walls clad in asbestos-infused marinite panels. Finally, workers lugged in suitcases stuffed with the sorts of flammable baggage that passengers usually kept in their staterooms—clothing, perfume, cologne, books, and magazines. Testers opened a suitcase and set it on fire.

Gibbs watched through the plate glass as the fire leapt from the suitcase and began spreading. The thermometer crept up as flames engulfed the entire room, reaching 1,300 degrees Fahrenheit. Reflections of the flames danced in his spectacle lenses. Finally the fire burned itself out.

The room was a mess. Marckwald's fabrics were scorched, and the marinite-clad aluminum bulkheads blackened. But the fire had been contained, and Gibbs was pleased with the performance of Marckwald's use of materials. "You can go into your stateroom and pile your luggage in the middle and set a match to it and when you come back in 3 hours it will have burned out," Gibbs said. "The room will be black with smoke, but not even the curtains will be burned," he added. "The test demonstrated," Gibbs concluded, "that the provisions for bulkheading followed in connection with the subject vessel appear to be satisfactory but no relaxation of insulation requirements was indicated."[16]

Then there were the pianos, which were to be custom-built for the ship by Steinway & Sons. Gibbs began pestering Steinway to build aluminum-framed pianos for the ship's lounges. Theodore Steinway, president of the august German-American manufacturing firm, put his foot down, not wanting his name on them. No metal piano, no matter how well designed, would have the same resonant tone as one made out of wood. After a flurry of letters and phone calls, an exasperated Steinway asked William Francis Gibbs to come to a demonstration. Steinway trotted out a gleaming concert grand of the type that would be used in the first-class ballroom, doused it with gasoline, and threw a match on it. The gasoline burned but the piano did not. Gibbs relented and allowed several wooden Steinway pianos to come aboard.[17]

This helped foster a popular boast about the ship: the only wood on board was to be found in the butcher's block and the pianos. This was not quite true. "There were a few other splinters here and there," Thomas Buermann admitted.[18] Among the "other splinters" was the Oregon pine that filled the bilge keels, and the lignum vitae lining the four propeller shafts. Native to Florida and the Bahamas, lignum vitae was often used for croquet mallets and cricket balls. Dense, self-lubricating, and resistant to rot, it had long been used to line propeller shafts. Despite all the advances in marine technology during World War II, no one had come up with a material that could match its durability. And since the bilge keels and propeller shaft linings would be soaked with seawater, Gibbs conceded that they couldn't catch fire.[19]

Now that the furnishings were approved, the shipyard workers began to outfit the interiors of the newly launched *United States.* Each day, as dozens of chairs, tables, and other custom-crafted furniture arrived at the outfitting pier, shipyard workers fastened big sheets of fireproof marinite board over the bare metal bulkheads and mechanical systems. To Gibbs, marinite was a godsend compared to flammable wood partitions used on earlier liners like *Morro Castle* twenty years earlier. In the early 1950s, no one knew about the risks of carcinogenic dust from asbestos, so workers sawed and drilled into marinite board without facemasks.

As the interior partitions were completed, a team of America's finest modern artists arrived to install the work they had completed offsite. One reporter saw artist Austin Purves, former head of the Cooper Union School of Art and creator of murals for the 1939 New York World's Fair, working with joiners. Purves supervised the installation of all 265 aluminum wall sculptures of state birds and flowers, as well as the United States Lines eagle insignias for the first-class grand staircase. "If you are fortunate enough to walk up unobserved," wrote David Yonan of the *Newport News Daily Press,* "you would see what appeared to be five joiners instead of four. The fifth one is Purves. He is dressed like them, at ease in a thin blue work shirt, faded by many washings, and ordinary jeans. Hatless and a cigarette in his mouth, he is pitching in, working up a sweat." To keep the screw heads invisible, workers centered the sculptures on the front side of the marinite panel and then screwed them in on the reverse side, pulling them flush against the surface. Because ordinary screwdrivers would ruin the fragile sculptures, the shipyard created special tools out of hard aluminum strips.[20]

Elsewhere off-site, dozens of subcontractors in all forty-eight states rushed to meet their own deadlines. The Ellison Bronze Company of Falconer, New York, for example, constructed two forty-five-foot-long bars for the ship's public rooms, as well as mirrors, window frames, and display cases for the first-class shopping arcade. The local newspaper, the *Jamestown Post-Journal,* boasted that the first-class bar was "the largest piece of fine metal fabricating ever undertaken in the Jamestown area," and that Ellison employees were working sixty-hour weeks to complete all of their work before the scheduled May 14 trial run.[21]

If the furnishings were modern and fireproof, William Francis's layout of the passenger accommodations was thoroughly traditional. There had been some talk about making the *United States* a two-class ship, then a novel concept. But in the end, the Gibbs brothers and General Franklin decided on a traditional three-class arrangement; tourist class, Gibbs argued, could be used to carry enlisted servicemen and their dependents to and from Europe.[22] As on earlier transatlantic steamers, first class was situated in the mid-portion of the vessel. This was where the least motion was felt, removed as it was from the propeller vibration at the stern and the heaving and pitching at the bow. Cabin class was located in the stern section, and tourist in the forward section.

Class segregation was strictly enforced aboard *United States* and ships like her, a holdover from the Victorian period. The only shared public spaces were the movie theater and the swimming pool, which were used by first- and cabin-class passengers at different times of the day. Separate doors and elevators ensured that there would be no mingling. Tourist-class passengers, segregated in the least desirable parts of the ship, had no access to the gym, but they did have their own movie theater.

Upon embarkation, a first-class passenger would walk up the gangway from the pier and onto the promenade deck, so named because of its twin strolling areas that ran along the sides of the ship. Each of these promenades was about four hundred feet long and enclosed by glass to protect walkers from the elements. During the voyage, they were lined with rows of red aluminum deck chairs. The sun deck, one deck above, had an open-air promenade that ran uninterrupted around the superstructure's perimeter under the lifeboats; the bridge and sports decks above had shuffleboard and deck tennis courts. Pushing through a set of double glass and steel doors, the passenger entered the promenade deck foyer, with its sweeping grand staircase and bank of elevators. A large aluminum eagle, the emblem of the United States Lines, sculpted by Austen Purves Jr., crowned the landing of the grand staircase.

Walking forward from the promenade deck foyer, the passenger would enter the "Observation Lounge," so named because of its thirty-two floor-to-ceiling windows that gave passengers sweeping views of the ocean rushing by. The H-shaped room spanned the entire 101-foot width of the ship. Despite its massive floor plan, the lounge was confined to a single deck in height, a fire safety limitation dictated by Navy design standards. Older vessels had two- or three-deck-high ceilings; high ceilings, overstuffed couches, and dark paneling created the feel of a grand hotel in London, Paris, or Berlin. Even the medium-sized *America* of 1940 had a two-deck-high lounge, adorned with lush murals and gleaming brass.

On *United States,* Marckwald had to make the low-ceiling space feel welcoming and soothing, not in any way claustrophobic. The room would be the most used area during the day, where first-class passengers would read magazines and books from the library, as well as play bridge and canasta. To make the Observation Lounge more inviting, Marckwald chose rich greens and soothing blues for the color scheme, to give the room a bright and cheerful appearance. The carpet was sea-green, and the chairs and couches were covered in blue cloth. She asked artist Raymond Wendell to create two massive murals out of gesso that would represent the underwater topography of the Atlantic Ocean. These murals, highlighted in gold leaf, would be placed on the room's two curved bulkheads. She also called for rows of tall floor lamps to be bolted to the floor. The lamps would bathe the room in a soft, glowing light in the evening hours, during which the picture window blinds would be closed.[23] To open up the space further, Marckwald used dozens of glass-topped card tables, as well as fourteen small, handcrafted round tables placed strategically around the room—each glass top was in a different shade of blue and adorned with small white stars.[24]

After admiring the space, the passenger would then turn around, walk aft through the stair foyer, and find himself in the ballroom. To appoint the large and luminous room, Marckwald commissioned Charles Gilbert to create a series of glass panels that would divide the space into sections. Each panel was etched with images of sea creatures and plants, highlighted with gold and silver leafing. When lit from above and below, the panels were stunning backdrops for men and women in evening dress. Marckwald capped the space with a shimmering, white-and-gold dome, which cast light on the dance floor and gave the space a two-deck height at its center. Banquettes and red barrel-backed chairs provided the seating around the dance floor. A bar and lounging area, located at the aft end of the room, was somewhat removed from the music and din of the ballroom. The kidney-shaped bar was forty-five feet long and one of the largest afloat. Here a white-jacketed bartender would shake cocktails and pour champagne into the wee hours of the morning.

Leaving the ballroom and continuing his walk toward the stern, the passenger would pass one of two special, intimate retreats for first-class passengers. The first was the Navajo Lounge, a small cocktail bar overlooking the starboard side enclosed promenade. Midnight blue in color scheme, the lounge would be named for murals by artist Peter Ostuni depicting sacred Navajo sand paintings, which would be backlit during the evening hours. Rather than using canvas, Ostuni painted the images in enamel on a metal base. The lighting, coloring, and artwork gave the Navajo Lounge a sleek, luminous look, especially at night.

The second refuge, opposite the Navajo Lounge on the port side, was a private restaurant. It could seat fifty of the ship's most distinguished passengers, who wanted to avoid the gawking crowds and autograph seekers in the main dining room, located several decks below. Like the

Navajo Lounge, it was also midnight blue in color scheme, with rich red curtains and chairs. It also featured seven large windows overlooking the enclosed promenade, as well as backlit crystal sculptures on the walls. To give his crystal light fixtures even more sparkle, artist Charles Lin Tissot mounted them on highly polished aluminum sheets.[25] Like its cocktail bar counterpart, the small space radiated urban sophistication and exclusivity.

Following the aroma of cognac and cigar smoke, the passenger continued aft to the most masculine space on board: the first-class smoking room. Traditionally, this room aboard a large transatlantic liner was a wood-paneled refuge, stuffed with leather armchairs, green baize gaming tables, baronial paneling, even a few animal heads. A working fireplace might add to the atmosphere. Marckwald, who derided what she called the traditional "elk horn" style, could not panel the walls with oak or mahogany, nor could she hang its windows with heavy velvet curtains. Plus she abhorred "muddy" colors. On *United States,* she had the smoke room walls painted jet black and covered the windows (which overlooked the enclosed promenade) with plaid drapery. For the room's centerpiece, William King created a simple aluminum Mercator projection of the world for the forward bulkhead. Small clocks positioned above each meridian showed the times of day around the world. A well-stocked bar was placed on the opposite side of the room. Chairs and couches upholstered in red leather—fireproofed, of course—completed the clubby look. It was not quite New York's University Club or London's Royal Automobile Club, but it was close enough considering the restrictions.

The first-class public room nearest the stern on the promenade deck was the movie theater. By placing the large room at the rear, Gibbs & Cox eliminated the typical specialty restaurant or outdoor café that held this space on older ships. Conferring with interior architects Eggers & Higgins, Gibbs concluded that American travelers would much rather watch first-run movies than have another sitting area. It was meant to be the finest cinema afloat: "Since the ship is expected to carry many stage and screen stars, as well as discriminating patrons of the arts, between Europe and America," a shipyard agent asserted, "all theater equipment and accoutrements were chosen to rival the finest motion picture installations in any of the world's large metropolitan centers."[26] Unlike makeshift cinemas aboard other liners, the space was as close as possible to its land-based equivalents: a sloping floor for better viewing, two aisles, and the best audio and visual equipment then available. As on *Normandie, United States*' theater had a stage for lectures and modest live performances.

The first-class dining room on A Deck (four decks below the other large first-class public rooms) was also placed near the ship's center of gravity to minimize motion. Vast in size, it spanned the full width of the ship. But as with the other public rooms, Marckwald wanted to make sure that the room remained spacious, airy, and welcoming rather than cluttering it up with ornamentation. The vaulted center section was two decks in height, but because of Navy structural requirements, the remainder of the space would still look chopped-up compared to first-class dining rooms aboard older liners. The room's stanchions, pillars, and size restrictions limited seating capacity to about four hundred, which meant that the 894 first-class passengers would have to dine in two sittings, grouped in tables for two, four, six, and eight diners. Having two meal seatings was a major drawback compared to the *Queens*, which could accommodate all their first-class passengers at once, allowing them to dine at their leisure.

Marckwald also made the room seem as spacious as possible by using bright and cheery colors. She accentuated the space with simplified classical cornices, and called for lots of indirect lighting, especially in the ceiling. There were no chandeliers or traditional sconces. The plush chairs would be upholstered in vibrant red fabric that complemented the oyster-colored walls and snow-white table linens, while the curtains were woven out of Dynel fabric and threads of real silver and gold. Spotlights were nestled in the corners and cornices, as well as below the sculptures, to bring them into relief.[27] As on *America,* a musicians' gallery was placed on the second level so that a Meyer Davis Orchestra could serenade passengers during dinner with light classics and popular show tunes.

First-class dining rooms also tended to have a large central decorative element that loomed over the captain's table. On *Normandie,* it was a towering, gilded statue of Peace (La Paix). On *Queen Mary,* it was a great Art Deco mural depicting the Atlantic Ocean, in which a miniature model of the ship would trace the vessel's actual progress along her route. Marckwald was determined to have such a work of art as the visual focus of the first-class dining room. She asked sculptress Gwen Lux to produce a piece that would be hung at the forward end of the room. Lux's creation, titled *Expressions of Freedom,* would consist of four stylized figures representing the "Four Freedoms." Each figure would stand just over four feet high. If cast in plaster, each figure would weigh about two hundred pounds.

William Francis Gibbs ordered her to find something lighter.

Lux proposed to mold her four figures out of "foam glass." Consisting of crushed glass bound together by carbon or limestone, the material was most commonly used as thermal insulation or a cork substitute, not a sculpture medium. Lux promised that her sculpture would be not only fireproof, but also extremely lightweight, with each figure weighing only forty pounds. They were so light, in fact, they could float. Lux also produced forty-eight foam glass state seals that would be placed around the perimeter of the dining room.[28]

First-class staterooms were spread out on the sun, upper, and main decks. Out of 1,962 total passenger berths, first class would get almost half: 894. General Franklin hoped these top-tier berths would be popular with the wealthy: movie stars, socialites, business tycoons, and European royalty. The postwar boom and the depletion of first-class berths during the war led Franklin and Frederic Gibbs to predict that there would be more than enough people who lived on New York's Park Avenue, Philadelphia's Main Line, and Chicago's Lake Shore Drive willing to pay $585 per person (equal to about $4,800 in 2012) for a standard, outside first-class cabin with a tub bath and trunk room during the 1952 peak summer season. First-class fares began at $370 for a basic inside stateroom, and a two-bedroom special suite went for $930 and up.

Although restrained in décor, the first-class cabins were extremely spacious compared to those in older ships—the largest double stateroom boasted 364 square feet of living space.[29] The wall treatments were light in color, usually cream or sky blue. Patterned and brightly colored curtains and bedspreads would give an airy cheeriness to the simple wall treatment and glass-and-aluminum furniture. Sleek aluminum lamps and sconces gave the staterooms a warm, yellow glow by night. In addition to individually controlled thermostats, electricians placed dimmers in each of the cabins—another touch of luxury unheard-of aboard other liners.

To please the ship's wealthiest passengers, Marckwald came up with special decorative schemes for fourteen first-class suites on the upper and main decks. The largest of them consisted of two bedrooms, two bathrooms, and an extremely large sitting room. The suites were so spacious that a reclusive millionaire could spend an entire five-day crossing in comfort, privately entertaining a select group of friends without having to set foot in any of the public rooms. A bellboy could deliver full meals, direct from the first-class kitchen, at a moment's notice.[30]

Yet compared to older ships, the suites aboard *United States* were relatively simple in layout, a concession to a more informal, American way of postwar living. This attitude was reflected in the names of the suites, which came from the American natural and nautical decorative motifs Marckwald had woven into the bedspreads, draperies, and upholstery: the "Seashell Suite," the "Butterfly Suite," the "Tree Suite," and most exquisite of all, the "Duck Suite."

Like most transatlantic liners of her time, *United States* had no outdoor pool—the indoor one was located deep in the stern section of the vessel and used by first- and cabin-class passengers at different times of the day. Those wanting to use the pool would take an elevator down to C Deck. The pool was not overly large, but the walls were decorated cheerfully by J. Scott Williams, who created enamel signal flags that spelled out "Come on in, the water's fine!" The pool itself was constructed of monel, a corrosion-resistant nickel alloy used for the spike of New York's Chrysler Building. A good-sized gymnasium, complete with stationary bicycles, punching bags, and weights, was located a deck above.

Located in the stern of the ship, cabin class was less spacious than first, but definitely not cheap. At a minimum of $230 (or about $1,900 in 2012) per person one way during the 1952 summer season, the 524 berths in cabin class were pitched toward upper-middle-class professionals and their families, as well as the privately wealthy who wanted to avoid the formality and high visibility of first class.

Cabin class enjoyed the use of an open-air promenade and gaming area on the promenade deck fantail, with a view of the wake churning behind the ship. In warm weather, a temporary outdoor swimming pool could be set up on top of the raised roof of the lounge below. Descending one level below to Upper Deck, a passenger would enter the cabin-class lounge, a pillared space flanked by two enclosed promenades. This room boasted comfortable black and red armchairs, set in arrangements for reading or card games. The center portion of the room had a raised ceiling, which gave its dance floor a sense of space. A large red carpet covered the floor during the day, and the walls had a golden hue.

One deck below on Main Deck was the cabin-class smoking room, essentially a smaller version of its first-class counterpart. Colorful, harlequin checkered curtains shaded the windows, and Lewis E. York's mural of seventeenth- and eighteenth-century American glassware and pottery hung above the bar. Two putty-green leather banquettes ran the length of the room, providing additional seating. It was intimate and clubby in feeling, a good space for late night cocktails and card games, although a slight rumbling from the propellers below might shake highballs and ashtrays perched on the glass tables. Descending the main staircase one more level, the passenger would reach the cabin class dining room on A Deck. Sharing a galley with the first-class dining room, the space spanned the full width of the ship, and had midnight-blue walls adorned with aluminum line sculptures depicting the season and constellations by Michael Lantz and Seymour Lipton. The chairs were covered in green leather and the tables set for groups of two, four, and six diners. Many who saw the cabin-class dining room for the first time thought it a "friendlier" place than its first-class counterpart.[31]

The décor in cabin-class staterooms resembled that in first class, although the rooms were much smaller. In addition to twin beds, two additional bunks could be pulled down from the ceiling, a setup ideal for families traveling together. All cabin-class staterooms would have private bathrooms, unheard-of in comparable staterooms aboard *Queen Mary* or *Queen Elizabeth*. Washbasins were located in the cabin itself.

Tourist class, located in the forward part of the ship and accommodating 544 passengers, was utilitarian in décor. With one-way fares priced about $175 per person (about $1,500 in 2012 terms), tourist class was almost as expensive as cabin class, but within reach of budget travelers, students, young families, and immigrants from Europe. Tourist class aboard *United States* was much more spacious and airy than the foul steerage areas of the great liners of fifty years before. But because of its location far forward, it was still subject to the heaving and pitching that had made life miserable for millions of European immigrants.

Public rooms included a small movie theater and a lounge on promenade deck, both of which had gray-blue walls and bright curtains. The lounge, which featured large windows overlooking the bow, had a small dance floor in the center, but dancing was made a bit unsteady by the upward sheer—or slope—of the deck. Two decks down on Main Deck was the tourist-class smoking room, which was somewhat cheerier than the lounge because of its bright red leather chairs and barstools. The dining room, located on A Deck, had beige walls decorated with eighteenth-century nautical motifs and was furnished with maroon curtains and chairs. Tables were set for groups of four or six, an improvement over long "boardinghouse" tables found in tourist class aboard older ships.

The tourist promenade areas included a small patch of deck between the bow breakwater and the curved base of the superstructure, as well as a couple of narrow strips on the bridge deck. Because of their proximity to the bow, both tourist promenade areas were windy and uninviting in all but the best weather.

To allow the shipping company to adjust to market demand, the plans also called for gates that could be moved up and down stateroom corridors. They were known as "crash gates" because a passenger could knock them down during an emergency, giving everyone regardless of class quick access to the lifeboats. Forty years earlier, a maze of locked gates kept hundreds of *Titanic*'s steerage passengers trapped below until it was too late. This would not happen on *United States.* These gates were also adjustable. During the high summer travel season, cabin and tourist class berths could be turned into first-class ones to meet demand, and vice versa during the stormy winter months. Regardless of class, cabins would be equipped with two luxuries that would have been unimaginable before the war. One was a telephone, with a connection controlled from a central switchboard manned around the clock. And they could be used for more than ordering a martini at three in the morning or nagging a family member in the next stateroom about being late for dinner; ship-to-shore phones, an expensive novelty when first installed aboard *Leviathan* in 1929, were now a reality aboard *United States.* By using the ship's radio transmitter, a passenger could make a call to anywhere in the world. The second luxury was individually controlled air-conditioning, adjusted by using a thermostat next to the cabin door. Gone were the days of stifling summer crossings and primitive forced-air ventilation. The trade-off, however, was that cabin portholes on *United States* were sealed shut; if a passenger wanted to breathe some sea air, a steward had to open the porthole for him.

As the interior neared completion, many of the artists complained that the decoration of the ship was a "great headache," and grumbled at the use of nontraditional materials such as foam glass.[32] Many who came aboard to observe the progress felt that the interiors of the new superliner, although modern and sleek, felt institutional and cold compared to those of older ships.

Marckwald, although proud of her work aboard *United States,* was also somewhat defensive and qualified about the results. "The *United States* is a ship, not an ancient inn with oaken beams and plaster walls," she shot back at one critic. "The best we can say is that the ship's decor is modern, American, that it is functional, and that color plays a most important part."[33]

As Marckwald supervised the interior work, Gibbs's longtime electrical engineer, Norman Zippler, and his team oversaw the installation of the ship's electrical and navigation system. To avoid electrical fires and short circuits, Zippler used many of the lessons he learned when rehabilitating *Leviathan.* If a bomb or torpedo hit the ship, and one part of the electrical system failed, alternate feeders would reduce overall voltage load. All lighting and power had emergency backup systems. The double electrical system was arranged in such a way that "the failure of one branch circuit or feeder will not leave any living space in darkness."

The bridge, located at the curved, forward end of the ship's superstructure, was the nerve center of the ship's operations. Unlike *Queen Mary*'s bridge, there was no wood paneling or polished brass trim. Everything was painted battle gray, including the two small ship's wheels, which required, according to Granville Parkinson of Gibbs & Cox, "little more effort than that required to turn the steering wheel of an automobile." Should the master decide, the ship could be steered by gyro-pilot (known by sailors as the "Metal Mike") on a fixed course, with no help from the helmsman.

If the ship were damaged or a fire broke out, the officer on watch could shut some or all watertight and fire doors using a panel located in the wheelhouse. The same officer could use a public address system to give evacuation orders to passengers and crew. There would be no confusion as there was aboard *Titanic, Lusitania,* or *Morro Castle,* none of which ships had a public address system or adequate emergency sirens.

And unlike *Lusitania,* which lost electrical power after a single torpedo strike, trapping passengers in jammed elevators and pitch-black corridors, *United States* was equipped with an emergency generating plant that would kick in if the main power plant were disabled. "The emergency diesel-engine-driven generators will start automatically upon the failure of the ship's service power," Parkinson noted, "but in the interim, while these units are coming up to speed and cutting in on the line, the emergency source is from storage batteries and direct current motor driven alternating current generators."

Radar, pioneered by the British navy during the early months of

World War II and refined by the Americans, allowed the captain to "bring the *United States* through harbor approaches which are enshrouded in fog or darkness, snow or rain, and bring it to its pier in absolute safety and on schedule." The radar system consisted of two netlike antennae swiveling atop the ship's single mast. "The powerful transmitters and receivers," Parkinson wrote, "operating alternatively, can supply to the screens in the wheelhouse an uninterrupted diagram of all objects within a large area surrounding the ship . . . the navigator knows at all times exactly how close he is to land, another ship, or any other object within range."[34] Radar was a navigational aid that Captain Smith of the *Titanic* would have found useful as he drove his ship through the Atlantic at full speed. But then again, the veteran White Star commodore ignored iceberg warnings received by wireless. In 1912, radio communications between ships consisted of spark-generated dots and dashes, not voice transmissions, and wireless operators would switch off their sets before turning in for the night. Warnings about navigational hazards were delivered at the convenience of the operators, whose main priority was sending messages from paying passengers to shore stations. By the early 1950s, however, radio was no longer a novelty, but an integral part of life safety at sea. By international law, radio sets aboard passenger liners had to be manned twenty-four hours a day, and incoming messages about navigation hazards took precedence over outgoing "wish you were here" cables from passengers.

A few months before *United States*' first trials, Gibbs decided to ensure the secrecy of the ship's advanced technology by making it a national security matter, even during commercial operation. He added some provisions to the sales contract that the United States Lines would sign when it took possession of the ship. One of them read that "plans and data concerning this ship are considered to be confidential and classified, as would be the case of a combatant naval ship. The plans and data shall not be released without the approval of the Federal Maritime Board." The leaking of such "confidential or classified" information could subject the owners, its agents, and representatives "to criminal liability under the laws of the United States." The United States Lines was obligated to fire or refuse to hire anyone whom the Federal Maritime Board or the Navy Department deemed a risk for "espionage, sabotage, or subversive activity."[35]

Gibbs's "Big Ship" was now protected not just from Soviet spies, but from commercial and professional competitors as well.

TURN HER UP

On May 14, 1952, just as the sun rose over the cranes and sheds of Newport News, the engines of *United States* rumbled to life, and smoke began to drift from her two red, white, and blue funnels. Parts of her interiors were still unfinished, and because she carried only enough fuel and provisions for a few days, she rode relatively high in the water. Several days earlier, Gibbs & Cox's chief engineer, Walter Bachman, had supervised a series of rigorous tests on the ship's engines. *United States* remained at the dock, her four manganese bronze propellers removed to keep the ship from swaying back and forth. All aboard seemed in order, and the propellers were remounted. She was ready to embark on the test that could establish America's postwar maritime greatness.

William Francis moved into first-class suite U-81, put away his threadbare suit, donned his coveralls and fedora, and headed up to the bridge.

Just before departure, yard president J. B. Woodward Jr. told the *Newport News Daily Press:* "If the people of the country could just understand what a remarkable ship this is, what lies behind the painted bulkheads in safety and stability, and what she will mean to the country in carrying power and speed, they would truly be proud of this American achievement."[1]

Public expectations were high, stoked by the fact that the great ship's speed achieved during the trials would be kept secret. Aboard ship, Maritime Administration chairman Admiral Cochrane told reporters that the ship "would go faster than any other of comparable size afloat." But when asked if the ship could make 40 knots, Cochrane declined to say anything.[2]

At 9 A.M. sharp, Commodore Manning ordered "slow astern," and *United States,* packed with 1,699 guests and crew, backed away from her pier and headed slowly down the James River. At the Old Point Comfort Lighthouse, Manning ordered the chief engineer to increase speed. The ship responded instantly, barely making a ripple or wake. Finally, at 11:45 A.M., after passing the Chesapeake Lighthouse, Manning ordered an increase of the turbine throttle to 100,000 shaft horsepower, and the helmsmen to steer *United States* on a due east course. The trial site lay eighty-five miles away. As the ship passed the breakwater, a photographer snapped a picture of William Francis and Frederic Gibbs atop the wheelhouse. In the black-and-white photo, Frederic leans against the rail and looks resolutely out to sea, the sunlight bathing his face. William Francis sits with his long legs akimbo. His right arm and black fedora hat shield most of his face as he peers warily at the photographer.[3]

As *United States* headed toward her destination at a steady 25 knots, the wind began to pick up, whipping off the tops of the wave crests into spray and whistling over the funnels. Despite the weather conditions, the ship remained remarkably steady, just as William Francis and Bachman had predicted in their model tests. The Gibbs brothers, Vincent Astor, and General Franklin went to the bridge to stand with Commodore Manning. They saw the prow cut through the waves, sending spray spattering against the windows. To clear the panes, officers on watch had to hand crank the wipers.

Grim-faced, Vincent Astor hunched over a set of charts and watched the movement of the gauges.

No outsiders were allowed to see what the gauges read. The *New York Herald Tribune* told readers that William Francis Gibbs was "passing the entire voyage watching the engines and seeing that no unauthorized person gets a peek at readings."[4]

It was not a luxury cruise. The ship's head chef, Otto Bismarck, was supervising a New York City boot camp preparing a legion of cooks, pastry chefs, and waiters for the ship's maiden voyage. Meanwhile, service in the first- and cabin-class dining rooms was hardly up to snuff. In the first-class dining room, a reporter from the *New Yorker* munched on fried chicken prepared by an interim chef known as "Clamchowder Jack." "Waiters mostly riggers," the *New Yorker* writer complained, "not very continental in manner."[5] The cooks managed some creativity: at lunch, Gibbs, Franklin, and others found loaves of bread shaped like an alligator and a sea turtle sitting on their table in the first-class dining room.[6]

As the sun shone down on the choppy Atlantic, Walter Hamshar of the *New York Herald Tribune* went looking for a deck chair. He was disappointed; the aluminum-frame chairs had not yet been brought aboard. Getting a drink also frustrated him. "The first-class smoking room and bar made life pleasanter for the guests who came for the ride," he wrote in his dispatch, "but there was considerable congestion that is not likely to occur when the liner's other public rooms and bars are opened during regular service."[7]

Slightly before 3 P.M., *United States* arrived at the trial site. Manning ran the ship on five-mile courses at low speed for the next six hours, against and with the wind. Everything went smoothly.

At the trial site, the crew of the Coast Guard cutter *Conifer* watched *United States* make her initial passes, purple rays of twilight streaking her white superstructure and two finned funnels. Lashed to the cutter's deck were two Raydist buoys. When dropped into the ocean, these devices would send out electric impulses that would bounce off *United States* and measure her speed with great precision. As they prepared the buoys for launch, those aboard the tossing cutter were stunned at the sleek ship's beauty as she sliced through the stormy seas.

As night fell and the wind howled against the bridge windows, Manning decided to call off the high-speed trials until the next day; it was too rough for the cutter's crew to launch the Raydist buoys. In the meantime, Manning circled the ship around the trial site, keeping her in a holding pattern until dawn. The lights of the liner cast an eerie glow through the mist whipped up by the gale. On board, the guests either turned in early or gathered around the pianos in the public rooms. As the sea crashed against the hull, a guest in the cabin-class lounge stood up to recite Matthew Arnold's "Dover Beach": "The sea is calm tonight, the tide is full, the moon lies fair upon the straits. . . ."[8]

Of course, the sea was not calm, and by daybreak, the gale had still not let up. In fact, it had gotten worse. The wind was now gusting at over 45 knots, almost 60 miles per hour, but it was now or never. Manning decided to move his ship into position for the official builder's speed trials.

On the bridge, General Franklin, Woodward, and Admiral Cochrane could barely contain their boyish excitement. Manning looked calm. Astor wore his perpetually sour expression. William Francis retained his composure, despite his growing excitement. Cochrane could not help himself and insisted, "Let's find out what she can do!" Manning ordered Chief Engineer Bill Kaiser to throttle *United States*' four Westinghouse turbines to 158,000 shaft horsepower, as specified in the construction contract. A strapping, six-foot-two Virginian, Kaiser had started his career as a lowly machinist aboard *Leviathan* nearly thirty years before. He remembered the trials off the Florida Coast in May 1923, in which William Francis Gibbs claimed that *Leviathan* had beaten *Mauretania*'s speed record. That was for publicity's sake. This was for real. Kaiser was now in charge of the most powerful set of engines ever placed on a ship, designed and built by American workers and American industry, without question the best both had to offer. Kaiser put down the phone, responded to the bridge's signal "Full ahead," and ordered the boiler room to give him full steam pressure. The gauge began to rise slowly. And then the four turbines began to roar, shaking the grating beneath his feet.

United States made thirteen runs along the five-mile course, with and against the wind. She was slicing through the churning ocean at an incredible speed, smoke pouring from her stacks, her bow kicking up great waves that blew onto her decks.

As the ship raced back and forth along the course, Franklin and Newport News executive vice president William Blewett moved to the warmth of the captain's cabin for a quiet chat.

"Jack, what do you think?" Blewett asked.

"We don't own her yet," Franklin said laughing, his face flushed with excitement. Not until the official trials in June were over would United States Lines take possession of the completed ship.

"Damn you," Blewett answered, as the ocean tore by the ship's windows. Manning had opened her engines up to the contract-specified power level, but Franklin knew there was significantly more in reserve. The high winds and pounding waves might put too much stress on the engines and propellers if she went full blower.

"Well, if you need advice about weather," Franklin told Blewett, "there's a man just above you on the bridge, Captain Manning, who can supply the facts."

Blewett sent a crewman up to the bridge.

He came back to the captain's cabin, and delivered Manning's response.

"You turn her up. I'll tell you when to shut her down!"

Manning then phoned Kaiser in the engine room and told him it was time to work up to her Navy-rating full power: 241,000 shaft horsepower, or a gear-crunching 200 tons of thrust on each propeller shaft. The steam pressure in the turbines would approach 1,000 pounds per square inch and would be heated to about 875 degrees Fahrenheit. Any leak would mean instant death for the entire engine room crew.

Kaiser watched the gauges carefully, clenching a cigar stub between his teeth. The two high-pressure turbine rotors, studded with thousands of blades, slowly built up to 5,800 revolutions per minute. The steam then fed into the two low-pressure turbines, which spun at 3,200 rpm. Reduction gears, which connected the four shafts to the turbines, then had to slow the eighteen-foot propellers down to 180 rpm. Never had a set of delicate reduction gears reduced the revolutions to the propellers by a ratio of thirty to one. Shafts had fractured and propeller blades fell off on many other liners. As Kaiser's men slowly inched her up, the ship's stern began to vibrate from the terrific force of the four propellers thrashing through the stormy seas.

United States was now thundering through the gale at over 34 knots.[9] Congressmen, executives, and reporters up in the first-class observation lounge noticed that their drinks were now trembling on the cocktail tables. Out on the decks, guests leaned over the railings and guessed they were going as fast as 35 knots, but could not find out for sure. The bridge was off-limits, and no one from the shipyard, the United States Lines, or Gibbs & Cox would answer any questions.

After fifteen minutes of building up to Navy power, Kaiser saw the emergency indicator lights flashing on the main control panel.

He picked up the phone to the bridge.

In the captain's cabin, one of Kaiser's overall-clad men appeared.

"The lights for the reduction gears in the after engine room are flashing," he told Franklin and Blewett. "We need more oil on the shafts. You'll have to shut her down."

Slowly, *United States* glided to a stop. As she wallowed in the rough seas, Manning, Franklin, and Gibbs conferred about what to do next.

After much discussion, the team decided to put off the maximum power run until the official trials, scheduled for early June.

The cutter's crew picked up the two Raydist buoys, and the captain signaled Manning that they too were setting a course back to Newport News.

As *United States* sailed home, reporters played a guessing game about the ship's speed. Up on deck, one reporter joked to an acquaintance that the ship had made 50 knots, an impossible feat.[10]

"My!" was all he could say.

Suddenly, a great gust of wind blasted across the fantail. The reporter's notes tore loose from his hands, flew over the railings, and fluttered into the boiling sea, followed by several hats and a wallet.[11]

Down in the cabin-class dining room, George Horne of the *New York Times* had more than speed on his mind as he picked at Clamchowder Jack's fried chicken. Turning his eyes toward the room's visual centerpiece, a backlit sculpture depicting allegories of the seasons and constellations, he spied artist Seymour Lipton's exceptionally well-endowed representation of Taurus the Bull.

Horne stormed out of the dining room. Lipton's work, he roared, was a gross violation of common decency.[12]

The liner returned to her shipyard berth in Newport News on the evening of May 16, brilliantly illuminated against the night sky.

All over the nation, but especially in New York and Washington, people picked up newspapers on May 17, eager to read about the trials. "In the last three days the *United States* has undergone rigorous tests," Horne wrote in the *Times.* "Into a 35-mile-an-hour wind she raced at a speed above 30 knots. The sea was moderately choppy, but she barely quavered and not one of the 1,700 persons reported seasickness."[13]

When asked by Horne about the ship's chances at capturing the Blue Riband, Admiral Cochrane replied tersely that the trials had "confirmed our conviction that the *United States* is the fastest and finest liner in the world."[14] Cochrane also admitted that "reduction gear bearings in two of the four engines gave indication of overheating, a circumstance that frequently happens in new machinery."[15]

Over the next week, *United States'* inboard reduction gears would be adjusted. The gears' bearings were slightly out of line, something easily fixed. With luck the weather would be better for the official trials in June.

Shipyard workers also made another adjustment, of a nonmechanical nature, to the ship. Not long after the trials, Horne found a package on his desk at the *Times.* When he tore off the wrapping paper, he found himself holding a stylized aluminum bull's penis, mounted on a mahogany plaque. Originally belonging to Taurus the Bull, it was a gift from William Francis Gibbs, with thanks for protecting the moral and other sensibilities of the traveling public.[16]

"In my business, if I didn't have a sense of humor I would have been dead long ago," Gibbs once said.[17]

Like many people who work around the sea and ships, William Francis Gibbs was superstitious. He would never sit at a table for thirteen. He would never write a letter with thirteen paragraphs.[18] He also firmly believed that it was bad luck for a woman to come aboard before the maiden voyage. That ironbound superstition held firm for both the May

14 builders trials and June 9 official trials: No women allowed. Not naval engineer Elaine Kaplan, who had designed the propellers. Not even his wife, Vera, who had to be content with a quick tour before departure. She joined more than twenty thousand other visitors, most of whom were locals who, according to the *Newport News Times-Herald,* were the "friends and relatives of men who had put in long hours in her building."[19]

Vera's tour was more private and thorough than one given to the average shipyard worker's wife. "On the ship were scenes of great activity," she wrote in her diary. "A great many men were scrubbing floors, or polishing this or that. Most of the furniture was in already. The movie theater was a gem; soft gray, with fluted walls, the chairs were most comfortable."

Vera also peeked inside the private dining room on the promenade deck, with its dark blue walls, red curtains, and crystal glass sculptures by Charles Lin Tissot. "One room had black walls with brilliant diamond stars of illuminated glass," she observed. "It was as though they were magnificent snowflakes, but no, they were heavenly bodies." Then there was the vast first-class ballroom, which she described as a "triumph," with its "glass screens, pure crystal, engraved with the most exquisite designs."[20]

Before leaving the ship, she strolled into the first-class dining room. Only five hundred guests would sail on these second trials, and they would find the dining setup much more sophisticated and polished than on the first. The tables were fully set with gleaming crystal, silverware, and plates adorned with black stars and the eagle seal of the United States Lines. Beneath the Gwen Lux sculpture *Expressions of Freedom,* Vera saw a large, painted sugar model of the ship, three feet long, placed on the buffet.[21]

On June 9, at 6:38 A.M., the ship pulled away from her pier and sailed down the James River for a second time. After a morning of calibration tests and compass adjustments, *United States* made a series of normal runs until eleven that evening. The reduction gears did not overheat, and Gibbs said that he was pleased, and that the "ship had exceeded his expectations in almost every respect."[22]

But only those on the bridge and the engine room knew what her true top speed was. The people aboard were left to guess by watching the water rush by. Was she making 35 knots, 40 knots . . . even 50 knots? No one could be sure.

Commodore Harry Manning kept his enthusiasm in check. The next morning, the seas were calm and the wind was gusting gently, perfect weather for Manning to run two punishing tests. The first was the "crash astern test," in which the engines were suddenly thrown into reverse while the ship was plowing ahead at full speed. The second was the "crash ahead test," in which the engines were thrown into forward motion while backing at full speed. The purpose of these maneuvers was not just to test the machinery, but to find out how long it took for the ship to come to a complete stop. Finally, after a day that would have shredded the machinery of most other big liners, Manning announced he was satisfied.[23] So did Gibbs.

When *United States* pulled into her fitting-out berth that evening, it was jammed with cheering shipyard workers. On board, a couple of huge brooms, spray-painted silver, were lashed to the top of the aluminum radar mast, symbolic of a "clean sweep." It meant that *United States* had broken all sea speed records during her second set of trials. Again reporters tried to find out what speeds the ship had achieved, but her designer and his coterie remained as tight-lipped as ever.

Vera Gibbs was waiting for her husband behind the security rope. As darkness fell and the ship was illuminated, Vera did not see a mass of steel and aluminum, but "a fairy palace . . . a shiny presence that had suddenly just happened," whose sleek form was surrounded by latticed cranes and shipyard lights twinkling behind them.[24]

Nearby, Vera spotted a small boy gazing at the liner. "I think she is the most beautiful thing I have ever seen!" the six-year-old shouted.

The main gangway doors swung open, and Vera watched people getting off the ship. "The first person I saw come off," Vera wrote, "was Admiral Lee, then W.F.'s secretaries, Mr. Kelly and Mr. Connelly, and then, after a while, W.F. and Frederic Gibbs. Everyone coming off more than cheerful.

"When W.F. came off he had to wait for somebody," Vera went on.

"[H]e perched himself on a post right by the side of the ship, and there he sat with his knees up to his chin, just looking at his ship—He was inside the lines, so we could not speak to him. Therefore, he was left in peace, just looking at the big and beautiful thing he created."[25]

This was the moment Gibbs had lived for since he was a small boy gazing up at the hull of *St. Louis* on the stocks at Cramp Shipyard. *From that day forward I dedicated my life to ships,* he had said. *I have never regretted it.*

"We all looked at her as a child gazes at a Christmas tree," Vera wrote, "full of wonder."[26]

CREWING UP THE BIG SHIP

A great ocean liner like *United States,* William Francis Gibbs once said, was a giant musical instrument, "with a thousand each playing a part."

As the liner neared completion, the United States Lines began to handpick a thousand men and women to crew up its new superliner for her maiden voyage.

There were three departments: deck, engine, and steward. Finding qualified men to fill slots for the first two was relatively easy. There were plenty of qualified and unemployed sailors looking for work, many of whom served in the merchant marine and the Navy during the war.

The stewards department, by far the largest and most visible to passengers, was another matter. The United States Lines lacked a consistent tradition of grand service; its passenger ships had nearly all been medium-sized and hence less popular with the social set. Even *Leviathan*'s passengers complained that service did not match the ship's grandeur. Cunard's *Queen Mary* and *Queen Elizabeth* boasted staff that cut their teeth on the company's legendary liners of the past, most notably the now-vanished *Mauretania, Berengaria,* and *Aquitania.* The same was true of ships of the French Line. As a result, many Americans preferred to travel on European liners, because they thought their own countrymen lacked the manners of their British, French, and German counterparts. In part because of the existing passenger prejudice, the United States Lines made a decision during the 1930s to hire a large number of naturalized German immigrant stewards and cooks. This backfired in 1940, when several of them aboard *America* were caught stealing blueprints that showed the location of her gun mountings. After the incident, the company stopped hiring German-Americans, but by 1952 it resumed recruiting them for the stewards department. A few other Europeans were taken on for the crew roster, all eager for the higher pay and longer vacations offered by American union contracts.

One German-American on the payroll was the much-feared chief chef Otto Bismarck (no relation to the German chancellor), who had started out scrubbing pots and pans aboard *Leviathan* thirty years earlier. The cooks who worked under him dreaded the sight of his tall white chef's hat floating above the hustle of *America*'s first-class galley. As befitting his name, Bismarck ran his kitchens like a Prussian drill sergeant, blowing a whistle when it was time to hose down the ranges. Yet he produced results: the German chef's culinary creations rivaled the haute cuisine on any ship afloat, including that of the French Line's chic *Ile de France.*

For added cachet, the company contracted out the personnel for the three orchestras (one for each class) to Meyer Davis, who supplied musicians for society functions up and down the East Coast. Of course, the largest and most polished ensemble would entertain first-class passengers, many of whom were familiar with his effervescent dance music. "What we provide is an atmosphere of orchestrated pulse which works on people in a subliminal way," Davis said of his music. "Under its influence I've seen shy debs and severe dowagers kick off their shoes and raise some wholesome hell."[1]

After the company had picked the crew, both old hands and new hires would go through an intense six-week training program. Here they would learn how to run the *United States* in the grand style of the Continental ships without sacrificing American friendliness. The staff would serve passengers in a manner that was "attentive but never conspicuous."[2]

For the eight hundred or so stewards, stewardesses, bellboys, bartenders, waiters, chefs, musicians, and other personnel needed to crew up *United States,* a job aboard a big superliner meant higher tips from wealthy passengers, known as "bloods." The new ship would also have a two-week rather than three-week turnaround, meaning more frequent family visits and shore leave. Finally, the crew quarters would be completely air-conditioned, a huge perk during summer crossings. *America* was only partially air-conditioned, and its crew quarters were rank during the peak summer travel season.

Some things wouldn't change, however. The hours would be long, and the stewards department would be on call twenty-four hours a day, whether to deliver a meal or drink to a finicky celebrity in the middle of the night, or scrub acres of linoleum floors, polish the metalwork, dust sculptures, wipe glass panels, or operate one of the many elevators. The worst was dealing with seasick, drunk, and sometimes belligerent passengers.

But above all, Gibbs insisted that his ship was to be cleaner than its European counterparts. Pristine was the word. And the crew quarters would be more spacious and comfortable than any other afloat. During the ship's construction, Big Joe Curran, head of the National Maritime Union, declared to General Franklin that the ship's crew quarters were "a substantial improvement over any quarters on any vessel in the past," and that "I can assure you they are satisfactory and she should prove to be a happy ship."[3]

One of the lucky men hired to work on *United States* was Assistant Bell Captain Bill Krudener, whose uncle Harry Tate had also helped found the National Maritime Union with Big Joe Curran. Born in 1928, Krudener grew up in Drexel Hill, Pennsylvania, near Philadelphia. When he graduated from high school, he accepted an offer from his uncle and decided to go to sea. In 1949 after getting his union book, he also got a new commission: bellboy aboard the newly refurbished United States Lines flagship *America.* Krudener, stocky and dark, had a charming smile and demeanor that made him popular with the passengers. Among the people he once served was Truman's secretary of state Dean Acheson, who showed Krudener exactly how he wanted his martinis mixed. During the crossing, Krudener helped with Acheson's parties held in his suite. Eleanor Roosevelt was once a guest, along with a couple of members of Congress. The fast pace and quick wit of the conversation amazed Krudener.

Krudener enjoyed life at sea. When *America* was docked at Le Havre, he and Chief Steward Herman Müller were standing at the cabin-class gangway as some college girls came aboard to return to New York. They had traveled the other way a month earlier.

"Oh Bill, so nice to see you again!" they said, throwing their arms around him. "Here's our cabin number, you know where to find us. Let's have some champagne."

"They are my cousins," Krudener said with a smile as the girls skipped down the corridor.

"How many cousins do you have?" Müller asked. "Everyone on this ship seems to be your cousin!"

Fraternization between crew members and passengers was strictly forbidden. But Müller couldn't fire Krudener, who was a hard worker and nephew of Harry Tate.

In 1952, Krudener and other members of *America*'s stewards department got a notice from United States Lines: those who wanted could transfer to the company's new flagship *United States.* As assistant bell captain aboard *United States,* Krudener would be working under the watchful eye of Chief Steward Müller, another German-American who brought to his work both precision and an iron fist. Sloppy service could ruin a liner's reputation, especially among the fickle ones in first class, who were accustomed to getting anything they wanted, at any time of the day or night.

One who chose to transfer was *America*'s assistant chief steward, a tall, stony-faced Swede named Andrew Malmsea, who supervised the special six-week "boot camp" at the U.S. Maritime Service Station in Sheepshead Bay, Brooklyn. The facility boasted a fully equipped galley, a simulated main dining room, mock-up cabins, as well as classrooms. The stakes were high. Since the maiden voyage would be a very high-profile event, there could be no complaints about sloppiness, confusion, or disorder. In addition to General Franklin and Vincent Astor, the first-class passenger list included Margaret Truman, the president's daughter.

Krudener didn't like living six weeks in barracks. "We knew how to do this stuff already," he said, like making beds the "right" way. "Everything was explained about how they wanted things done." But at least he could visit relatives in nearby Flatbush. He also liked his new uniform: a navy cutaway coat highlighted by epaulets, gold braid, and gray lapels; close-fitting blue-gray pants with twin red stripes running down each leg; then there was a starched white shirt, and a black bow tie.[4]

On June 6, 1952, members of the press descended on Sheepshead Bay to get a taste of what passenger life would be like aboard *United States.* A reporter from the *New York Times* was impressed. In addition to the food being "excellent," he wrote that "matches were always ready whenever a cigarette was taken out, water glasses were filled, and a continuous supply of butter attested to the attentiveness of the stewards." While the waiters were trained to properly serve Chef Bismarck's meals, the bellboys were taught how to mix drinks and "even tie a fumbly-fingered passenger's bow tie."[5]

The galley staff also learned how to operate a new cooking contraption known as a RadaRange. "A housewife's dream," according to one newspaper, the primitive microwave could cook a steak to well-done in a few minutes. But the device still had bugs. "Chef Otto Bismarck invented a sauce that made steaks look charcoal-broiled," recalled one crew member, because "without it, the steak did not look appetizing."[6]

In the end, Krudener was glad that he received additional training, but he couldn't stand Malmsea. Chief Steward Müller's assistant always addressed bellboys as "Bells" rather than by their first name. "Snob," Krudener recalled. "He just had this look that made you want to kick him in the ass."

Commodore Manning didn't show up at the boot camp to rehearse what he would do on board—which was a lineup and inspection of the stewards department before the start of each voyage. Pants had to be creased and pressed, fingernails clean, and bow ties straight. Krudener knew Manning from his three years as assistant bell captain on *America.* "He was like a little Napoleon," he said about the commodore. "He was one of those people who wanted to show their authority over the littlest thing."

On June 17, Krudener joined five hundred other Sheepshead Bay graduates at Penn Station to board a special train to Newport News. "It was one big wild party!" Krudener recalled of the trip. "They had to rebuild the train cars after we left!"[7]

The train pulled directly up to the outfitting pier. The sight of the ship's two finned funnels towering above the rooftops and gleaming in the June sunlight took the crew's breath away.

"It was just so massive, you couldn't believe it," deck steward Jim Green remembered. "The *America* looked like a tugboat compared to *United States.*"[8]

After stepping off the train, Krudener and the rest of the stewards departments clambered up the gangways to drop off their gear in their cabins and began to set up the ship for the two-day delivery voyage to New York. Food had to be loaded on board, sheets and linens cleaned, upholstery and carpets vacuumed, and liquor cabinets stocked.

Krudener believed that African-Americans were treated reasonably well by the National Maritime Union, and he experienced no real racial tensions aboard ship. But many of the ship's African-American crew decided it was best to stay on *United States* rather than venture into Newport News after dark. On the outfitting pier were two water fountains, one labeled "White" and the other "Colored," a clear indication that the Virginia town was part of the segregated South.

During the next few weeks, as workers added the last touches of exterior paint, the United States Lines outfitted the ship's passenger areas for the maiden voyage, a task comparable to furnishing New York's Waldorf-Astoria. The steward's manifest included 2,200 aprons, 900 cook's caps, 3,000 laundry bags, 7,000 passenger and 3,000 crew bedspreads, 125,000 pieces of chinaware, 57,000 pieces of glassware, 170,380 napkins, 10,000 coat hangers, 6,000 pillows, 44,000 bed sheets, and seven caskets for the ship's morgue, a grim necessity on all modern transatlantic liners.[9] Most of the passenger items were emblazoned with the black eagle insignia of the United States Lines, first introduced by Clement Griscom for its predecessor company, the American Line, in the 1890s.

There would also be 1,200 "freeloaders," as Krudener called them, on board for the big party on the delivery voyage from Newport News to New York: admirals, generals, congressmen, senators, shipyard executives, and others who begged and cajoled for a berth, preferably one in first class. Walter Jones, *United States'* publicity director, had also invited more than 160 reporters, an offer not easy for any member of the press to refuse. True to Gibbs's superstition, because this sailing preceded the maiden voyage, no women were invited. Indeed, there were no women passengers at all; the only two aboard were members of the crew. Women would not be allowed as passengers until the liner's official maiden voyage.

"It was some party," sniffed Assistant Bell Captain Krudener.[10]

Two trains carrying guests left New York's Penn Station on June 21 in the mid-morning and arrived in Newport News that evening. Among those who boarded *United States* for the run up to New York was a harried-looking Charles Sawyer, President Harry Truman's secretary of commerce.

Sawyer was happy to get away from his desk. Even as the public eagerly anticipated the maiden voyage, he was waging an intense battle with President Truman to keep the ship's precarious financial arrangement from collapsing. Whether the arrangement would hold was still to be determined.

TRUMAN ON THE ATTACK

In June 1952, when the new superliner had completed her trial runs, the contract between the U.S. Maritime Administration and the United States Lines had been in place for over two years. President Truman and his comptroller general, Lindsay Warren, continued to insist that the owners of *United States* had cheated the American taxpayers out of millions of dollars. The original agreement, Warren asserted, was "riddled with impropriety."[1] Moreover, the president's top naval aide warned Truman that if the government sold the ship to the company at the original $28 million, it would result in "serious criticism" of the administration.[2]

But also by June 1952, the Truman administration was collapsing under the weight of corruption charges and the stalemate in Korea. Truman's firing of the popular general Douglas MacArthur proved to be a political blunder, even if the president had correctly asserted the constitutional principle of civilian authority over the military. The Democrats were facing huge losses in both houses of Congress. It also appeared that Eisenhower, Franklin's old commanding officer from the 301st Tank Battalion, was going to win the White House.

With a backdrop of bad poll numbers, Truman's fight to get several million dollars back from the United States Lines grew more vitriolic.

For his part, Secretary of Commerce Charles Sawyer worried that if Truman and the comptroller general got their way, the United States Lines would walk away from the new superliner and the government would be stuck operating her. Sawyer felt that the two men were trying to turn *United States,* a ship that was already a huge hit with the American public, into a political football. In a year of civic turmoil—of McCarthyism and Cold War worries—the completion of *United States* was good news for the country. Why not just leave the liner alone?

Three months before the trials, the general counsels of the Department of Commerce and the Maritime Administration had concluded that "a suit by the Government to set aside the sales contract or a defense by the Government predicated on a claim of illegality would not be sustained by the courts." Sawyer wrote Lindsay Warren that "the contract abovementioned between the Government and the U.S. Lines Co. is a valid and binding contract on both parties," and that there had been "no suggestion that improper influences were exerted on any member of the Commission or that false representations were made by U.S. Lines Co."[3] The commerce secretary hoped that Warren would drop the matter.

He didn't. In April, Warren wrote to Sawyer saying that "I will continue to employ every power of this office to prevent subsidy grants and national defense allowances which exceed those by law."[4]

Sawyer tried one more time to convince Warren that what was past was past, and the public was happy with the ship. On May 6, Sawyer wrote to Warren saying, "It seems to me more important for the United States to carry out its contract already entered into than the Government be saddled with a ship it cannot run and a lawsuit which it cannot win."[5]

Even so, to the delight of the president, Lindsay Warren developed a line of legal reasoning of his own. The General Accounting Office, not the Department of Commerce, he said, had the final legal say over the matter. Most of the $25 million in national defense features, he went on, had not been spent to further the interests of the American military, but to further a commercial quest for the Blue Riband by the United States

Lines.[6] Warren wanted the GAO to renegotiate the contract with the United States Lines. The new purchase price, Warren insisted, should be $38 million, half the liner's total cost and $10 million more than the original contract price.

Sawyer then made a direct appeal to the president to stop Warren. Citing Maritime Board chair Admiral Edward Cochrane, Sawyer said that going after the United States Lines by cutting off their subsidies demonstrated an "immoral and unwarranted attitude" on the government's part, and was not in line with American concepts of "fair play nor the traditions which have governed the relationship between the Government and its citizens since our republic was founded."[7]

Cochrane had tipped off General Franklin about what Truman and Warren were up to, and Franklin came down to Washington to meet with the admiral and the Federal Maritime Board. "We could not agree to reopening of the contract and to pay more for the ship," he wrote to Cochrane on June 6. "This decision has been ratified by the Board of Directors of the United States Lines."

Franklin was deeply bothered by the controversy. It was more than just the possibility of forking over another $10 million to the government. "It is a pity that she makes her debut under a cloud of controversy," he said in a letter to Cochrane. "From every standpoint of national prestige this vessel should be delivered under our contract."[8]

Three days later, Secretary of Commerce Sawyer, who probably read General Franklin's letter to Admiral Cochrane, wrote the president, saying that "there is no course open except for Admiral Cochrane to go through with the contract on June 20th in accordance with its terms."[9]

President Truman refused. "As you know," he wrote to Sawyer, "the maintenance of a strong merchant marine has always been of the greatest importance to me. . . . The construction of the S.S. *United States* has been followed with especial interest by me." He then challenged the United States Lines to do what he felt was the right thing, or else face the consequences. "I sincerely trust," he continued, "that the officers of the United States Lines Company will promptly accede to the renewed request for an exploratory conference without prejudice to any rights either party may have under the contract."[10]

On June 20, one day before *United States'* delivery voyage from Newport News to New York, Secretary of Commerce Sawyer defied the president and simply ignored Warren's claim to jurisdiction: he handed the ship over to the United States Lines under the original contract terms. William Francis Gibbs and Commodore Manning attended the formal transfer ceremony. Sawyer then packed his suitcase, headed to Washington's Union Station, and boarded the train for Newport News. The two-day voyage to New York would no doubt give his nerves a rest. There would be fresh sea air, cocktails, shuffleboard, and lounging in a deck chair. He planned to ignore any incoming telegrams or ship-to-shore phone calls from the comptroller general.

Just before leaving for his train, Sawyer fired off a letter to Congressman John F. Shelley, chairman of the House Merchant Marine and Fisheries Committee. Sawyer accused Warren of trying to give the impression to the public that he was a "knight in white armor . . . defending the taxpayers from some nefarious plot to which I and the Maritime Board are parties." In reality, Warren was merely doing some "Monday morning quarter-backing," a practice that "may be a pleasant pastime for sports writers but it cannot become a rule of conduct for officials of the Government of the United States."[11]

"Mr. Warren has by his statement in effect dared to turn this ship over to the United States Lines pursuant to this contract," the letter concluded. "Well, I turned it over. He claimed that it was in his power to tell me that I could not go through with this contract. 'Upon what meat doth this our Caesar feed, That he has grown great?' "[12]

Chairman Shelley agreed, responding that Warren's attacks on the Secretary of Commerce were "wholly gratuitous."[13]

But Truman hit the roof when he found out about the sale. He sent a letter to Attorney General James P. McGranery the very day Sawyer handed over the ship to Franklin. The letter was released to the press and copies sent to Warren, Cochrane, and Sawyer.

The president asserted that he had been a firm supporter of the construction of *United States* to promote international commerce and national defense, but the United States Lines, he argued, had benefited from these "construction subsidies, tax benefits, and other privileges accorded by the Merchant Marine Act" for too long, and special legislation to amend excessive subsidies was needed to bring the company to heel.[14]

He then told the press, "I deplore this attitude on the part of the company . . . the government is not helpless in this matter."[15]

A happy Warren read the president's letter and wrote back immediately. "I am pleased and in strong accord with your letter of June 20, 1952," he chirped. "I shall be happy to confer with the Attorney General at any time he wishes."[16]

Sawyer did not respond. By the twenty-first, he was aboard *United States,* bound for New York. Just before moving into his stateroom, he again said to a Newport News reporter that the contract was "binding and unchangeable," adding, "No other passenger ship ever built is so beautiful, so fast, so safe, so useful."[17]

But the argument was not over. Later in June, Attorney General McGranery would issue the Justice Department's opinion: the government would punish the United States Lines by withholding two operating subsidy payments, not to exceed $10,000,000, the amount the White House said the United States Lines owed the government. Truman liked McGranery's opinion.[18]

General Franklin told the press that the line welcomed an inquiry, and that the United States Lines was "ready and willing to cooperate fully with the Attorney General."[19] But he could not back down. He knew that without these subsidy agreements as written in the original contract, his company could not afford to run *United States* as expected by the Navy and by a fickle traveling public.

As the Big Ship slipped out of Hampton Roads on June 21, 1952, William Francis Gibbs barricaded himself in his suite. He had left his name off the passenger list and planned to evade all strangers who wanted to congratulate him. For him the delivery voyage was not a time for cocktails, but for double-checking the ship's mechanical systems and giving tours to a few, high-ranking officials. Vera, who had been with her husband in Newport News, went by train back to New York. She would join the ship and her husband on July 3 for the maiden voyage. Gala occasions were her element, and she looked forward to cocktails and laughter in the luminous Navajo lounge. She knew her husband would be on the bridge with Commodore Manning for the entire first crossing. Gibbs would be there because he suspected Manning might push the ship too fast, against the company's clear instruction.

One important player decided not to take part in the upcoming maiden voyage, with its social events and media circus. Frederic Gibbs planned to stay at home in his bachelor flat and follow the newspaper reports. Besides, someone had to run the firm, and Frederic was the only person his brother really trusted.

THE NEW SEA QUEEN ARRIVES

As *United States* entered New York harbor for the first time on June 23, 1952, she was greeted by a cacophony of hooting tugs and spraying fireboats. Planes and helicopters carrying newsreel cameramen whizzed over the tops of her funnels and her radar mast.

From the skyscraper at 21 West Street, a giant banner hung from the windows: WELCOME *UNITED STATES,* GIBBS & COX, INC."[1] Not since 1946, when the *Queens* sailed into the harbor packed with thousands of victorious troops, did the people of New York turn out in such huge numbers to greet an ocean liner.

Thousands in the crowd watching the ship come in were not content just to look at *United States* from afar—they *had* to get on that ship, if only for a couple of hours.

On June 28 at 9 A.M., the ship was opened to the public. Admission was one dollar, with the proceeds going to the Travelers Aid Society. Working the huge crowds was a teenager hawking pictures of the ship at ten cents apiece. Hot dog and ice cream vendors staked out positions along the queue, turning the sidewalk into what was described as a "summery midway."[2]

From the ship's upper decks, William Francis Gibbs looked down at the crowd with delight. "The flood began early in the morning and continued throughout the day," he recalled. "Thousands of people came. It required the inspector to call for the reserves on three different occasions—the mounted police—and by the time the day was well on, the arrangements in front of Pier 86 looked like a Fifth Avenue parade. . . ."[3]

Down on the West Side Highway, the line shuffled slowly forward, the crowds murmuring with anticipation. "I wanted to see that boat," said Mildred Tross, one of nearly twenty-five thousand visitors waiting in a queue that stretched for eight blocks. "I wanted to see it. It's the only way I'll ever get to see it." The sun beat down on the moving crowd as it inched toward Forty-Sixth Street, people's eyes trained upward on the red, white, and blue funnels looming over pier catwalks. "Midwesterners who never expect to cross the ocean," the *New York Times* reported, "and people from all over stirred by the drama of a great ship converged yesterday on the liner *United States,* alongside Pier 86."[4]

A taxi driver passing by the new liner pointed at her and proudly declared to his passenger: "You know this is the *United States.* This ship was built here. It's our ship."[5]

Deputy Chief Inspector William McQuade, the New York police officer in charge of crowd control, felt absolutely overwhelmed. Maiden arrivals at Luxury Liner Row always drew lots of curious people. But previous debuts paled compared to the crush of visitors clamoring for a chance to get a close look at the Big Ship. The men, women, and children snaking toward the *United States* were tangible proof of what the *Times* called "the drawing power of the new sea queen."[6] The sight of the big American flag fluttering from her sternpost made patriotic hearts happy, and the veil of mystery created by Gibbs about the ship had its desired effect with the general public, which had no interest in the $10 million subsidy dispute between General Franklin and President Truman.

As he stood by the gangway doors, Assistant Bell Captain Krudener knew visiting day would make extra work for the cleaning crew, but he was glad to be in New York and rid of the freeloaders. The two-day voyage from Newport News had been chaotic. "These guys," he said, "had all the booze they wanted." In an age when men did not walk out of their cabins without wearing a coat and tie, and rarely went in public without a hat, Krudener found his charges walking around the first-class public rooms in T-shirts and shorts. During the trip, one of the first-class elevators had broken down between decks, trapping several drunken passengers inside. To keep them happy, Krudener threw three bottles of scotch into the elevator car as the crew fixed the problem.

"They didn't care if they ever came out," he recalled.[7]

On the day of the liner's arrival in New York, the crowd on board was not so rowdy. Still, gawking people tracked mud all over the sea-green carpet of the observation lounge and the black linoleum of Purser's Square that the crew had scrubbed and waxed to a high shine. Someone knocked out the plug to a water main on one of the aft decks, soaking many visitors. The slow-moving line outside came to a halt as the crew replugged the main and stopped the geyser. Despite the delays, more than nineteen thousand people toured the Big Ship that day.

At 5:30 P.M., John Brennan of the United States Lines announced to the three thousand people still waiting on the street that the ship was now closed to the public. An enormous groan came up from the crowd.

Brennan promised the disappointed people who already paid for admission that they would get their money back. As they moved away from the pier, they glanced over their shoulders at the great liner that loomed over them.[8]

Despite the great show, not everything had been on display. All throughout the day, the doors to the engine rooms and bridge remained locked and secured by guards. Gibbs did not want any Russian spies or—worse, perhaps—rival naval architects who might try their hand at the escapade he himself had pulled at *Normandie*'s maiden arrival in New York.

The onboard parties went on for nearly two weeks, as did the loading and provisioning. "We worked every day," Krudener recalled. "They were showing off the ship to all the celebrities." Although he collapsed in his bunk exhausted every night, the assistant bell captain still felt that "it was quite a wild, good time."

Deck steward Jim Green was awed by the New York and Washington elite who swarmed aboard the ship. Every day there was one glittering luncheon and dinner party after another for shipping executives, bankers, and politicians. Before one gala affair, Green passed the first-class dining room and saw an enormous ice sculpture in the form of an eagle, perched above a buffet stacked with caviar, aspic, and cheeses.

The parties went on late into the night as the Big Ship rested at Pier 86, her twinkling lights reflected on the Hudson River. Her public rooms echoed with laughter and clinking glasses as the nation's rich and powerful enjoyed themselves.

William Francis Gibbs still loathed these sorts of events. They were unpredictable, loud, and full of conversation for which he had no patience. He loved being the font of all knowledge at his office, or speaking at professional dinners, but he hated being the center of attention in other settings, even on board his dream ship.

The staff of Gibbs & Cox knew this full well, so they decided to throw him a party that he would like. At 4 P.M. on July 1, the *United States* design team filed into the tourist-class theater. The team, which included Elaine Kaplan, contributed money to purchase a special gift for him.

Matthew Forrest, the hull designer who had worked with William Francis since *Malolo* entered service in 1927, made the opening remarks from theater stage.

The gift sat on a pedestal, wrapped in cloth and adorned with striped ribbons and bunting. Its recipient stood to the left of the stage, his white hands clasped over his threadbare black suit jacket.

"Mr. Gibbs," Forrest began, "we hope this is not a complete surprise to you. This gathering is not intended to hide your light under a bushel, but it is merely the desire of old associates who have been with you from ten to over thirty years, to have a moment with you by themselves, without benefit of publicity, and we are gathered here and represent the Staff as a whole and are mindful of those who are not with us today who have left us in the past, and who could not be here."[9]

The statue was unveiled. It was a scaled-down replica of the ancient

Greek statue called "The Winged Victory of Samothrace." A winged female alighting on the prow of a warship, it honored a victory at sea by the ancient democratic city-state of Athens.

As "a symbol of achievement," the staff felt, "it is considered especially suitable for this occasion."

Forrest presented his boss with a scroll signed by everyone who worked on *United States* from design to completion:

> Naval architect, marine engineer, and brilliant executive who has given impetus to the beginning of a new epoch in the creative art of ship design and construction . . .
>
> Throughout the years he has followed closely the history of disasters at sea and devoted unceasing efforts to the study of ship design to promote greater security and safety for the sea traveler. . . .
>
> To William Francis Gibbs, we pay homage for this, his greatest achievement. . . .
>
> We, his associates, by the presentation of this symbol, honor his advancement in his profession and hold dear our association with him and the example he has set in leadership that has inspired and prepared us to collaborate with him on even greater ventures in the future.

To deafening applause, Gibbs took the podium, and then motioned for Frederic to join him. Hundreds of eyes trained on his gaunt frame. His staff wondered what this reticent man would say.

"The fact is," Gibbs began, "that this ship is the product of many minds, much enthusiasm, of people who with head and hand have joined in a great project and thus presented to their fellow citizens an example of something which has been denied these fellow citizens for 100 years."

He then apologized to all "their wives and their sweethearts for keeping them in the office long after they should have left; for making them work on Saturday, for complaining bitterly about their stupidity, because none of them is stupid—and for doing all those things that a taskmaster does in driving the Egyptians to build a pyramid."[10]

That same day, General Franklin asked for a special meeting in the United States Lines boardroom at One Broadway. It was forty-eight hours before sailing day. Among the directors present was Vincent Astor. "The atmosphere was electric for in the back of our minds was the idea of the Blue Riband Trophy," Franklin recalled. He ordered his personal secretary and a notary public to record the meeting.

Two guests arrived in the boardroom and took their seats at the long, gleaming table. First was Commodore Manning, trim and resplendent in his blue uniform, his steel-blue eyes alert and ablaze. Second was Chief Engineer Bill Kaiser, lumbering and red-faced.

Manning hated being called away from his ship. And he dreaded the prospect of these men telling him what to do. He held his tongue.

Franklin called the board meeting to order. "I think this is a very fast ship," he said, "and she can break the record, east and west. Captain Manning, what do you think?"

"I agree, sir," Manning replied.

"Well, how many revolutions will you need?"

"One hundred and sixty-five, sir," Manning responded.

Franklin's tone then hardened. "I want to make one thing perfectly clear. The complete safety of the passengers and the ship is number one priority. You will shut her down if you encounter heavy weather or fog."

Manning noticed the secretary was busy typing, and the notary was ready to certify an agreement.

"I instructed that the document be removed from the company vault and publicized for the official record if, in the event, the ship went down," Franklin wrote in his memoirs. "Regardless of the trophy, I could not allow the U.S. Lines to be placed in an irresponsible position."[11] Astor and Franklin had not forgotten the *Titanic* disaster, which had taken place almost exactly forty years earlier.

As much as he respected Manning as a sailor, William Francis Gibbs had misgivings. Earlier he had urged the company to make its position clear. "I warned them that in my opinion no attempt should be made to operate at high speed until they had had several crossings at moderate speed and that the crew had been thoroughly broken in."

But ship and passenger safety was only one reason Gibbs didn't want the record broken in dramatic fashion. For the tight-lipped creator of *United States*, it was a matter of national security: no one must know how fast the ship could really go. The previous year, at lunch with United States Lines vice president John Brennan, he made clear just what he *didn't* want to happen on the maiden voyage. "I said they could beat it by a reasonable amount, or make the average 32 knots," Gibbs wrote, "but that they should under no circumstances indicate this to the British by performance what the power of this ship is."[12]

Against all odds, the ship's speed stayed secret. The newspapers kept on guessing, the public speculating. Once Commodore Manning took command of the ship, however, Gibbs worried that the famously independent captain would do what he wanted.

Despite his nervousness, Gibbs did his best to keep his mouth shut. Speaking for his firm, William Francis said that "I have no connection with the operation of this ship, except occasionally when my advice is asked."[13]

Finally, after two weeks of parties, tours, and photo ops, the Big Ship was ready for her first paying passengers, all eagerly awaiting one of the most thrilling trips in Atlantic history.

A VERY FAST LADY

Boarding began at 9 A.M. sharp on July 3. In front of Pier 86, there was a crush of honking taxis. Earlier, stevedores had secured the three passenger gangways—first, cabin, and tourist—to the ship's side. Conveyor belts carried stacks of steamer trunks into the cargo holds; each passenger was allowed twenty-five cubic feet of trunk space, a generous allowance, especially when compared to the almost nothing that passengers could take on the new, cramped transatlantic turboprop planes.

For Bismarck's bakers, the day had started at four in the morning, as they began to make the dough for hundreds of bread loaves and dinner rolls. Down below, Kaiser's engineering team began to fire up the eight boilers to get up a full head of steam. Lazy wisps of smoke started drifting from the two finned stacks. On the bridge, the quartermaster began filling out the deck log, which recorded all navigational data for the ship's voyage. It was one of four such logs maintained by the ship's officers. There was also the official log, listing passengers, crew, and cargo, which had to be filed with the U.S. Coast Guard. The bell book recorded all signals given from bridge to engine room. Finally, there was the highly confidential engineer's log, which recorded engine performance data.[1]

Glancing out the bridge windows, Manning could see the first rays of dawn bathe the skyscrapers of Manhattan in pink and orange. A steaming cup of coffee rested beside the log book as he penciled in "Winds: SW-3, Weather: Clear; longshoreman aboard; resumed loading #3, 4, 5 hatches."

Asked by reporters what his hopes were for the maiden voyage, Manning answered, "I've been instructed to keep to schedule. After all, the main thing is a safe passage."[2]

Hoping for last-minute cancellations, fifty candidates showed up at Pier 86, luggage in tow and passports in hand. All of them were turned away.

Just before passengers started boarding, the entire stewards department lined up along the first class enclosed promenade for inspection. Epaulets, gold braid, and brass buttons glinted in the morning sunlight. The ship's executive officer, Frederick Fender, came down from the bridge to look for crooked bow ties and scuffed shoes. As he did, he reminded the crew what an historic occasion the voyage represented, and how the United States Lines expected the best from everyone.

General John and Emily Franklin had Suite U-92-94. Their twenty-two-year-old daughter, Laura, had U-90, her own stateroom next door. As a little girl, she had traveled aboard *Manhattan* and *Washington* on her father's business trips, and once found herself on Vincent Astor's knee. "He always wore a blue cap aboard the ship," she recalled.[3]

As soon as she boarded the new ship, Laura knew this would be an experience nothing like sailing on the stuffy old *Manhattan* and *Washington*. *United States* was sleek, modern, and fresh. As she walked through the public rooms, she took in the aroma of fruit baskets, Chanel No. 5, and champagne. The strains of Meyer Davis's saxophones and violins drifted in and out of earshot. All around, elegantly dressed people filled the air with expectant chatter and laughter. Laura overheard two words pop up in many conversations: Blue Riband.

"You just felt the excitement the minute you put your foot on her," she recalled. "We seemed to know from the start she was out to break the record. . . ."[4]

When she opened the door to her parents' suite, Laura saw two huge bedrooms, a large living room, and three private bathrooms. The suite even had its own trunk room. Unlike the dark, wood-paneled suites on the *Queens,* it was airy, spacious, and well lit. Once in the suite, Laura scanned the first-class passenger list. On board were Fritz Reiner, conductor of the Metropolitan Opera; Sara Roosevelt, FDR's granddaughter and a cousin of the Astors; and General David Sarnoff, president of RCA, the Radio Corporation of America. Next door to the Franklins were Vincent Astor and his wife, Minnie.* They occupied Suite U-87-89, the already-famous "Duck" Suite named for the pink pearl murals that decorated its walls.

Margaret Truman (daughter of the president, who was, even at that moment, still opposed to the government's role in the liner's construction and operation) was staying one deck below the Franklins in Cabin M-55. She shared a room with her friend Drucie Horton, married daughter of Treasury secretary John W. Snyder. Margaret told the press that she was going on a six-week vacation to Europe, and was traveling on *United States* not as the president's daughter, but just as a tourist.

General Franklin, who was anxious about the comfort of his most important passenger, asked his daughter if she could go down the corridor to check on how Miss Truman was doing.

"I went over to the cabin, which was full of people, all very excited," Laura recalled of her visit to M-55. She introduced herself to Margaret, who "was very informal, lovely."

Amid the chattering visitors was an unassuming, middle-aged lady, clad in a raincoat. She was standing by herself, holding a drink.

Laura walked over to her and said, "I'm Laura Franklin."

"I'm Bess Truman," the woman replied.

"You could have knocked me over with a feather!" Laura recalled.

---

* Vincent and Minnie Astor divorced shortly after the maiden voyage. In 1953, Vincent married Brooke Russell, who inherited the bulk of his fortune and became a world-famous philanthropist, giving away millions to cultural and humanitarian causes. Brooke Astor died in 2007 at age 105. Two years later, her son Anthony Marshall was convicted of plundering her $198 million estate.

"I was talking to the First Lady, with no secret service people around. She was very sweet, very modest, and I had a lovely talk with her."

Laura asked if the first lady needed anything.

"No I have my bourbon here, thank you very much," Bess Truman replied.[5]

Bess and Secretary Snyder both came aboard to see their daughters off. All four posed at the rail for the *New York Herald Tribune* photographer, big smiles on their faces. This time the Secret Service appeared in force. George Horne observed them "brushing aside eager amateur photographers and women eager for a look at the President's wife and daughter."[6]

Margaret Truman also took the time to talk to a few of the musicians in the first-class orchestra. "She had come out of her way to tell me how she'd enjoyed the music," one of them recalled, "and she was just a lovely person."[7]

Unlike the trials and delivery voyage, Gibbs invited many of the important women who helped decorate the liner on her maiden voyage: Dorothy Marckwald and Anne Urquhart, along with sculptress Gwen Lux. Elaine Kaplan was not on board—her husband had booked a trip to Canada, much to Elaine's chagrin. However, she complied and went on vacation with him instead of going on the maiden voyage. Admiral Edward Cochrane was not making the trip, either.

Down the hall from the Franklins, Don Iddon, New York columnist for the London *Daily Mail,* was assigned a single first-class cabin on U-deck and given free passage. "A modest little nook," he lamented, "but it is big enough for a single traveler and is air-conditioned—as is all the ship."[8]

Known as "Britain's Walter Winchell," Iddon entertained his British readers with a "hodgepodge of gossip, press agentry and political hip-shooting," all under the umbrella of "Rule, Britannia." Iddon, who described himself as a "terrific egoist," became the main source of what was happening in America for a huge swath of the British public. "[H]e leaves the impression," *Time* wrote, "that most Americans guide their lives by astrology, gorge themselves on thick steaks, give their daughters

$10,000 debuts, and are all ready to jump into aluminum pajamas and lead-foil brassieres at the first hint of atomic attack."[9]

Iddon did not like what he found on the first-class passenger list—he was disappointed by the lack of "big names." There were lots of wealthy and successful businessmen, he said, as well as politicians and Navy brass. Margaret Truman was a charming person, a talented writer and musician. But where were the movie stars and the European royalty who made crossings interesting for a gossip columnist?

He then wandered the public rooms. "It is ultra-modern," he said of the big American liner, "a big chromium, air-conditioned, streamlined Park Avenue apartment house afloat." He added, "The *Queens* are more opulent, richer, more spacious and gracious, more dignified, and possibly a little bit old fashioned."[10]

As the passengers and guests streamed aboard, William Francis's excitement grew, but so did his anxiety. Once she left the pier, the ship would be in the care of Commodore Manning. And Manning did not like asking for advice.

At 10:45 A.M., all loading of the freight and baggage was complete, and the harbor pilot arrived on the bridge. He would steer the ship as far as Ambrose Lightship, and then disembark and return to shore by tender. By 11:50 A.M., the last passengers had scrambled aboard, and the stewards had finished shooing the last eight thousand bon voyage revelers off the ship.

Gibbs & Cox engineer Sidney Malmquist checked on the engine room as *United States* prepared to back away from her pier. The engineers, dressed in their white overalls, stood at attention, looking at the gauges and making last-minute adjustments as the machinery around them whirred. Nearby, the firemen starting lighting additional burners in the boilers, to provide an extra head of steam upon clearing New York harbor. When the bridge rang the "Standby!" signal, Malmquist watched as the crew scrambled to their stations.

Just after noon, Commodore Manning sounded three husky bellows from the whistles, followed by a long blast, and ordered "Slow astern." Down in the engine room, Malmquist observed that "the whispering of steam flowing to the propulsion units can now be heard. An air of alertness still prevails in the machinery spaces although an interested observer can detect a partial smile, a sense of relaxed tension on the part of the operating personnel."[11]

The tugs *Barbara Moran* and *Nancy Moran* began pushing *United States* back into the Hudson River. The liner's four propellers slowly churned up the water around the stern, sending up billows of river mud. Twenty minutes later, the bow of *United States* was pointed downstream, and the tugs slipped their lines and let her free.

The Big Ship's 1,660 passengers leaned over the railings, wildly cheering to their friends and relatives on the pier head, which was draped in red, white, and blue bunting. Confetti and steamers continued to rain down from her upper decks. Shouts of "Bon voyage!" and "Happy landings!" rang from pier to ship.[12]

Meyer Davis's men lined up on deck and played "Anchors Aweigh," but had a hard time being heard above the din of whistle salutes as the ship pulled back from the pier and out into the Hudson. By 12:45 P.M., still dripping streamers from her promenade deck and railings, *United States* glided past the gleaming skyscrapers of lower Manhattan, then the Statue of Liberty. Those standing on the deck as the ship passed 21 West Street noticed that Gibbs employees had hung three giant banners from the windows:

GOOD LUCK **_UNITED STATES_** **GIBBS & COX**

Once past Ambrose Lightship, at 2:20 P.M., the harbor pilot got off the ship. Manning ordered the telegraphs rung to "full ahead." Down below, Kaiser opened up the steam valves and the turbines spun faster. Malmquist watched as Chief Engineer Kaiser, cigar in his mouth, barked out instructions to his subordinates. "Above these muffled voices," he observed, "the quiet rhythm of the machinery plant can be heard."[13]

Deck steward Jim Green, who had been on his feet since daybreak, felt the decks rumble beneath him as she picked up speed. "It was like a greyhound," he said.

Although there was excitement among the crew about breaking the speed record, there was also the fear of what might happen if the brand-new ship were pushed too hard, as she picked up speed and raced toward Europe.

At 3:30 P.M., Manning ordered all hands to their emergency boat stations. Boat drills had become standard practice at sea since the *Titanic* disaster, yet passengers were merely "encouraged" to come. Hundreds of passengers donned their bright orange "Mae West" lifejackets (sailors joked they made their wearers look like the busty movie star) and lined up along the promenade deck. But for the crew, boat drill was mandatory. Commodore Manning descended from the bridge for inspection, not saying a single word to anyone.

By 5 P.M., *United States* had reached 30 knots. Manning told the press that "the plan was to move up the speed gradually."

By nightfall, the ship's four propellers were spinning at 160 revolutions per minute, pushing *United States* through the Atlantic at around *Queen Mary*'s regular service speed. Yet aside from a bit of shaking in the stern, the passengers could barely sense any effort coming from the machinery at work. By 7 P.M., the first-class dining room echoed with the clink of silverware, crystal, china, and the murmur of conversation, occasionally punctuated by a ripple of laughter. Passengers had to choose between an early and a late seating, except for the VIPs, like Margaret Truman; they could either eat in the fifty-seat private restaurant on the promenade deck, or in the center section of the first-class dining room, which boasted a two-deck-high vaulted ceiling and always had open seating. The "semi-millionaires," as Assistant Bell Captain Krudener described the "ordinary" first-class passengers, were relegated to the sides of the main dining room. These areas were only one deck high and did not boast the same visibility as the center section.

Like a British butler, Assistant Chief Steward Malmsea determined the seating chart based on a passenger's status. According to Krudener, a passenger eager for a prestigious table might slip Malmsea a twenty- or fifty-dollar bill to help the prestige commandant make a decision.[14]

As they opened up their menus, first-class passengers were pleased by the lavish selection of entrées: Braised Roulade of Beef "Forestier,"

Lobster à la Newburg en Cassolette, California Asparagus with Hollandaise Sauce, Long Island Duckling, and Tenderloin Steak with Sauce Bernaise and Mushroom Sauté.[15] If a passenger did not want anything from the menu, made-to-order dishes, including kosher ones, could be prepared with advance notice.

After dinner, Don Iddon watched dozens of passengers cluster around the first-class cocktail bar, its shelves stacked high with crystal glassware. The Meyer Davis Orchestra struck up their signature show tune medleys, featuring favorites by Cole Porter, George Gershwin, and Jerome Kern. The music grew more up-tempo, the laughter louder, and as the band moved from quick to slow numbers, and from major to minor, the dancers were awed as the ballroom's lighting shifted from dim to bright.[16] The ship's sleek décor and Park Avenue–style ambience had started to grow on Iddon. The ballroom, packed to the walls with dancers, was "palatial," and other public rooms resembled posh nightclubs, such as "the Rainbow Room and the Starlight Roof in New York and the American Bar at the Savoy."[17]

Manning, dressed in his dark greatcoat, walked into the ballroom. The commodore chatted with a few passengers, and then asked a few of the women to dance. The introverted Manning proved to be adept at the tango.[18] He then quietly slipped back to the bridge.[19]

Those who did not want to dance retreated to one of many small seating nooks, hidden by illuminated, etched glass panels of swirling sea creatures and plants. For those wanting something quieter, there was demitasse coffee and sherry being served in the smoking room.

Although they had received weeks of training to prepare for the voyage, the stewards department scrambled to keep up with the work. In the middle of the night, a single steward had to shine all the shoes left out in the corridors. The laundry workers had to press hundreds of passenger items each day, as well as thousands of sets of bed linens and steward's jackets. The clattering galleys, their ranges and counters lined with dozens of chefs, had to cook eight thousand meals daily. When the kitchen closed, uneaten piles of steaks, fish, and other delicacies sat on the galley counters and the buffet tables. Bismarck blew his whistle, letting everyone know it was time to hose down the ranges. But before his workers began to clean up, dozens of Krudener's staff descended on the galley like locusts. They grabbed the unclaimed filet mignons, stuffed them into tin pots filled with rice, and ran back to the crew lounges to scarf them down. Only the leftover lobster tails were off-limits. They would be turned into lobster salad for the following day's lunch. The stewards department also could munch on leftover frog's legs that came from a big tureen in the first-class smoking room.[20]

During a few hours' rest, the serving staff would retire to their cubbyhole lounges and cabins, hang up their jackets, pinafores, chef's hats, and bow ties, and enjoy a game of cards before turning in. The crew menu was basic but hearty: cream of tomato soup, fried filet of sole, and roast leg of lamb. They could also watch a movie or listen to popular music on portable record players. Gambling among the crew was a chronic problem, especially among the nonstop poker players. Most of the staff would be up by 5 A.M. A few engineers would remain on watch through the night.

By the end of the first day, deck steward Jim Green was so tired he could barely stand up. He was already miffed that the commodore had docked his first day's pay for sleeping through a lifeboat drill. "As a deck steward, I was working 7 days a week, 13 hours a day," Green recalled. "You'd work from 8 A.M. to 12 noon. You'd then be off from noon to 3 P.M., then on again from 3 to 5 P.M., then on again from 7 P.M. to 10 P.M. In addition to your normal chores, you'd have a lot of extra duties to do."

The Meyer Davis musicians' schedule was also demanding, especially for the first-class orchestra. "You'd usually play a tea concert," one musician recalled, "then about 6:30 o'clock would be dinner . . . you'd probably play till about 9 o'clock. You'd start again about 10 P.M. and play for whatever time would be appropriate, probably about 2 hours."[21]

Despite their best efforts, at least one passenger complained about the service. "The fastest passenger ship in the world has not got the fastest service," sniffed Iddon. "The Cunard's have. The system of pressing a button for bellboys does not seem to work very well, and some stewards have too much to do at peak rush hours."[22]

Friday, July 4 dawned cold and foggy. Yet Manning did not slow the liner down in the crowded summer C-track shipping lane, the northernmost and shortest of the three routes. In fact, he ordered Kaiser to increase steam.

Passengers who got up early in the morning found fresh copies of the *Ocean Press* in their staterooms, its pages full of United Press stories radioed to the ship only a few hours earlier. As they sipped their coffee, they read that in Vienna, U.N. secretary-general Trygve Lie had called the Korean War not "a war about a small territory but . . . a fight for the general principles of the U.N." In Chicago, Senator Robert Taft, Republican of Ohio—a bitter foe of President Truman—continued to seek his party's presidential nomination, despite General Eisenhower's popularity. In baseball, the New York Giants had beaten the Brooklyn Dodgers 4–3. The *Ocean Press* also reported that the United States Lines received 6,000 applications for 1,000 spaces atop the Ocean Liner Terminal in Southampton for the anticipated July 8 arrival, and that 2,500 spaces on top of two other piers were snatched up in half an hour.

For those looking for some exercise, the *Ocean Press* informed first-class passengers that the gym and swimming pool opened at 7 A.M., and that outdoor games such as shuffleboard and deck tennis would be set up at 10:30 A.M., weather permitting. The "Ship's Notices" section urged passengers to "always hold onto safety ropes, hand rails, or secured furniture when crossing open lobbies, through public rooms, or in the Dining Room." It also urged its readers that "to avoid injuries to fingers, do not attempt to open or close ports," and that women passengers "are advised to wear low heel shoes."[23]

At twelve noon, the press corps trooped into the observation lounge and took out their notepads. Manning, General Franklin, and William Francis Gibbs filed into the room and took their seats in the green leather chairs.

Iddon wrote that the commodore was a "handsome, very alert man and he can be tough." He was less impressed with "red-faced, massive" General Franklin, "who did not say much about the ship except: 'We're doing all right pretty good. I'll say we're doing alright.' "

Manning said that for her first full day at sea, *United States* had averaged 34.11 knots and had clocked 696 nautical miles.

While he might be rude to the crew and even passengers, Manning loved talking to reporters. "He likes to score off the Press with mild sarcasm and astringent comments," Iddon noted, "but I found him amiable and there is no doubt he is wonderfully proud of his wonder ship."[24]

When asked by reporters about what he thought of the ship, Gibbs kept a deadpan face, but he could barely contain his enthusiasm. "This ship is the product of explosive power—American industry," he said. "The *United States* moves as if it were jet-propelled. She is a very fast lady indeed."[25]

It was too breezy outside for deck games, so many of the first-class passengers lounged in the deck chairs lining the enclosed promenades. They chatted with each other, read a book from the ship's library, or simply watched the ocean sweep by the windows. There was a bridge party in the observation lounge at 3:30 P.M. Those wanting to hear some music could go to the ballroom to hear a subset of the Meyer Davis men play light classical favorites at afternoon tea. The 1:45 P.M. feature in the movie theater was *Lovely to Look At,* starring Howard Keel and Kathryn Grayson. Before changing for dinner, passengers could also have a cocktail or two in the black and red refuge of the smoking room.

Anyone looking for adventure in other classes was disappointed when they read in the *Ocean Press* that day that "regulations prohibit your going into classes other than in which you are booked. Should you be observed, it is the duty of the staff to ask you to leave."[26]

"Go back up through the gates," Krudener would tell people he suspected were sneaking in from tourist class.[27]

Disregarding what he was told, Don Iddon journeyed into tourist class, where he asked a young British woman what she thought of *United States.* "It's all together different," she sniffed. "The *Queens* have dignity, royal dignity, you might say. They are stately and very British and majestic, you might say, Mr. Iddon."[28]

Yet some of the passengers in first class were unhappy, as the number of top-tier berths given to company executives and Gibbs & Cox staff had created a dance-card problem: there were too many men in first class. Cabin class, on the other hand, had too many women. Walter Hamshar of the *Herald Tribune* noted that "the shortage of dancing partners in the two classes will be adjusted by the time Independence Day festivities get under way," probably meaning that some lucky single women in cabin class got a coveted invitation to dance in the first-class ballroom.[29]

The Independence Day Dinner capped the first full day at sea, and custom dictated full evening wear: black tie and gowns in first class. The ship's hairdressers and manicurists struggled to keep up with demand. Everyone was in a festive mood despite the dreary weather.

Before putting on his dinner jacket and heading to the first-class dining room, George Horne of the *New York Times* sent a telegram update to his news desk, reporting that the liner had reached the land-based equivalent of 39.8 land miles an hour, and that "no merchant vessel has ever traveled so fast for a sustained period as far as records are known." Despite such speed, passengers could not believe "the minimum of vibration and the absence of a sense of speed" as the ship hurtled through the Atlantic.[30]

Commodore Manning remained quietly confident. When asked about his expectations for the next day's run, he simply responded: "Tomorrow is another day."[31]

Those who placed bets wondered if *United States* had broken *Queen Mary*'s 1938 one-day record of 738 nautical miles, or an average speed of 32.08 knots.

But on the next day, Saturday, July 5, the fog refused to lift. Even so, *United States* averaged 35.60 knots and clocked an astonishing 801 nautical miles, shattering the first of the British rival's records. She was traveling toward England so fast that the ship's clocks, controlled centrally from the bridge, were advanced 90 minutes that night rather than the usual 60—on transatlantic liners, this was a way of gradually adjusting passenger sleeping patterns to the ship's destination time zone. But a good number of his fellow travelers did not care how fast they were going. Horne reported that many passengers who "dined and danced in

Independence Day celebrations" were sleeping off their hangovers until late morning.[32]

For Manning, the sleep deprivation was beginning to put dark circles under his eyes. Yet he refused to leave the bridge, and kept drinking coffee. He hoped that the fog would lift the next day, since he knew that the ship had more power in reserve.

That day, Captain Harry Grattidge of the New York–bound *Queen Mary* sent Commodore Manning a telegram reading: "Welcome to the family of big liners on the Atlantic!" Manning glanced at it, and then turned his eyes back to the heaving gray horizon.[33]

The morning of July 6 showed overcast skies and gray seas, with a 20-knot headwind blowing from the east. But the fog had burned off. Manning asked Kaiser for more steam. The ship that day would be passing several big foreign-flagged liners heading in the opposite direction, including the French Line's *Liberté* (the former *Europa*) as well as the westbound *America*.

Down in the engine room, Kaiser watched the pressure gauge move higher as the whine of the four turbines grew shriller, almost reaching a shriek. "It wasn't only confidence in the *SS United States* we were gaining as we whipped into the home stretch," Kaiser said, "but also confidence in ourselves. The foreign flags had had a monopoly on this sea-queen business entirely too damned long. We could build ships to beat them, and we had men smart enough to run the ships too! Every 'man jack' down in those engine holds was fighting for something, just as Manning had fought and was fighting in the bridge."[34]

Those up early enough felt the ship shudder as she picked up even more speed. "That ship just took off," musician Walter Scott recalled when the fog lifted. "It was unbelievable to me the incredible speed it was traveling."

Some guessed she was exceeding 40 land miles per hour through the moderate swells.

Second Officer Asterio Alessandrelli took out his 16-millimeter camera and stepped out on the bridge. He leaned over the port wing and captured quick shots of the ship's wake foaming violently astern.

He then pointed his lens forward toward the bows: spray kicked up onto the foredeck as *United States* plowed ahead into the steel-gray Atlantic, studded with whitecaps.[35]

At 11 A.M., stewards served cups of hot chicken bouillon on tables or carts pushed along rows of deck chairs. After the fog finally lifted, the hardier passengers braved the winds howling across her upper decks, but it was still too windy to play shuffleboard, quoits, or deck tennis. A few played Ping-Pong in the shelter of the enclosed promenade.

At the noon press conference, the reporters managed to get a few words out of Gibbs. "My expectations are rather high, and the ship is running them hard," he said tersely, then headed back to the bridge with Manning.[36]

America appeared on the horizon heading in the opposite direction, making a steady 22 knots. As the two running mates passed abeam of each other, they exchanged raucous salutes from their whistles. Within a few minutes, *America* had vanished astern.

But another westbound liner was closing in on *United States* from the opposite direction, and judging by the image on the bridge radar screen, she was a big one. Late on the afternoon of July 6, as darkness closed in, three red and black funnels appeared in the distance, belching clouds of black smoke. As she grew closer, passengers could see her prow was kicking up a huge, cresting bow wave.

When the announcement came over *Queen Mary*'s public address system that the ship about to pass was the new American vessel, passengers watching a movie in her first-class cinema got out of their seats and ran to the starboard rails.[37]

Queen Mary was making about 29 knots as she swept past *United States* at 5 P.M. The two ships had a combined speed of over 70 land miles an hour.

Manning let loose a long bellow from his whistles, *Queen Mary*'s Captain Grattidge responded, and the two ships cut loose with a spine-tingling bass chord that sounded over the Atlantic for miles. The Union Jack dipped from the *Queen*'s mainmast in salute, followed by the Stars and Stripes from *United States*' radar mast. Within minutes, *Queen*

Mary's cruiser stern disappeared into the twilight, and the passengers on both ships left the railings and headed below. It was time to dress for dinner.

It was Gala Dinner Night on *United States,* a traditional ritual for the last evening at sea. All three dining rooms were decked out with streamers and balloons, as were the passengers in their party hats and finest clothes. It had been announced over the public address system that *United States* would pass Bishop Rock at about six in the morning. Most of the passengers planned to stay up through the night and continue to party, a prospect that the already overworked crew dreaded.

Outside, the overcast skies continued to darken. Thunderhead clouds grew larger on the horizon, and the winds freshened from the northeast. *United States* was steaming into a full gale, with winds gusting up to 60 mile per hour. As visibility deteriorated, the radar went on the fritz. Manning became so frustrated that he pushed the repair technician aside and shoved his finger into the device. The shock flung the commodore against the bridge bulkhead, but amazingly he was uninjured.[38]

As he stood on the bridge watching the prow rise and fall, William Francis may have recalled that it was the same kind of weather that kept *Mauretania* from taking the Blue Riband on her maiden voyage in 1907. That Manning was still pouring on the steam through fog and gale-force winds must have made Gibbs's blood boil. Not only that, but the commodore was disobeying company orders that the record was not to be beaten by a big margin. Barring an accident, it looked like the margin would be huge.

As the ship began to sway ominously from side to side, Assistant Bell Captain Krudener, dressed in his full uniform of cutaway and epaulets, stood like a sentry next to the big swinging doors leading to the first-class dining room, pulling them open each time he saw a couple sweep down the grand staircase. Up in the musicians' gallery, the white-jacketed players continued to serenade the diners.

Commodore Manning entered the dining room and took his seat at the captain's table. Sitting on his right side was the guest of honor, Margaret Truman. He had reason to be proud. He announced to the table that on July 6, *United States* had broken the record set the day before, clocking 814 miles and tearing through the rough Atlantic at 36.17 knots.

Also at the captain's table were William Francis and Vera Gibbs, sculptress Gwen Lux, Newport News Shipbuilding's William Blewett, and Gibbs & Cox chief engineer Walter Bachman. "Since Mr. Gibbs asked me to sit at his table with Mrs. Gibbs and a young woman who is traveling with her," Bachman wrote Elaine Kaplan, "sundry misguided people think perhaps I am somebody. As a result, I have met a large proportion of the celebrities on the ship."[39]

During dinner, Gibbs asked Blewett if he knew what Miss Lux's sculpture hanging above them represented. Blewett, who knew more about engineering than art, said that he did not. Gravely, Gibbs explained the meaning of the sculpture's German name, and proceeded to explain in great detail how the figures represented the four seasons.

Gibbs's erudition impressed the dinner guests, especially Blewett.

Lux could barely contain her laughter. "Mr. Gibbs was making it all up," the artist told the table. "The four figures represent the four freedoms, and he knows it perfectly well!"[40]

In his letter to Kaplan, Bachman also quoted something written by Don Iddon, but not published: "This has been the maiden voyage, but *United States* has behaved like no maiden. She is a very fast lady, a woman of the world, sleek, sophisticated, and maybe a little ruthless."[41]

Bachman's letter, to be sent ashore as soon as *United States* reached Le Havre, was marked "Airmail."

The party moved on to the lounges and ballrooms. Horse racing started at 9:15 in the first class ballroom; the ponies were aluminum, moved by bellboys across the dance floor following rolls of the dice, which moved some horses faster than others. For those who wanted the silver screen, the theater offered *She's Working Her Way Through College,* starring Virginia Mayo and Ronald Reagan.[42]

The gala began at 10:30 P.M. Revelers in party hats drank champagne and spun around to the strains of the orchestra, whose tunes probably included Cole Porter's exuberant "Ridin' High." In the wee hours of the morning, Laura Franklin, who was "full of champagne," headed up a conga line consisting of Margaret Truman, Drucie Snyder, and a few newspaper reporters. After circling the ballroom, they paraded down the enclosed promenade deck, then up several flights of stairs to the bridge.

The two women, braving the pelting rain and ignoring the "do not enter" signs, found Commodore Harry Manning, William Francis Gibbs, and a clutch of officers on the bridge, their faces illuminated by the pale light of the ship's compass.

When Manning saw the conga line burst into the bridge house, he bellowed, "What are you doing here?"

He then saw Margaret Truman.

"Miss Truman, we are about to break the record," he said quietly. "Would you like to put your finger on the wheel?"

Margaret wrapped her fingers around the spokes for a few moments.

"Let Laura do it," Margaret said, looking at Commodore Manning. The president's daughter let go of the wheel and moved aside. Laura put her fingers on the spokes.

A few seconds of this was too much for Manning, even if it was General Franklin's daughter.

"Please leave!" he hissed in Laura's ear.

The two women retreated to the ballroom, with the revelers in tow. Laura then went down to her parents' suite to wake up her father.

"Daddy, we're about to break the record!" she shouted.

General Franklin rubbed his eyes and looked out the porthole at the ocean racing past. The ship was going fast. *Too fast,* he thought. He started cursing. Manning was defying company orders to slow down in bad weather.[43]

By then, hundreds of passengers had packed the first-class ballroom, champagne glasses in hand and tension building. The captain promised to sound the whistle if the record were broken. At 6:16 A.M., Manning and the officers spotted the flashing Bishop Rock lighthouse, perched on an outcrop, through the fog. As she swept by the stone beacon that marked the end of the eastbound Blue Riband course, her three whistles let loose a great bellow that shook the entire ship. Within seconds, the whistle blasts were followed by popping champagne corks, music from three orchestras, and celebratory cheers and singing from 1,660 passengers. The party spilled out onto the promenades. People sang and danced as the musicians marched up and down the decks playing "The Star-Spangled Banner."

The stewards didn't cheer quite as loudly: "The crew could have cared less," deck steward Jim Green recalled. "They were so tired."[44]

Margaret Truman was with Commodore Manning on the bridge. Both had their eyes fixed on the clock. "Congratulations!" she said, her face beaming. They shook hands, and Manning walked away, exhausted.

United States' time for the eastbound voyage was 3 days, 10 hours, and 42 minutes. She had slashed the *Queen Mary*'s record time by an astounding ten hours. On this first run, steaming so fast that the Atlantic waters stripped most of the black paint off her bow, *United States* achieved an average speed of 35.59 knots, a full four knots faster than *Queen Mary*'s earlier record run.[45]

As the ship neared Le Havre, France, her first port of call, William Francis urged his colleagues that "we should let the ship speak for itself and not tell the British and French in vainglorious terms what we have accomplished."[46]

When *United States* entered the English Channel, Manning dropped the ship's speed to 20 knots. The channel was always crowded and dangerous, but the weather was clearing and vessels of all shapes and sizes were positioning themselves to get up close to the new Blue Riband holder of the Atlantic. Leaning over the rail as the ship sailed through the Le Havre breakwater, Laura Franklin saw dozens of boats, "all decorated with American flags, and signs saying 'Congratulations to the Big U.'" *United States* acknowledged them with "several blasts of her very unladylike ship's horn," Laura said.[47]

French tugs eased the new liner into her berth at 1:24 P.M. on July 7. Her continental passengers disembarked and boarded the Boat Train for Paris. More than five hundred French guests came aboard for breakfast the next morning, at which General Franklin was to speak. After several minutes of bad French, Franklin left for the smoking room and ordered a double scotch.

The bartender told a frazzled Franklin it was one of the best speeches he had ever given: "the French couldn't understand you and neither could the Americans."[48]

At 12:46 P.M. the next day, *United States* passed through the Le Havre breakwater for the two hour cross-channel trip to Southampton, England.

"The journey across the Channel was made on a beautiful sunlit day," Matthew Forrest wrote. "There was a brisk wind blowing too much for the dressing of the Ship with the I.C.S Code flags. . . . We crossed at a good rate of speed, possibly to give all the shoreside visitors who had not crossed with us a little idea of what the 'Big Ship' can do." When *United States* passed the Nab Lighthouse and entered "Southampton Water," Forrest, along with others standing at the rail, was amazed by "one of the greatest collections of small craft I had ever seen. . . . Sailing yachts, power cruisers, excursion boats and even outboards. All were crowded up to the gunwales with people. All gaily decorated with flags and blowing their horns and whistles for all they were worth."[49]

As the liner approached, streamers and confetti flew out of the windows of the Ocean Liner Terminal, as thousands of people lined the docks. The people of England, who had produced the great ships that inspired young Gibbs, stood in awe at the spectacular new American masterpiece.

George Horne cabled his account of the reception of the great vessel's arrival in Southampton to the *New York Times*. "This is a port that loves ships and England is a nation that loves them," he wrote. "The people sometimes considered reserved in their public demonstrations dropped the barriers and let everything go. Along the Solent, along the Hampshire countryside, along the beaches, on crowded piers, in open green fields, masses of people were standing to say to *United States*, 'Welcome, Welcome.' "[50]

Telegrams flooded into the radio room. Prime Minister Winston Churchill cabled the *United States:* "Congratulations on your wonderful achievement."

Gracious in defeat, Captain Grattidge of the *Queen Mary* wrote Manning: "Godspeed. Welcome to the Atlantic. Am sacking my chief engineer."

Manning wrote back: "I've still got more speed up my sleeve—we were just cruising."[51]

As the ship moved into her dock and the Southampton longshoremen secured her lines, those watching from the deck of *United States* were surprised by the warmth of the British reception, especially after their *Queen Mary* had just given up the Blue Riband. Not only that, the *Mary*'s running mate, *Queen Elizabeth,* had to leave a day ahead of schedule to make room for *United States.*

"As we approached," Forrest wrote, "we could see the various levels covered with people. They had even arranged for them to be on the roof of the 1,500 foot long building. There was a big banner which said 'WELCOME' and another one which said 'GOOD LUCK TO THE NEW SHIP.' "

James Black, a Scotsman who worked in the Gibbs & Cox model shop, stepped out onto the forward hatch, carrying a set of bagpipes under his arm. "At this sight," Forrest said, "those on the dock burst into cheers. The band on the dock played, the Ship's Orchestra on the Promenade Deck played, the Piper played and everyone cheered, so it was quite a noise that greeted the *United States.*"[52]

At this, the greatest moment in his life, William Francis Gibbs faced the newsreel cameras. "It's a great pleasure to be here this morning, on this ship," he told the world, "and it also gives me a feeling of humility in having the responsibility of the hopes and aspirations of my fellow citizens that have been embodied in this ship."[53]

For the failed lawyer from Philadelphia, now heralded as America's greatest naval architect, the moment couldn't have been sweeter. As he stood on the bridge of the finest ship in the world, William Francis Gibbs looked down at the cheering British crowds on the docks below. He then turned to look at the decks of the ship and saw hundreds of wildly waving passengers and crew. Finally, he looked up at the massive red, white, and blue funnels that towered over him and then at the Stars and Stripes that fluttered from the radar mast. On July 8, 1952, this reserved, introverted, and driven man could now proudly wave his hat to the cheering crowds and bask in the glory that was rightly his.

MASTERS OF VICTORY

When *United States* left Southampton on July 10, all on board were confident the ship would also take the westward speed record, and so the new crop of passengers, including comedian Milton Berle, were relatively subdued. Some lobbied General Franklin to lash two silver brooms to the ship's radar mast as a symbol of a "clean sweep." Franklin refused, but he did allow for a forty-foot-long blue pennant, symbolizing the Blue Riband, to be hoisted once the ship entered New York harbor.

Sometime during the maiden voyage, Vera Cravath Gibbs ducked out of a party and went back to her stateroom for some peace and quiet. It had been over twenty-five years since Vera, then a young divorcée and daughter of New York's most powerful lawyer, first spotted an awkward, grave-looking man standing by himself at a cocktail party. She at first found him odd, but soon discovered that the man had only one thing in life he wanted to do: to design the finest, fastest ship in the world. She had been part of every setback and triumph her husband experienced during their marriage: the scrapping of *Leviathan,* the war-profiteering charges in 1944, the months of perfecting Design 12201, the disputes with the Truman administration, the launching, and, finally, the spectacular, history-making maiden voyage.

"The trip of trips was now drawing to a close," Vera wrote in her diary. "I look back on the weeks, months, and years that W.F. spent on the S.S. *United States.* I wonder how his enthusiasm remained undiminished. The series of disappointments he had to face, the political battles he had to face, all those went on for so long. Those aggravations kept repeating themselves, with slight variations, over and over again. What I always wondered was why the wellspring of W.F.'s enthusiasm didn't dry up. I am reminded of what Edmund Burke wrote: 'The nerve that never relaxes, thought that never wanders, the purpose that never wavers, these are the masters of victory.' "[1]

United States passed the Ambrose Lightship at the entrance to New York harbor at 4:29 P.M. on July 14. Even though prevailing winds and currents made westbound crossings more difficult and slower than eastbound ones, *United States* still averaged 34.51 knots, making the trip in 3 days, 12 hours, and 12 minutes—fully 9 hours and 36 minutes faster than *Queen Mary*'s best westbound time.

But her passengers were to spend another night on board. Manning dropped anchor off Staten Island. The harbor pilot was to take her through the Verrazano Narrows first thing the next morning and bring her in to Pier 86 to a wakened New York City.

By 7 A.M. on July 15, it seemed that all of New York had turned out to watch *United States* come home. On a hot, hazy summer morning, tens of thousands of people lined the Battery and the Hudson waterfront, cheering and waving as the Big Ship moved slowly by. "Dozens of small craft, tug boats, two Fire Department boats—each shooting up fifteen giant streams of water—and ferry boats set up a din for the returning mistress of the seas," Milton Lewis of the *Herald Tribune* wrote, "and to each one the *United States* returned a salute." The deep roar from the liner's 120 separate salutes was so loud that passengers covered their ears.[2]

A photographer standing on the pier took a picture of Commodore Manning in his resplendent summer whites, standing on the starboard wing of the bridge as he gave orders to his docking pilot. Facing the photographer dead on, his left hand is raised, index finger pointed up—the way of signaling a docking pilot to go "dead slow ahead" or "stop" so that the ship would line up with the pier markings.[3] General Franklin, in a dark suit and Panama hat, stands next to Manning, leaning on the bridge railing. His eyes are trained on the commodore. Gibbs stands toward the back, near the pilothouse. He appears to be caught in mid-sentence, his mouth slightly open, his expression agitated. Not until her lines were made fast and the gangways secured would her designer ease up.

By 9:12 A.M., *United States* had swung into her berth at Pier 86 and her lines made secure. Voyage Number One had officially ended. The 1,650 passengers began to disembark and collect their luggage. The crew looked forward to a few days' leave before provisioning up for Voyage Number Two.

"In 100 years, we have been able to do this just once," Manning said to reporters gathered in his cabin. "It should wake our people up and make them realize what a ship like this means—particularly her qualities as a troop transport."

Chief Engineer Bill Kaiser emerged from the engine room and told the press that the Big Ship had enough power "in reserve" to take on any rival, but for future crossings the ship would be throttled down to an economical 30-knot speed.[4]

There was a ticker-tape parade on July 18—New York's traditional way to salute its heroes in war and peace. Paper streamers flew down from office windows in the tall buildings along lower Broadway, as 150,000 spectators cheered the captain and crew rolling by in open limousines. The party moved into the Starlight Room at the Waldorf-Astoria, where Manning continued to hold center stage. "To me she is the most beautiful thing I have ever seen," he said, "far more beautiful than any woman I have ever known."[5]

The commodore, despite his brusque confidence, did not look well. He had lost ten pounds during the trip.[6]

William Francis was also exhausted. Sometime after returning to the Glass Menagerie at 21 West Street, Gibbs had one more thing to do. He sat down and wrote a note to Vera, expressing both his gratitude and guilt.

"I thought you did your part in adding to the gay spirit of the trip and it was a great pleasure to be with you on such a momentous occasion," he wrote. "I am afraid I could not be as entertaining as I would have liked, but it seemed to me that whenever I got settled down, something came up which made it necessary to put one's mind on the business at hand. Many thanks for your long continued patience and enthusiasm about this ship. I have many times thought you must get tired about being with a person who seemed to have but one thing in mind. However, you can agree, it was quite a big thing."[7]

On July 22, 1952, the crew of *United States* assembled in the first-class theater for an 11 A.M. staff meeting. Nearly one thousand officers, engineers, deckhands, chefs, pursers, stewards, stewardesses, and bellboys found their way into a space built to seat three hundred. From the stage, William Francis peered down at them through his spectacles, his long fingers curled around the edges of the aluminum lectern.

As the lights dimmed, Gibbs announced that he would present each one of them with a medal commemorating the record-breaking voyage: "not the kind of medal that one hangs on one's manly chest, but a larger one that you retain in a proper case as a memento of an occasion would be required to cover the thought I had in mind."[8]

Each three-inch medallion bore a bas-relief of *United States*, along with a map of the Atlantic showing the eastbound and westbound tracks, along with each of the voyage times:

EASTBOUND PASSAGE AMBROSE TO BISHOP ROCK
2942 MILES, 3 DAYS, 10 HOURS, 40 MINUTES
AVERAGE SPEED: 35.59 KNOTS
WESTBOUND PASSAGE BISHOP ROCK TO AMBROSE
2906 MILES, 3 DAYS, 12 HOURS, 12 MINUTES
AVERAGE SPEED: 34.51 KNOTS

Before presenting the crew with the medallions, however, Gibbs drew on one of his favorite analogies to describe his vision of how a great ship was more than an enormous machine, but rather had a life of her own.

"It is my good fortune, loving music, to have seen the rehearsals of a great symphonic orchestra," he said. "I have noticed with keen attention, that these great conductors take the greatest possible care in the training of the orchestra. They know how every particular instrument should be played, and the actual strength and softness of each tone.

"But here's an instrument—this ship—with a thousand each playing a part; and when we think that the first time that this ship—this instrument—was ever operated by the crew that took it out on its maiden voyage—it left Newport News one afternoon, and was in New York the next morning. The next time it went out to sea, it performed as never a ship performed before."[9]

His voice then softened. "I want to confess to you," Gibbs continued, "that I have a great affection for this ship. It's a large object, but I think my affection is coextensive with its size, and likewise I have an affection for every man who has made possible and makes possible day by day, the extraordinary success of this ship so far."[10]

Gibbs then slipped a wooden ruler from his suit pocket and rapped his knuckles with it. Knowing there was no wood on board to knock on, the crew roared with laughter. "It may seem sentimental, and I'm not noted for sentimentality, but I feel today that I owe you all a tremendous debt of gratitude. I wish you well, and may God be with us, and may God be with this ship."[11]

A thousand men and women leapt to their feet and filled the theater with applause and cheers. Gibbs folded his speech, walked off the stage, and sat down.

The following day at noon, *United States* pulled away from her berth at Pier 86 to begin Voyage Number Two. Standing on the pier, her creator watched five tugs push her back into the Hudson River.

Commodore Manning and the harbor pilot turned her bow to the narrows, and then the Big Ship gradually vanished into the summer haze. Turning his back to the empty pier, William Francis Gibbs got back into a car ready to take him back to 21 West Street.

Manning's association with the great ship proved to be short-lived.

He went too far one last time. On the return leg of her second voyage, *United States* left France within minutes of the Cunard liner *Queen Elizabeth*. Both liners were booked to capacity. Manning spotted the *Queen* just off Bishop's Rock. Sailors' etiquette dictated that if Manning was to pass the British ship, he would have to do it at a good distance to avoid any appearance of racing. Manning didn't. He set a course toward the British ship at a steady speed of 31 knots, closed in on the *Queen*, and passed her in plain sight.

United States arrived in New York at 2:55 A.M. on August 4, 1952. *Queen Elizabeth* arrived 11 hours later, maintaining her scheduled service speed of 28 knots.

Reporters were all over Manning, asking if there had been a race. "There wasn't any race," the commodore snorted. "We just raced away from her."

Queen Elizabeth's master, commodore, George E. Cove, claimed that his ship's arrival time had been decided months in advance. "I have no knowledge of any race," Cove said coolly. "Commodore Manning and I merely exchanged greetings off Bishop Rock."[12]

Yet soon after docking, Manning received a stern phone call from the United States Lines office—just hours after completing the second voyage, he was sacked. There had been whispers about how Gibbs was constantly feuding with Manning over how to run the Big Ship.

At a short ceremony on *United States*' bridge, the ship's officers presented Manning with a ship's clock as a retirement gift, and the fiery sailor then left for his home in New Jersey. Aside from commanding *United States* for a few relief trips, his career at sea was over.

Manning was replaced by *America*'s former captain, John Anderson, who was much more popular with his crews and deemed more reliable by Gibbs and the United States Lines.

During the high summer season of 1952, *United States* sailed at capacity and took her place as the most popular liner on the Atlantic. But Truman stubbornly continued to fight the original terms of sale, and threatened to cut off the ship's operational subsidies. Finally, after Dwight Eisenhower's election in November 1952, Truman's special naval aide Robert Dennison prepared to pass the file on "The Case of the United States Lines Company" to the next administration. The ship's funding would remain unresolved until May 1954, when the United States Lines agreed to a revalued purchase price for *United States* of $34,850,000. That was about $7 million more than the initial June 1952 purchase price—less than the original $10 million demanded by Truman's comptroller general, Lindsay Warren, but enough to satisfy all parties. In addition, the government would no longer withhold the ship's operating subsidies. This cash stream was safe for the time being.[13]

For William Francis Gibbs, the ship was never the result of a partnership between private business and the government. For him, not money but the ship's design and construction created *United States,* and for him, creativity came out of a team, but never a bureaucracy. "This great ship that lies up at Pier 86," he said after returning to his office, "is the embodiment to me, of the force and power of a free people working with individual initiative, coming together and cooperating to produce a result which is far beyond the capacity of any one agent or any one individual to produce by himself or cause to be produced by himself."[14] When asked why *United States* was so perfect, he answered flatly, "The reasons we managed to construct that ship was that the government left us to ourselves, putting no regulations in the way. So we designed the *ideal ship.*"[15]

Indeed, for the next decade, William Francis Gibbs's "ideal ship" would lead a charmed career, the ship of choice for the rich and powerful, universally acclaimed as the most beautiful and modern liner afloat, an icon recognizable to millions of Americans. Stunned by the success of his Big Ship, her designer developed a fanatical devotion to his finest creation, making sure that she remained as perfect in service as she was on his drawing board. *United States,* in fact, became an extension of himself, his ideals, and also the medium in which he reached out to others.

Yet it was also clear that *United States* was an achievement he could never duplicate; by the 1950s, the public's fascination with the superliner as a technological paradigm was nearing its end. It was a sea change that Gibbs would avoid confronting for as long as he could.

THE HALCYON YEARS

Once every two weeks, William Francis Gibbs made an early morning excursion before going to the office. He had his usual 6 A.M. breakfast of coffee and Uneeda biscuits in the apartment at 945 Fifth Avenue, as a chauffeur-driven black Cadillac idled out front. After Gibbs got in, the chrome-edged beast roared down Fifth Avenue, the foliage of Central Park zipping past the windows. Turning west near City Hall, the car sped across the Brooklyn Bridge, and through the sleeping streets of Brooklyn. At the foot of the Verrazano Narrows, the driver pulled off the road, and William Francis got out.

Around him were drab two-story row houses, their windows dark. The lanky figure strode past them toward the lapping waters of the Lower Bay. His brown eyes, squinting through wire-rim glasses, peered through the fog. Salty bay breezes whipped his gray trench coat. All around was the noise of foghorns, buoys, and screeching seagulls.

His ears strained for one very distinct sound.

Then a rich, mellow, deep bass note shook the morning air. A half-mile away, *United States* slowly emerged from the haze, signal lights blinking and smoke drifting from her two finned stacks. Veiled by the morning fog, she glided slowly through the narrows toward the shadowy skyscrapers of Manhattan.

All but a few of the passengers would be asleep in their cabins. A few small, dark figures stood at the railings, catching a glimpse of the New World. A crew member might lean over the bow rail, holding a lighted cigarette. He would flick it overboard, and Gibbs would see a tiny dot of orange arc toward the quenching waters below. Then looking up to the bridge windows, he would see the lights glowing.

United States pushed on through the fog toward her Hudson River pier, a large American flag fluttering from her sternpost. She would be docking at Pier 86 in a few hours. Gibbs turned and walked back to the Cadillac, the soles of his cracked shoes meeting the gravel. The glow of the early morning sun now spread across the Atlantic horizon. William Francis Gibbs had seen his Big Ship come in.

By 7 A.M. another round of coffee and biscuits awaited Gibbs in his office at 21 West Street. Here he scanned the shipping news in the *New York Times* and clipped out articles that interested him. One, dated March 12, 1954, read: "The 83,000 ton Cunard liner *Queen Elizabeth* reported on her arrival here today that two passengers had suffered broken bones, social events had been caneceled [*sic*] and interior decorations were smashed in the vessel's stormiest Atlantic crossing of the winter."[1]

William Francis Gibbs remained a charismatic figure, one whose monumental work ethic served as a benchmark for his subordinates. Even if they found him cantankerous or demanding, employees could never say he was lazy. Chief Engineer Walter Bachman wrote that even as he aged, Gibbs was a "natural leader, who inspired great loyalty in his staff and confidence and cooperation in those with whom he did business."[2]

Women, who would face workplace barriers for decades to come, found openings at Gibbs & Cox. When in the presence of women he did not know, Gibbs was gallant but aloof. A kiss on the cheek would lead to him shouting, "For God's sake, stop that!"[3] He still had little time for women he considered less than accomplished, no matter how charming they might be.

But for women he believed had achieved something, he was

unfailingly generous and open. When Elaine Kaplan told Gibbs that she and her husband had decided to have a family, Gibbs offered a nanny, a car service, anything to get her to stay on after her child was born. Kaplan respectfully refused, and left Gibbs & Cox. Yet for Kaplan, *United States* remained her "first baby." A model of the ship was displayed prominently in her home.

"Aside from having children," her daughter Susan Caccavale recalled, "her time at Gibbs & Cox was the happiest time of her life." The Big Ship, whose innovative propellers Kaplan had designed, would be a lifelong source of professional pride. "In those days, a woman's identity was to stay at home and have children," Caccavale said. "If you were to ask my mother her identity, she would say a nautical mechanical engineer, and then a mother."[4]

Gibbs's loyalty to the employees he valued extended to both sexes. In stark contrast to most employers of the time, William Francis Gibbs based his hires not on background, but ability. African-Americans worked alongside their white colleagues.[5] Bachman said that Gibbs "took a close personal interest in his employees, and quietly, often anonymously, assisted those in need." When an employee died and left no family, the boss quietly paid the funeral expenses.[6] If Gibbs overheard that an employee needed a new bicycle or suit, one of the best might mysteriously appear by his or her drafting table. If an employee or employee's spouse got sick, flowers would always arrive by the bedside, with a handwritten note from Gibbs.[7]

William Francis also attended a huge number of weddings and funerals. He popped up at churches and synagogues all over New York City, usually sitting in the back row. And few things made him happier than to be the master of ceremonies at an employee's wedding reception or birthday party. Raising his glass high, he would make his signature toast: "To all you want. Doubled. Good health, and the Big Ship!" He would then empty his glass into the glass of someone sitting next to him.[8]

Gibbs had every right to toast his Big Ship, and he was also pleased by *United States'* passenger lists, which nearly always included royalty like

Queen Frederica of Greece and the Duke and Duchess of Windsor; movie stars like Marilyn Monroe, Jack Benny, Burt Lancaster, Cary Grant, Marlon Brando, and Judy Garland; business tycoons like Walt Disney. Salvador Dali was a passenger; so was Mahalia Jackson. During one crossing, Duke Ellington and his orchestra provided the evening's entertainment in the first-class ballroom, along with Broadway star and fellow passenger Ethel Merman. Walt Disney was so impressed with the ship that he gave her a starring role in his comedy *Bon Voyage!* The movie, about the European travels of an Indiana family, starred Fred MacMurray and Jane Wyman, and most of the ship sequences were shot on board during an eastbound crossing.[9]

In port, crew members would see Gibbs roam the miles of corridors and wind his way through the machinery spaces. He tapped machinery, looked over gauges, stared at the propeller shafts, ran his hands over the aluminum railings, and even sniffed the air coming out of the air-conditioning ducts. And he kept copious notes on all aspects of the ship's performance.

When asked to explain his obsession, Gibbs barked defensively, "I'm interested. My God, who wouldn't be?"[10]

He kept close to operations even when the liner was out at sea. At 8 A.M. New York time, without fail every morning, the phone would ring in Chief Engineer Bill Kaiser's office. Commodore Anderson would come down from the bridge to join Kaiser, and a familiar, gruff voice would begin asking questions:

"How does it look ahead?"

"Is everything running to suit you?"

"Are the passengers happy?"[11]

Above Kaiser's desk was a maze of spinning dials and flashing lights showing revolutions per minute for each of the four propellers, as well as other measures of the ship's performance like oil consumption. But Gibbs refused to discuss the oil consumption rate on the phone. He was afraid that someone was listening in.

In fact, the Federal Communications Commission was tuning in, and told Gibbs that if he did not clean up his salty sailor language, it would take away his phone.[12]

Officially, Gibbs never traveled aboard his ship after its maiden voyage. But this was contradicted by at least one member of the crew. "That's not true!" Jim Green insisted. "I saw him pop up on numerous voyages. He would appear out of nowhere and in the oddest places, often the crew's quarters. He would ask us how the ship was doing."[13]

Despite her superior hull design, *United States* rolled just like any other transatlantic liner when big waves kicked up. Crewmen were still awed by the North Atlantic in winter, and even the hardiest ones got seasick. "Unbelievable waves," Bell Captain Bill Krudener said. "People don't realize how big they are."[14] During the worst storms, the motion got so bad that crew members who were berthed in the forward part of the ship would get tossed out of their bunks. Unlike the *Queens*, which were fitted with newfangled Denny-Brown stabilizer fins in the mid-1950s, *United States* relied only on her hull shape and bilge keels to stay steady in bad weather. While the two bigger British ships dropped with a sickening lurch when headed into the waves, *United States* would often slice right through them, sending clouds of spray flying onto the foredeck.

The ship also required a delicate touch from the helmsman in bad weather. Despite their beauty, Gibbs's beloved big finned stacks acted like sails and caused her stern to swing from side to side. Quartermaster Leslie Barton, who was in charge of steering the ship, described the liner as a "young colt . . . skittish, light, and always her own boss, but she would listen to reason."[15]

For passengers who had not found their "sea legs," the results were predictable. "Sawdust, Kelly!" Krudener would yell to an English porter when one of the passengers threw up while running from the dining room to the lavatory. The porter dropped a shovelful of sawdust on the vomit, and then swept the mess off the linoleum floor.

During one gala dinner, as Krudener stood by the doors to the first-class dining room, he felt the ship "cock itself." He looked at the three-hundred-odd diners in tuxedos and evening gowns, eating filet mignons and sipping expensive wines.

Krudener knew what was coming next: a freak wave. He grabbed one of the stanchions and planted his feet firmly on the floor.

With a roar, a mass of water struck the ship. Krudener held on as

United States lurched sickeningly over to one side. He watched people fall out of their chairs, dishes sail through the air, and waiters holding silver trays slip and tumble onto the floor.

The Big Ship then jerked back onto an even keel, sending people rolling in the other direction. Shouts and screams mingled with the din of smashing crockery and crystal.

"What a mess!" Krudener recalled. "All those people in their finest outfits, covered in gravy, ketchup, and wine." Miraculously, no one was hurt.

After that voyage, United States Lines installed "seatbelts" beneath the dining room chairs that anchored them to the floor.[16]

During a stormy September passage in 1956, Vera Cravath Gibbs sat in her first-class stateroom and wrote in her diary, "Right now, I am sitting in my room while our dear ship rolls," she wrote. "There is a heavy sea, and I am having my wish. Every previous time I have crossed on the S.S. *United States,* the water has been calm. Sunday evening was quite rough . . . ropes up, people slipping around."[17]

During that same voyage, Vera opened the door to the sun deck and only barely kept her footing as the wind slammed into her face. But no matter the weather, she made a pilgrimage to a special spot during every crossing. "I took a windy walk on the boat deck," she wrote, "and of course, went right forward to gaze at the Gibbs & Cox plaque, which I must visit on each trip."[18]

The gleaming aluminum builder's plaque hung just below the bridge. It had no decorative flourishes. The lettering was black and simple. Figures for tonnage, length, and record-breaking speeds were nowhere to be found. It read:

UNITED STATES
DESIGNERS: GIBBS & COX, INC.
BUILDERS: 1952 NEWPORT NEWS SHIPBUILDING AND DRY DOCK CO. NO. 488
MARITIME COMMISSION NO. 2917

Because her husband did not travel with her on her jaunts to Europe, Vera brought friends with her instead. One was Eugenia McCrary.

"Several times, Vera and I took what she called the 'family rowboat' over to Paris for 18 hours," she recalled, "catching the ship upon its return journey to New York. During the interval, we'd dine in some fabulous restaurant do some quick shopping, mainly for perfume and silk scarves, and attend the latest popular French play. I can easily say that traveling with such a close friend on the *SS United States* to go over for dinner remains one of the most glamorous and treasured moments of my life."[19]

Quick trips to Europe were common among first-class passengers, and Vera was just one of many members of New York society who used *United States* the way ordinary people used the Staten Island Ferry. They often brought their cars with them. At the start of each voyage, cranes lifted up to forty automobiles from the pier and dropped them into the hold. Most of them were luxury makes: Cadillacs, Chryslers, Mercedes-Benzes, Rolls-Royces.

Artist Cissie Levy of Philadelphia remembered that preparation for her honeymoon aboard *United States* meant packing several steamer trunks for all the clothing required for the five-day trip, including formal wear, because first-class passengers were expected to change what they wore several times a day. The Levys' traveling companions were *TV Guide* media tycoons Walter and Leonore Annenberg, close friends of her new husband's family.[20] The Duck Suite especially had to be booked well in advance, usually around the travel plans of its most famous devotees, the Duke and Duchess of Windsor, the star-crossed and controversial royal couple.

When the Windsors came aboard, they brought along one hundred and fifty Louis Vuitton steamer trunks and suitcases. (Their secretary would inventory their contents, which included many fabulous jewels, twice during the voyage.) Stewards replaced United States Lines' linens with Windsor monogrammed Porthault towels and sheets. Standard lightbulbs from lamps and ceiling fixtures were replaced with pink-hued ones, presumably because they were kinder to the duchess's complexion. Custom drapes were installed, leopard-skin rugs were thrown on the floor, and pictures in jewel-encrusted frames placed on the tables. Their suite was no longer fireproof, but it was now fit for royalty, and looked as if the Windsors lived there year-round. Not only that, but their two pug dogs were allowed to romp around the suite during the day.[21]

The prickly Wallis usually turned down Commodore Anderson's invitation to dinner. "Ship captains were inclined to pinch!" she once complained.[22]

Even so, the United States Lines offered the Windsors a half-price rate for the Duck Suite to keep them from defecting back to the *Queens*. The special deal benefited everybody: the Windsors saved money, and the United States Lines got great publicity for their ship—the Duke and Duchess gave a press conference aboard after every trip—even if the couple's demands made life difficult for the stewards department. Everybody wanted to travel on the same ship as the Windsors. And the duke was the only passenger allowed to visit the bridge whenever he pleased, nattily attired in a tweed jacket, plaid socks, and wild-patterned pants. There he would sip a scotch and regale the ship's officers with witty anecdotes. But he was no aristocratic bore; he actually amused officers with stories about his golf game.[23]

Gibbs understood the importance of the Windsors' loyalty to the Big Ship. Every time the Windsors boarded *United States* in New York, they received a gift from Mr. Gibbs. "Thank you so much for the beautiful roses," the duchess wrote the naval architect after one trip in 1957. "It was indeed kind of you to think of us. . . . She is a wonderful ship and you are justly proud of her. The Duke joins me in renewed thanks for the flowers and we send you our best wishes for a pleasant summer.[24]

Yet some celebrities found the formality on board disconcerting. During one trip, Marlon Brando sat down at the dinner table and glanced at the gleaming array of silverware laid out in front of him. He turned to his dinner companion, Laura Franklin, the daughter of the United States Lines president.

"Laura," he whispered sheepishly, "which fork is for the caviar?"

After the meal, Brando also asked Bell Captain Bill Krudener if he could borrow a guitar and find a spot away from gawkers and autograph hunters.

"The next thing I knew," Krudener remembered, "I had 500 guys in and around my cabin all playing and singing along with Marlon Brando."[25]

The crew knew which passengers treated them with respect and which looked right through them. On his first trip in 1955, Joe Rota got assigned to operating the elevator that ran between the main deck and the swimming pool. The first day out, when Rota opened the elevator door, actor Burt Lancaster and his two young sons were standing outside. Lancaster was on his way to Paris for the filming of *Trapeze*.

The actor looked so familiar to Rota that what came out of his mouth was an involuntary "Burt!"

Lancaster looked, in Rota's words, as if he were about to "step into a snake pit." He protectively put his arms around his son's shoulders and hesitated before getting in.

After dropping his boys off at the pool, Lancaster looked warily at the elevator and then started making for the stairs.

"Mr. Lancaster, can I talk to you for a moment?" Rota yelled across the pool.

Lancaster stopped and walked toward the elevator.

"This is only my second trip to sea," Rota said. "When the door opened, it was exactly like seeing an old friend, and I had grown up seeing your movies. I have to apologize."

Lancaster's face broke into his trademark big grin. "That's perfectly all right. I understand. Take me up."

When the ship docked in Le Havre, Lancaster picked Rota out from the line of bell staff to carry his luggage from the pier to the Paris boat train. The actor then gave the new elevator operator a handsome five-dollar tip.[26]

Similarly, on another trip, Bill Krudener found artist Salvador Dali and his wife, Gala, to be extremely courteous. One day, Dali took out a piece of paper and pressed it into Krudener's hands.

"This is for you," he said.

It was a signed pencil sketch of the Eiffel Tower and the Statue of Liberty, reaching across the Atlantic and shaking hands. Krudener took it right to his office, put it on his desk, and returned to work. When Krudener came back, he found his room cleaned, and the sketch gone.[27]

The crew was officially forbidden from "fraternizing" with passengers in all classes, but at least one of them broke the rule. During a crossing in 1957, deck steward Jim Green was walking on the sun deck when a ladies' kerchief fluttered down from the tourist-class promenade above. He looked up and saw a young woman saying, "It's mine."

Green clambered up and handed the kerchief to a beautiful twenty-seven-year-old named Frieda Kerstgens. Five years earlier, Frieda and her family had watched from the railings of *America* as the brand-new *United States* swept by on her record-breaking voyage. As she watched the beautiful ship disappear on the horizon, Frieda vowed that one day, she would take a trip on the superliner.

Jim Green and Frieda Kerstgens fell in love, and were soon married.

When asked if she dropped the kerchief on purpose many years later, Jim Green's answer was "She'll never tell." After which, she said, "No, I won't."[28]

Although popular with celebrities, *United States* also attracted many ordinary travelers, for whom the voyage was the trip of a lifetime. Cabin class was popular with prosperous professionals, as well as the privately rich who wanted to keep a low profile. Joseph and Judith Follmann, a couple from New Rochelle, New York, were typical of the prosperous rather than superwealthy clientele who sailed on *United States.* Joseph was an insurance executive, while Judith worked with the American National Standards Institute. On a trip to Paris in the 1950s, the Follmanns traveled first class to Europe, and cabin class back home. For Judith, first class was supremely elegant, a glimpse into a champagne-and-caviar world they could sample once in a while.

The passage back home in cabin class was much more sedate. "We played bridge all the way across the Atlantic," she recalled.[29]

In tourist class, American college students mingled with European budget travelers and immigrants. Public rooms were stark, staterooms cramped, and the lack of private bathrooms made this part of the ship feel like a college dormitory. But for young people on board, tourist class was the liveliest part of the ship, much more fun than bingo games with a bunch of old people in suits and tuxedos.

Frank Nolan, a nineteen-year-old student at Notre Dame, booked a tourist-class berth on the way to a study abroad program at the University of Innsbrück in Austria. There were more than thirty people in his group of college students, all of whom were "on the look-out for a good time the moment we set sail." Since days were largely unscheduled, most of them spent their time drinking beer in the smoking room until their cash ran out. Sometimes, "sugar daddies" would sidle up to the bar and buy the young college women drinks. Passengers gathered around the piano, alcohol and cigarettes in hand, for late night sing-alongs. Even in tourist class, Nolan could ring a steward to deliver a club sandwich and bottle of beer to his stateroom at 3 A.M. The gala dinner, which took place on his birthday, was better than he expected: crab salad, pâté de foie gras, stuffed Cornish game hen. He also tried his hand at skeet shooting—the clay pigeons were flung from the fantail, and passengers fired at them from the railings as they flew over the ship's wake. And despite the wind, Nolan spent hours walking the decks, contemplating the vastness of the North Atlantic. "We were always discovering new things and meeting new people," Nolan recalled. "We were not conscious of time during the voyage."[30]

For immigrants in tourist class, *United States* was often their first taste of America. One was ten-year-old Kurt Wich, whose family fled war-devastated Germany for a new life in Philadelphia. At the gala dinner of their February 1953 crossing, immigrants and students alike were dressed in party hats and blowing noisemakers. Kurt's father, who had served time in an American POW camp, pointed out the African-American waiters in the tourist-class dining room. "They are celebrating the birthday of the president who gave them their freedom."[31]

A more recent president than Lincoln had left political office virtually penniless three years earlier, and might have booked tourist class if not for a windfall. After leaving the White House, the highly unpopular Harry Truman went back to Independence, Missouri, and took up residence in a white frame Victorian house that belonged to his wife, Bess. With no personal wealth, Truman was forced to make do on a scant Army pension of $111.96 per month, barely enough to pay the bills, let alone the cheapest stateroom aboard *United States*. But then the ex-president received an offer from *Life* magazine for $670,000 to write his memoirs.

When they were published, Harry Truman and Bess decided to sail to Europe in style. They had other reasons to celebrate: their daughter Margaret had just married *New York Times* editor Clifton Daniel Jr.

The *United States* left New York with the Trumans aboard on May 11, 1956. The last time Truman had been at sea was eleven years earlier, when he sailed aboard the cruiser USS *Augusta* to the Potsdam Conference. This was a different kind of trip. The sea was smooth, his suite spacious and comfortable, and the service impeccable.

The first morning at sea, radio bellboy Joe Rota knocked on Truman's stateroom door.

Bess Truman opened the door. "Yes, come in," she told Rota.

He saw the former president enjoying breakfast in the suite's sitting room.

"I'm Joe Rota," he said. "I'm the radio bellman. I'll be handling all the radio telegrams, and I'll be glad to bring them to you in the morning."

"Where are you from, Joseph?" Truman asked, peering through his thick glasses.

"New Jersey," Rota responded.

Truman leaned in closer. "Are you a Democrat or a Republican?"

"Mr. President," he answered, "everyone's a Democrat in this room."

Truman chuckled. "You should be working in Washington."

While on his daily rounds delivering radiograms, Rota would see the president pacing the promenade deck arm in arm with Bess, his eyes trained on the gray ocean rushing past the windows.

When he ran into Rota, he always asked the same question: "So how are things in New Jersey, Joseph?"

"The same, Mr. President," he would answer. "Unfortunately the same."[32]

The Trumans' trip was a public relations coup for the United States Lines. When the former president disembarked in Le Havre, the newsreel cameramen filmed him coming down the gangplank and greeting the French honor guard. The great ship loomed behind him as he held his hat to his chest and walked down the pier. The bitter controversy surrounding the ship's construction had faded into a distant memory—even Truman recognized what a great national achievement *United States* was, and the place she held in the hearts of the American people.

The mid-1950s was a halcyon era for *United States*—she was consistently sailing at or near capacity—but there had been one ominous incident during that high-water mark year of 1956, a tragedy that not only validated Gibbs's obsession with safety, but also gave the American public the chance to watch a big liner sink on television. The Atlantic was as treacherous as ever, and as ever, punished negligence.

On July 25, 1956, the three-year-old Italian liner *Andrea Doria,* carrying 1,706 passengers and crew, was on the final stretch of a routine voyage from Genoa to New York. Although not in the same superliner class as *United States* and the *Queens*, the beautiful 697-foot-long flagship of the Italian Line was a symbol of the rebirth of postwar Italy. Top-flight cuisine and service made her popular among wealthy Americans.

As *Andrea Doria* approached Nantucket near the end of her crossing, Captain Piero Calamai spotted a fog bank and slowed his ship down. Even with radar to guide him, Calamai remained on edge. These were dangerous waters. Three decades earlier, in the same location and weather conditions, a Norwegian freighter had rammed the new Matson liner *Malolo* amidships, with her designer William Francis Gibbs aboard.

Andrea Doria's captain had good reason to worry about collisions. For all her glamour and popularity, *Andrea Doria* was unstable in rough seas, due not only to poor design, but to sloppy, company-mandated ballasting. Finished only a year after *United States, Andrea Doria* was designed to stay afloat with only 2 of her 11 compartments open to the sea. The bigger *United States,* on the other hand, could stay afloat with 5 of her 20 compartments flooded. Earlier that year, an alliance of American shipbuilders and owners lobbied for a relaxation of strict compartmentalization requirements. If the Italians could get away with a "two compartment ship," why couldn't the Americans?[33]

At 11:15 P.M., Captain Calamai of *Andrea Doria* spotted a bank of lights quickly approaching his ship's starboard side. He ordered the helmsman to take evasive action by turning to port, but it was too late.

The small Swedish liner *Stockholm,* built with a prow reinforced to crush Baltic Sea ice, smashed into *Andrea Doria*'s starboard flank. Both ships had advanced radar systems, but their commanding officers appear to have misinterpreted the other ship's courses, steering *toward* rather than *away* from each other. The sound of ripping steel and smashing wood was followed by the screams of dozens of passengers being crushed to death. As *Stockholm* drifted astern, the mauled *Andrea Doria* immediately heeled over to starboard, and her lights flickered out. With a 20-degree list, most of her lifeboats could not be lowered.

Passengers and crew aboard *Andrea Doria* knew immediately that the ship was doomed. "They had swarmed out of their staterooms and up from the lower decks," a *New York Times* reporter wrote. "At no time did they panic. But the list made the footing perilous. On the boat deck the passengers nestled against the high side of the superstructure to await their rescue."

Eleven hours after the collision, as helicopters bearing TV cameramen circled overhead, *Andrea Doria* rolled onto her side and sank. The collision killed forty-six people on *Andrea Doria,* five on *Stockholm,* which remained afloat despite a crumpled bow. Miraculously, 1,661 of *Andrea Doria*'s passengers and crew, including Philadelphia mayor Richardson Dilworth, escaped in lifeboats. Most survivors were picked up by the French Line's *Ile de France.*[34]

After the disaster, one of the lobbyists hoping for a relaxation of American maritime safety standards called William Francis Gibbs, who had refused to support them.

"Gibbs," he sighed, "you're always right."[35]

Yet to the millions in American households who watched *Andrea Doria* capsize on their television screens, the lesson was different. The disaster was their first television image of a big liner going down. They had read about the sinking of *Titanic* and *Lusitania.* They may have seen illustrations, out of artists' imaginations, of what happened to these long-ago ships. They may even have seen newsreels of the burning of *Morro Castle,* a disaster then two decades in the past.

But all that was history. This was today. The *Andrea Doria* disaster proved a devastating blow to *United States* and other great passenger ships. The horror of the images on the new medium reinforced the accusations of sloppy navigation and improper maintenance by the shipping lines. Suddenly luxury liners did not seem much safer than those cramped turboprop planes now making noisy, bumpy flights across the Atlantic. And a new and faster aircraft, the jet, would soon be leaving its contrails across the skies.

TROUBLE ASHORE

As the 1950s proceeded and the Big Ship prospered, the "happy prospect" of designing a sister ship to *United States* kept William Francis Gibbs working long hours into his seventies.[1]

By all outward appearances, Gibbs & Cox, with a staff of one thousand, did not seem to be resting on its laurels after designing *United States*. *Fortune* magazine reporter Richard Austin Smith, writing about Gibbs in 1957, toured the great engineering firm just as it was finishing the design of two new medium-sized liners for the Grace Line: the 14,000-ton *Santa Paula* and *Santa Rosa*. Both looked like miniature versions of *United States,* except with only one funnel. Overall, Gibbs & Cox was grossing between $7 million and $8 million and clearing a profit of $400,000 each year from its Navy contracts.[2]

But Smith noticed that there were rumblings of discontent among the staff, especially younger designers who felt frustrated by what they saw as the traditionalism of the firm. Some in the Navy agreed. Although Gibbs & Cox was working on new designs for destroyers, it was well behind its competitors in developing newer types of craft such as submarines and aircraft carriers. It was an era when the Navy was looking hard at nuclear energy to drive powerful new ships and submarines. In the vanguard were the big shipyards; Newport News Shipbuilding was designing an atomic aircraft carrier and an atomic cruiser. Top Navy brass began to feel that Gibbs & Cox was "trailing the field, with only a joint General Electric nuclear contract to show."[3]

Commercial shipping, meanwhile, was also undergoing a revolution. Until the mid-1950s, cargo had to be offloaded from trains and trucks, hauled to the pier, and lowered into cargo holds. Shipping entrepreneur Malcom McLean's Sea-Land company changed that by developing a logistical breakthrough called containerization. This was a method by which a sealed container could be lifted directly from trucks and trains and lowered onto a vessel's deck, and vice versa. Containerization saved shippers a lot of time and money, in large part because it eliminated the need for armies of longshoremen. Frederic Gibbs's idea to build fully integrated sea-land terminals—revolutionary in 1915—was becoming a reality fifty years later. Not in Montauk, but across from Manhattan on the New Jersey side of New York harbor, with its vast stretches of open land and easy access to new interstate highways.

Containerization would in time make all older cargo ships obsolete and spell the doom for old ports like New York, with their cramped finger piers and the shortage of land for trucks to park. But Gibbs & Cox was not designing container ships. Its biggest commercial clients, the United States Lines and the Grace Line, were still operating traditional cargo vessels. The rival firm of J. J. Henry & Company, also located at 21 West Street, got the Sea-Land contracts instead. Gibbs & Cox, once famous for radical ideas like high-pressure, high-temperature steam, had grown conservative.

And then there was the matter of the proposed running mate to *United States*. Many of the younger men at Gibbs & Cox wanted the new *America* to be a complete departure from her sister ship—as they told *Fortune's* Smith, something "strikingly new, not a copy."[4]

Across the Atlantic, they saw competition coming. The French Line had just laid the keel of a new liner to replace *Ile de France* and *Liberté,* both approaching their thirtieth year. Estimated to weigh in at 66,000 tons, the sleek, streamlined *France* would stretch 1,035 feet in length, making her the longest ship in the world, and bigger than *United States.* Although she borrowed some design details from Yourkevich's

Normandie, France would be a completely new ship. And she would have only two classes: first and tourist. The French Line hinted that their new superliner might try to take the Blue Riband away from *United States.*

Designs for the proposed running mate for *United States* became a flashpoint of contention at 21 West Street. Gibbs was standing firm: why improve upon perfection? The new *America* would be basically the same as *United States.*

There were also problems with the structure of the firm's ownership. It was a partnership controlled entirely by William Francis and Frederic Gibbs, who owned all of the company's stock. They paid themselves relatively modest salaries and never declared a dividend on their shares. Some of the senior engineers wanted part of the "crock of gold" that they were convinced the Gibbs brothers had squirreled away, and they wanted the old man to distribute some of the stock and the unpaid dividends hidden within the equity to senior members of the staff.[5]

William Francis would have none of it. The company would remain his and his brother's alone. When asked who was next in line when he retired, he would look down at his middle coat button and mutter, "A committee is studying it."[6]

In private, he was more candid about his future. "When I retire, I'll be dead," he told a friend.[7]

When not designing or drumming up business, Gibbs took his friends to the theater, opera, and Broadway musicals. He also loved the circus, and whenever Barnum and Bailey's came to New York, he would rent a box in Madison Square Garden and take friends to marvel at the trapeze artists, jugglers, and lion tamers. When going out on the town, Gibbs often shed his shabby suits and cracked shoes for full evening dress: white tie, tails, polished patent leather pumps, and red-lined opera cape. He especially enjoyed the musical *My Fair Lady*, based on George Bernard Shaw's *Pygmalion.* So much in fact that he saw the show eighteen times.[8]

Gibbs was more than an observer of the theater world. While his wife, Vera, sat on the board of the Metropolitan Opera Guild, he served as chairman of the Neighborhood Playhouse Theater. Fellow board member Wynn Handman enjoyed Gibbs's presence at board meetings, where the naval architect spoke up with "authority, lubricated with humor." The actor and architect became close friends.

An even earlier theater friendship was the one William Francis formed with actress Katharine Cornell. From the day they met, the two had formed an instant connection. Like the naval architect, Cornell accepted nothing less than the absolute best in her work, never forgetting, said the *New York Times,* "the day or the city where she had a 'perfect performance.' "[9]

Cornell grew to know Gibbs as few outsiders did. In the early 1950s, they made a ritual of meeting each other for lunches where he could open up and relax. She went with him to Yankee home games, where Gibbs let loose whenever a New York player hit a home run. "He became so enthusiastic," she laughed. "He loved it."

The actress recognized the performer in Gibbs. In her view, he used his unpleasantness as a dramatic act to achieve the results he wanted. "His wonderful dour expression—it was just a pose on his part," she said. "I don't know whether it arose from shyness; but it was absolutely marvelous . . . people were so enchanted to see him smile."

She also knew that William Francis was a lonely man, despite his family and influential friends. "He sought perfection, that's why he cut himself off," Cornell said of him. "He was very economical with his friends. He was very controversial, very concentrated, so concentrated on his work that he had very little time just to chitchat. If he liked you, he was very helpful in anything you were interested in, and he was always very generous."[10]

He reveled in the praise from other leading ladies of the stage. "Dear Francis—You're my ideal!" wrote actress Annie Jackson on a photograph that hung prominently in his New York apartment.[11]

Although bad at cocktail parties, the onetime Harvard recluse found that he had a knack as a public speaker. His deadpan demeanor and dry wit won over audiences at dinners all over New York City. At the 1960

Thomas Edison Foundation Mass Media Awards dinner, hundreds of guests, including Elizabeth Taylor, sat in the Waldorf-Astoria's gleaming Sert Room. The proceedings were dreary. Ahead of them, and still to be endured, was the presentation of the awards at the end of the event.

After he was introduced, Gibbs left his table, mounted the podium, and stared out across a sea of vacant faces.

"Ladies and gentlemen," he declared, "I have had some very sad experiences. Tonight, I think, is much the worst."

Everyone in the room began to laugh, then laugh louder, and then to applaud.

"I honor scientists," Gibbs told the audience before presenting the awards. "I honor educators because I can never aspire to be either. My sole claim to distinction is being a jack of all trades and master of none."[12]

Despite his professional success and involvement in the arts, observers could not help wonder about his private life.

"I do believe that you love the *United States* more than your wife," a reporter said to William Francis Gibbs.

"You are a thousand percent correct," he replied.[13]

Vera Gibbs, present during the exchange, did not show hurt. She had spent a life acknowledging with a smile her husband's obsession with work, and she kept any personal thoughts to herself, especially around the press. But by the mid-1950s, things worked best when they were apart: he in his office, and she shuttling back and forth across the Atlantic on the "Family Rowboat." In her diary, she noted how her husband would send her messages each day: "I had a nice telegram from William Francis." But she did not sit around waiting. "Each day I received messages from him, either through the Captain or the Chief Engineer. I finally sent one to him, in which I said that the only flaw in the trip was social complications. I always seem to be trying to wedge in one more cocktail party."[14]

By this time, their three children were grown and on their own.

Known by her friends as a woman of "infinite patience" who possessed a "large reserve of humor," Vera Gibbs built a life of striking independence, tempered with her concern for the welfare of her stubborn husband.[15] Personal friends were entertained at the 945 Fifth Avenue apartment, whose living room boasted two large likenesses facing each other: an oil painting of Vera's father, the lawyer Paul Cravath, and a bronze bust of her husband, William Francis Gibbs.

Now in his early seventies, Gibbs had long been accepted by society, and not just through Vera. Despite his seeming remoteness, numerous cultural, social, and business leaders counted him a friend. He had been a member of the University, Century, Broad Street, India House, Piping Rock, and New York Yacht clubs.[16] Still, he seemed restless and decided to build a seaside retreat. In the late 1950s, during destroyer trials off Rockport, Massachusetts, he saw a beautiful rock outcropping, about an acre in size, jutting into the Atlantic Ocean. In 1960, William Francis and Vera Gibbs purchased the outcropping as the spot for a new summer house.

What emerged from the naval architect's drafting board was a rough-hewn structure built of gray granite, its walls anchored deep into the soil. Like the Big Ship, almost no wood was used in its construction. A verse from the gospel of Matthew was chiseled into the stone mantelpiece: "Therefore whosoever heareth these sayings of mine, and doeth them, I will liken him to a wise man, which built his house upon a rock; And the floods came, and the winds blew, and beat upon that house, and it fell not: for it was founded upon a rock."[17]

The house proved to be just that, described as "snug and comfortable even in the worst weather."[18]

Gibbs also purchased a large motor yacht. True to form, *Weather* was square, plain, and sturdy.

Vera encouraged his enthusiasm for the new house. Far from the social obligations of New York and Locust Valley, the busiest sounds were the cries of the gulls and the pounding of surf.

But the diversion did not last. Within a few years, Vera found herself alone at Rockport during the summer, with Gibbs making only occasional visits.[19] She occupied herself by reading and going for long, frigid swims around the point.

Her husband also avoided her pre-opera parties in New York. "Will doesn't care much for the opera," she told an acquaintance of her son Adrian.

When at the opera, she had strong opinions, even as her hearing deteriorated and she had to wear a hearing aid. "I can't stand that soprano!" she once shouted loud enough for the entire house to hear. During intermission, she told one guest she was off to her private locker for a swig of whiskey.[20]

In New York, the couple saw each other at breakfast or in the late evening. William Francis did not leave his office for lunch, but ate at his desk. When he dined out, he always ordered the same thing: pork chops, canned peaches, and a mix of chocolate and vanilla ice cream. He would wash it down with his only alcoholic indulgence: half a whiskey old-fashioned. He cared little about food, in the words of his friend Frank Braynard, but rather "the quality of the service and the atmosphere of the dining room."[21]

He did not keep late nights during the week, however, usually coming home and getting to bed around 9:30 P.M. At this point in their lives, the couple kept separate bedrooms. Vera's was stuffed to the gunwales with books. Her husband's was full of mementos: a silver tennis trophy that he won with his brother, Frederic; his earliest ship drawings; a framed Lincoln quotation: "The darkness of the quiet past is inadequate to the stormy present." There was also a photograph of 1733 Walnut Street, his childhood home, taken when it was one of Philadelphia's grandest mansions.[22]

"Before he's asleep, I come in and there he is lying in bed looking at the pictures he carries," Vera said. "It's always either his model fire engine or the *United States*."[23] He also stubbornly held on to another machine. In the garage of their Long Island estate he kept a 1903 Mercedes—the same car that he and Frederic had raced in their youth, when it was the ultimate symbol of modernity, speed, and status.

Gibbs also felt his reclusive brother Frederic needed to have some fun, and encouraged him in buying an updated version of their childhood car: a red Ferrari.

"The expensive sound of a Ferrari getting away," Gibbs said. "A dreadful racket."[24]

Gibbs did not lavish the same care on his children that he did on his machines. Neither of his sons ever showed any interest in becoming naval architects.

After serving as an Army lieutenant during World War II, his stepson, Adrian Larkin, took a job at the International Institute of Education in New York and also kept a house in Pride's Crossing, Massachusetts, a fashionable town not far from Rockport. He probably wanted to be near his mother.

The elder, Francis (known as Frank), moved to Colorado, where he honed his skills as a singer, drummer, and radio announcer. He eventually came back to Rockport, Massachusetts, and then moved to Maine.

Perhaps resenting the attention his father lavished on the Big Ship, Frank kept his father's achievements hidden from his daughter Susan, born in 1962. "I had no idea he had even achieved fame." Susan Gibbs said of the grandfather she had met only a few times as a young child.[25] For Susan, both her paternal grandparents were remote and formal figures, relics of a fast-waning, old-order East Coast establishment.

According to his wife, Paula, Frank Gibbs would proudly point out that the tourist guidebook to Rockport, Massachusetts, had an entry on the antics of his mother's rambunctious dog Luther, but said nothing about his father, William Francis Gibbs, who she knew was a "designer of ships like the SS *United States,* the fastest passenger liner ever built in the history of the world."[26]

Gibbs's other son, Christopher, also fled the social world of his parents, taking a job as a schoolteacher in Apple Valley, California. His father once tried to reach out to him by sending a framed copy of the poem "If" by Rudyard Kipling. "I have found it to be not too complex to understand, but very specific in his message, as well as in its many implications," Christopher wrote his mother about the gift. One stanza must have conveyed to him what William Francis Gibbs expected of his son, and perhaps of himself:

Figure 40. The control panel in one of the ship's two engine rooms. Unlike other liners, no passenger tours were permitted of any of the machinery spaces aboard *United States*. *The Frank O. Braynard Collection.*

Figure 41. Gibbs watches the dry dock flood the day before *United States*' christening. *The Mariners' Museum.*

Figure 42. General Franklin addresses a crowd of twenty thousand at the *United States* christening ceremony, June 23, 1951. *The Mariners' Museum.*

Figure 43. The newly christened *United States* is towed to the fitting-out pier, where over the next year her interior appointments will be installed. *The Mariners' Museum.*

Figure 44. The Duck Suite aboard *United States.* It boasted two bedrooms, a living room, and three bathrooms. The most luxurious suite on board, it was occupied four times a year by the Duke and Duchess of Windsor. *The Mariners' Museum.*

Figure 45. The first-class observation lounge aboard *United States*. The concave gesso and gold-leaf mural depicts the underwater topography of the North Atlantic Ocean. *The Mark Perry Collection.*

Figure 46. The first-class ballroom aboard *United States,* with its domed ceiling, circular dance floor, gold-leaf walls, and etched glass panels depicting undersea life. *The Mark Perry Collection.*

Figure 47. The first-class dining room aboard *United States,* which seated four hundred passengers at a time and had a two-deck-high center section. *The Mark Perry Collection.*

Figure 48. The cabin-class dining room aboard *United States,* with aluminum line sculptures by Seymour Lipton. *The Mark Perry Collection.*

Figure 49. The tourist-class dining room aboard *United States,* located in the forward section of the ship. *The Mark Perry Collection.*

Figure 50. A four-person cabin-class stateroom aboard *United States,* with one of the upper berths pulled down. *The Mark Perry Collection.*

Figure 51. Sculptress Gwen Lux installs *Expressions of Freedom* above *United States'* first-class dining room buffet. *The Mariners' Museum.*

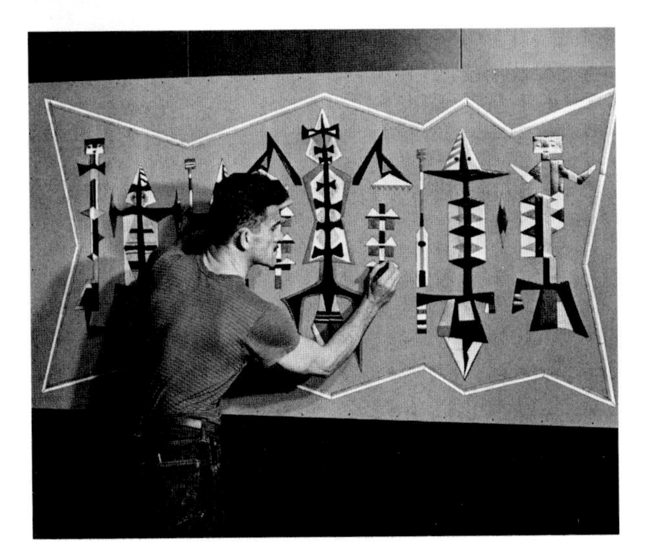

Figure 52. Peter Ostuni paints the murals of the Navajo cocktail lounge aboard *United States. The Mariners' Museum.*

Figure 53. Newport News shipyard workers leaving *United States* as she nears completion. More than three thousand people worked on the ship from February 1950 to June 1952. They included painters, riveters, welders, pipe fitters, electricians, upholsterers, and artists. William Francis Gibbs was constantly on-site, making sure that the ship was being built to his exacting specifications. *The Mariners' Museum.*

Figure 54. William Francis Gibbs (sitting) and Frederic Gibbs atop the bridge as *United States* steams out of Newport News for her trial runs, May 14, 1952. *The Frank O. Braynard Collection.*

Figure 55. Lunch in the first-class dining room during the first trial run. From left to right: Vincent Astor, General John Franklin, J. B. Woodward Jr., Admiral Edward Cochrane, William Francis Gibbs, William Blewett Jr., and an unknown officer. *The Frank O. Braynard Collection.*

Figure 56. *United States* at speed during her May 14–15, 1952, trial runs, as viewed from the Coast Guard cutter *Conifer. The Mariners' Museum.*

Figure 57. Maiden arrival of *United States* in New York, June 23, 1952. *The Frank O. Braynard Collection.*

Figure 58. Sailing day at Pier 86, at the foot of Forty-Sixth Street, New York. *United States'* upper decks can be seen through the pier shed windows. Passengers boarded at separate gangways for first, cabin, and tourist classes. The baggage allowance consisted of all the baggage they could carry plus twenty-five cubic feet of steamer trunk space, per person. Passengers with pets checked them with the ship's kennel master. *The Mariners' Museum.*

Figure 59. A bartender mixes cocktails in the first-class ballroom. *Courtesy of British Pathé.*

Figure 60. Passengers reading in the first-class observation lounge during the maiden voyage. *British Pathé.*

Figure 61. Cabin-class passengers play deck tennis on the maiden voyage. *British Pathé.*

Figure 62. View from the port bridge wing as the ship tears through a vicious gale at more than 36 knots. *British Pathé.*

Figure 63. *United States* blasting past Bishop Rock, England, in the early morning of July 7, 1952, breaking *Queen Mary*'s fourteen-year record by ten hours. Crossing time: 3 days, 10 hours, 40 minutes, at an average speed of 35.59 knots. *The Frank O. Braynard Collection.*

Figure 64. Margaret Truman (daughter of President Truman) and Commodore Harry Manning shaking hands on the bridge of *United States* at 6:16 A.M., July 7, 1952, the moment the ship broke the eastbound transatlantic speed record. *British Pathé.*

Figure 65. William Francis Gibbs during an interview with the British press on July 8, 1952. "It's a great pleasure to be here this morning, on this ship," he told the world, "and it also gives me a feeling of humility in having the responsibility of the hopes and aspirations of my fellow citizens that have been embodied in this ship." *British Pathé.*

MAIDEN VOYAGE
QUADRUPLE SCREW TURBINE STEAMER
UNITED STATES
Commodore Harry Manning
Captain, U. S. N. R.

Abstract of Log **Voyage 1, Westbound**
From SOUTHAMPTON to NEW YORK, via LE HAVRE

Left Nab Tower, 5:00 p.m., BST, July 10, 1952 Arrived Havre L. V., 7:24 p.m., BST, July 10, 1952
Distance, Nab Tower to Le Havre L. V.: 75 miles
Steaming Time: 2 Hours, 24 Minutes — Average Speed: 31.25 Knots

DATE	LAT. N.	LONG. W.	NAUT. MILES	SPEED	WIND	REMARKS
July 11						Departure Havre L. V., 2:00 a.m., BST
" 11	49-49	08-49	341	34.10	W-5	Abeam Bishop Rock 9:17 a.m., BST
" 12	48-10	31-35	902	36.08	W-3	Moderate Sea
" 13	42-56	50-43	865	33.92	Var-2	Light fog; speed reduced
" 14	40-26	69-51	872	34.19	Var-1	Smoth Sea, Hazy
" 14			175			Arrived Ambrose L. V., 4:29 p.m., EDT

Passage, Bishop Rock to Ambrose L. V.: 2906 Miles
3 Days, 12 Hours, 12 Minutes — Average Speed: 34.51 Knots

Total Distance, Le Havre to New York: 3155 Miles
Steaming Time: 3 Days, 19 Hours, 29 Minutes — Average Speed: 34.48 Knots

NOTE: A Nautical Mile is approximately 15 per cent longer than a Statute or Land Mile

These passages are world records. It is the first time in a century that an American ship has captured the Blue Ribbon of the North Atlantic for both East and West passages. The United States Lines is rightfully proud of the achievement. We believe you are too.

Figure 66. Westbound maiden voyage log, from Southampton, England, to New York City. Steaming time of 3 days 12 hours, 12 minutes, average speed 34.51 knots. *Courtesy of Bill Krudener.*

Figure 67. Bell Captain Bill Krudener and entertainer Bob Hope at the entrance to the first-class dining room. *Courtesy of Bill Krudener.*

Figure 68. Vera Cravath Gibbs (right) and friend Eugenia McCrary dressed for an evening in first class. Vera took frequent trips aboard the "family rowboat" while her husband remained ashore. *Courtesy of Susan L. Gibbs.*

Figure 69. John Wayne chats with Commodore Anderson on the bridge during a blustery winter crossing in early 1956. *Courtesy Joe Rota.*

Figure 70. Cooks in Chef Bismarck's galley. *Courtesy of Joe Rota.*

Figure 71. Actress Kim Novak poses next to the ship. *Courtesy of Joe Rota.*

Figure 72. Artist Salvador Dali (right) and his wife, Gala, in the first-class ballroom. *Courtesy of Joe Rota.*

Figure 73. *United States* arrives at Pier 86 in New York on a bitterly cold winter day. The ship's hull is encrusted in ice following a stormy Atlantic crossing. *The Frank O. Braynard Collection.*

Figure 74. William Francis Gibbs watches his Big Ship come in. *Marvin Koner/Corbis.*

Figure 75. William Francis Gibbs in his seventies, contemplating New York harbor from the "Glass Menagerie." *Marvin Koner/Corbis.*

Figure 76. *United States* in June 2011. Rusting, stripped, and faded in the city of William Francis Gibbs's birth. Despite her rusted exterior and forty years of neglect, the ship remains in excellent structure condition, testament to the skill of her designer and builders. *Author's collection.*

Figure 77. The gutted first-class ballroom, long ago the domain of the rich and famous. Only the dance floor and the band stage have survived the 1984 furniture auction in Newport News and 1992 asbestos removal in Ukraine. The rest of the interior has been stripped down to the bare bulkheads. Only traces remain: a stray bar stool, a telephone switchboard in the "Times Square" crew area, and a torn dinner menu found in a crew compartment. *Author's collection.*

Figure 78. A crowd of more than five hundred people gathers in front of *United States* on July 1, 2010, to celebrate philanthropist H. F. "Gerry" Lenfest's $5.8 million gift to the SS United States Conservancy for the purchase and maintenance of the ship. The Stars and Stripes flutters from her radar mast for the first time in years. *Author's collection.*

If you can talk with crowds and keep your virtue,
Or walk with kings—nor lose the common touch,
If neither foes nor loving friends can hurt you;
If all men count with you, but none too much,
If you can fill the unforgiving minute
With sixty seconds' worth of distance run,
Yours is the Earth and everything that's in it,
And—which is more—you'll be a Man, my son!

Christopher hung the poem above his fireplace, "where it is always in view, as a reminder."

In 1966, Christopher went back to New York to visit his father, for what turned out to be the last time. "I have read and heard him described in terms of 'greatness,' " he wrote to his mother of the encounter, "and I think that when I had the chance to visit with him last summer, I came away with a feeling of what this 'greatness' referred to. A newly-found feeling, I may add."[27]

It seems that William Francis and Vera Gibbs, as trailblazing as they were in their professional and social lives, were coldly Victorian as parents raising children. The children, in turn, kept a distance from their father, his world, and his greatest achievement; nor did they become any part of their mother's in New York.

With his children grown and his wife busy, William Francis Gibbs found he needed something more than just work. As he grew older, the ascetic engineer devoted more of his time to St. Thomas Episcopal Church on Fifth Avenue. The parishioners of the Gothic church were high-church in their leanings, and William Francis grumbled that a "humble man" like himself had a hard time understanding all the rituals. No matter, sustaining the life of the church became his main diversion, and he enjoyed donning his formal outfit of striped trousers, cutaway morning coat, and homburg hat each Sunday.[28] As a vestryman at St. Thomas, he insisted that expensive paid pews be abolished, and he also removed the church's wooden doors, replacing them with ones made of glass that allowed pedestrians to look inside.[29]

It was through the Episcopal Church that Gibbs left another legacy, one that he hoped would last as long as the Big Ship. In 1960, Gibbs's old friend, actor Wynn Handman, approached William Francis with the idea of starting a new congregation for the arts community. Not only would it be an active church; it would also be the home of a new arts venue of Handman's called the American Place Theater. William Francis threw himself into the project, and the renovated St. Clement's Church, located only a few blocks from the Big Ship's berth at Pier 86, opened its doors in 1962. The church's guiding belief was "There will be no outcasts."[30]

Next to designing a sister ship for *United States,* Gibbs's deep involvement with the Episcopal Church brought him his greatest satisfaction in his later years. Yet as a new decade dawned, he could not help but notice that the public was increasingly traveling to Europe and back by passenger aircraft, something he had brushed off during his entire career.

Gradually, it became clear that *United States* would not signal a new era of American maritime dominance, but the swan song of a whole way of life on the North Atlantic, one to which William Francis Gibbs had dedicated his life.

And after half a century of intense commitment to the design of ships, William Francis had no choice but to hold on to the Big Ship even as the world changed around him.

"Once such a man as W.F.G. produces a masterpiece," one observer noted, "he becomes captive to it."[31]

The sister ship to *United States* would never be built. As the 1950s ended, the Big Ship was under attack by her unionized crew and by the jet airplane. It was a fight that even William Francis Gibbs could not win.

NO TIME TO SAY GOOD-BYE

During a stormy winter crossing in the late 1950s, Bell Captain Bill Krudener found himself tending to a seasick couple on their honeymoon. They never left their first-class cabin, nor did they stop vomiting. Krudener dutifully cleaned up the mess. As the couple stumbled down the gangplank, the green-faced husband, who had lost about fifteen pounds, glared at him in anger.

"Next time, we are flying!" he said.[1]

The man's words signaled a new, unstoppable trend that began on October 4, 1958, when the first commercial jet took off from London's Heathrow Airport and screeched onto the tarmac of New York's Idlewild six hours later. With the coming of the jet, travel writer John Malcolm Brinnin observed, "sea travel did not so much decline or diminish as plummet—almost out of sight."

For the first few years after the arrival of the transatlantic jet in the autumn of 1958, *United States* somehow continued to carry healthy passenger loads, including a large number of American military personnel traveling to bases in West Germany. "The ship held her passengers well, especially in the summer time," ship's photographer Joe Rota recalled. "But when the weather began to deteriorate around October, then we noticed more of a drop."[2]

To the old-timers in the ship business, the bleak days of the Great Depression returned when giants such as *Leviathan, Mauretania,* and *Olympic* sailed less than half full. But this was different. The world was prosperous, the stock market was soaring, and Americans had money.

In response to the crisis, the shipping companies decided to get rid of the older ships. In 1959, the first of the big old liners went to the scrappers when the French Line disposed of the thirty-one-year-old *Ile de France.* She was quickly followed by the old flagship *Liberté,* built in 1929 as the German greyhound *Europa.*[3] Both ships were replaced in 1962 by the giant *France,* at 1,035 feet the longest liner in the world and capable of sailing at 35 knots. But even the French Line conceded there would be no attempt to capture the Blue Riband from *United States.*

"Flying has cut down trans-Atlantic travel time so that there is no meaning to surface speed records," company president Jean Marie told the press, closing the door on more than 120 years of hard-won innovation, as generations of ship designers chased the Blue Riband.[4]

The Italian Line remained optimistic, commissioning three big new liners in the early 1960s: *Leonardo da Vinci, Michelangelo,* and *Raffaello.* All were suffused with Italian glamour, style, and heavy government subsidies, but operated in the red as soon as they entered service.

No one, it turned out, could overcome the sudden paradigm shift in transatlantic travel. Even Cunard, with its weekly express duo of the *Queen Elizabeth* and *Queen Mary,* began to stumble financially. By the early 1960s, the two great ships were losing millions of dollars every year. Maintenance cutbacks hurt them as well, transforming them from floating Ritz-Carltons to shabby-genteel hostelries. Their hulls became streaked in rust, brass work dull, luxurious lounges and staterooms dingy. Menus grew shorter and service surlier. "As their final decade drew to its end, the *Queens,* more often than not, were ghost ships of the Great Circle," John Malcolm Brinnin wrote. "It was then possible for a single solitary passenger to turn up for tea in the dim depths of the grand saloon and sit, magnificently alone, while a dozen white jacketed stewards stood around like sentries, alert to his command. . . ." The captain's receptions, populated by dwindling bands of doddering old men and aging dowagers, were "an occasion of uneasy gaiety and an index to despair."[5]

As passenger lists declined, labor troubles escalated, and more sailings were canceled or delayed by strikes. Joe Curran's National Maritime Union refused to allow the United States Lines to lay off crew members when the ship was lightly booked, especially in the winter. "Members felt that the shipping companies were taking advantage of the workers every place they could," a bellboy recalled, "and they were making a fortune, and if it weren't for the unions the ship couldn't sail. Often there was no work for the bell staff. But we had to be there."[6]

In June 1958, the ship's engineering officers, all members of the Marine Engineers Beneficial Association, walked off *United States* just before she was to set sail with a nearly full complement of passengers. The *New York Times* reported that the union demanded "a 'substantial' wage increase, job security, longer vacations, higher pensions, and minor changes in working rules." The dispute was resolved only after "what was described as a complete capitulation by the United States Lines in shipboard conferences that began at 10 A.M.—two hours before sailing time." *United States* sailed for Europe eight hours late. Passengers made the best of the delay by sunbathing in deck chairs.[7]

On October 1 of the same year, the members of the International Union of Masters, Mates and Pilots walked off the job, demanding longer vacations. The strike dragged on, and the union refused to budge.

On October 3, a fuming General Franklin called a press conference and announced that for the first time ever, the company would cancel a sailing of *United States*.

Franklin declared that the negotiations failed "because of impossible demands, including featherbedding."

"We are determined to fight it through," he vowed, even though, by company estimates, the strike cost the United States Lines $500,000 in lost revenue, as well as $700,000 in wages for the crew. Franklin, used to settling NMU disagreements over a scotch and soda with Big Joe Curran, found his powers of persuasion slipping away when faced with leaders of other maritime unions.[8]

The strike was eventually settled and the ship made her next crossing. But the unions felt they had won. If passengers were inconvenienced frequently enough, they figured, the United States Lines had no choice but to give in.

Because of her $12 million annual operating subsidy and her status as a naval reserve vessel, *United States* remained impeccably clean and well maintained, even as her passenger revenues fell. But the company's influence in Washington waned when Vincent Astor, the company's most powerful stockholder and director, died of a heart attack in 1959, leaving his third wife, Brooke, his entire fortune.

In 1960, the company's passenger business suffered another blow when the Department of Commerce ended the transportation of military personnel aboard the ships of the United States Lines. More and more troops would be taken by air to bases abroad.[9] The days of *United States* as an unofficial troopship were over, even though her federal operating subsidy remained safe for the time being. Scrambling to make up the lost revenue from the cancellation of the armed forces transport contract, the United States Lines drastically revised the Big Ship's sailing schedule. The company added Bremerhaven, Germany, to the ship's summer schedule, hoping to pick up more passengers, especially European immigrants. The Immigration Act of 1965, signed by President Lyndon Johnson, had opened America's borders to more people from Eastern Europe and Asia, reversing the discriminatory provisions of laws passed in the 1920s.

The year his company lost its military personnel transport contract, General Franklin stepped down as president of the United States Lines to become chairman of the board, appointing William B. Rand, the husband of his daughter Emily, in his place. But Rand's tenure was short-lived. The shipping dynasty that began with Clement Griscom and J. P. Morgan and was carried forward by Philip Franklin, Vincent Astor, and John Franklin ended in late 1967 when Walter Kidde & Company purchased a controlling stake in the company once known as the International Mercantile Marine.[10] The conglomerate had little interest in keeping *United States* in service if she continued to lose money.

By 1964, the United States Lines had decided that one of its two money-losing passenger ships had to go. *America* was sold to the Greek-owned Chandris Line for just $4.5 million. (Chandris would convert her into a one-class immigrant ship that sailed between Europe and Australia.) The sale left *United States* without a running mate, cutting into the schedule of sailings the United States Lines could provide the public. That same year, Commodore Anderson retired from *United States* and was replaced by his executive officer, Leroy Alexanderson, another graduate of the New York Maritime Academy. A large number of the ship's crew turned over shortly afterward, as Alexanderson had a reputation as an authoritarian.

As the 1960s wore on, the labor disputes grew more frequent. In January 1965, the New York longshoremen walked off the job. To keep *United States* sailing on schedule, a small army of United States Lines employees, from office boys to senior executives, trooped over to Pier 86. As the winds howled and the temperatures dropped into the twenties, men accustomed to office work loaded 32 automobiles, 13 tons of beer, 19 tons of vegetables, 3 tons of ice cream, 490 pieces of heavy luggage, and 3,500 items of hand baggage. Of course, the Duke and Duchess of Windsor traveled the way they always did, with their customary 83 pieces of baggage.[11]

Still, with the 1960s stock market boom putting money in people's pockets, the shipping lines believed that a market existed for travelers to destinations more exotic than Europe. Pleasure cruises, where the shipboard experience was half the "destination," were becoming popular. The United States Lines lobbied the Maritime Administration to let the company divert *United States* to cruises during the winter months, as the Cunard ships were already doing. After much grumbling, the Maritime Administration relented. By the mid-1960s, *United States* was sailing on winter jaunts to Bermuda, the Caribbean, and Rio de Janeiro.

The cruises were popular, but were not enough to keep the United States Lines from bleeding red ink. Meanwhile, as passenger lists slipped, operating expenses—especially the burning of 600 tons of fuel a day—remained significant. In 1966, the United States Lines lost a staggering $8 million. On one eastbound voyage, the Big Ship left

Europe carrying only 202 in first class, 263 in cabin class, and 296 in tourist class: 941 crew members looking after 761 passengers.[12]

Such dismal numbers became the norm. *United States,* the *Queens*, and the remaining big liners found themselves competing for the one traveler in twenty-five who still crossed the Atlantic by ship. Although consuming $12 million a year in government subsidies, *United States* was losing $4 million annually by the late 1960s. The government subsidy translated to $400 for every passenger ticket sold.[13]

The company even tried a rather sad new slogan to sell tickets: "Travel with the Unrushables" replaced "The Fastest Ship in the World."[14]

The liner's government subsidies were scheduled to end in 1977, when she turned twenty-five. However, there were rumblings in the press that neither the company nor the Maritime Administration wanted to wait that long to retire her. "I am as much a romantic as the next fellow, as an admirer of a truly great ship," the new president of the United States Lines, John P. McMullen, told the head of the Maritime Administration. "But I am also a practical shipping man."[15]

In the spring of 1967, the Cunard Line announced that *Queen Elizabeth* and *Queen Mary* would be retired and sold to the highest bidders. The two liners would be replaced by a single, modern ship that would spend half the year on the Atlantic and the other half cruising. Most guessed the old Cunard *Queens*, fading symbols of Britain's maritime might, would be sold to the wreckers of Taiwan or Japan.

About this time, the eighty-year-old William Francis Gibbs suddenly fell ill. His condition worsened, and by August he was confined to a bed at St. Luke's–Roosevelt Hospital. For the first time at 21 West Street, the stool at his drawing board was empty, his phone silent.

Shortly after noon on September 7, 1967, Commodore Alexanderson raised his right hand in salute as *United States* sailed past 21 West Street on the start of Voyage 353. The ship's other officers, assembled on the bridge, followed suit, and *United States*' three whistles let out a blast.

William Francis Gibbs had died the day before. The cause of death was congestive heart failure. He had just turned eighty-one.

As *United States* passed the Statue of Liberty, she was trailed by another outbound vessel. The Greek liner *Queen Frederica* was a small prewar relic, old and worn by service. As she steamed past 21 West Street, the captain of the ship once named *Malolo* gave a series of short, husky blasts to salute the man who had designed her back in 1927.

The funeral was held on September 11, at Madison Avenue Presbyterian Church, where William Francis and Vera Gibbs were married forty years before. The longtime rector of St. Thomas, Dr. Frederick Morris, wrote Vera Gibbs that he had a "real disappointment and hurt in William Francis' final attitude and decisions. But we are all determined to forget and to lay it to infirmity and illness. A goodly number of us will attend the funeral in a body."[16]

The church was packed with friends and employees for the simple service. Reverend John O. Mellin, a personal friend of William Francis Gibbs, delivered the eulogy, praising a man who "revealed to those around him a loyalty of perfection."[17]

Vera Gibbs employed her characteristic graciousness and composure to greet mourners at the viewing and service and to hear expressions of their sympathy. Son Francis Gibbs attended his father's funeral, but his brother Christopher decided to stay in California. He wrote a note to his mother, saying that his teaching load and his own fear of flying kept him from coming. "This reason, 'per se,' may sound rather strange and even callous in a way—but, after considerable meditation on my part as well as recalling my last year's visit with Dad, I came to the conclusion that he wanted me to carry on 'Full steam ahead,' to coin a phrase." Although he hinted at a reconciliation of sorts with his father, Christopher's letter to Vera still pointed to a strained relationship. "Although the loss of Dad has affected me in certain specific ways," he wrote. "I nonetheless feel that each of us continue to handle our lives as seems most suitable and appropriate. We must each reach our own decisions, whether right or wrong in the eyes of others."[18]

Many others wrote to Vera. Lewis Lapham, former president of the

Grace Line (and father of the editor of *Harper's*), wrote about the service that "all he was and stood for was there, strong and abundant."[19]

One condolence note, scrawled in script, came from the man once known as the greatest mariner in America. "The world has lost a great man," retired United States Lines commodore Harry Manning wrote. "Personally I had a great admiration for him. He cannot be replaced."

The naval architect would have most appreciated the sentiments of the Chief Engineer Bill Kaiser. "I miss the old so-and-so," he said, "asking every morning how things were going and how many revolutions we are making."[20]

On September 22, John J. McCloy, a onetime poor boy from Philadelphia and now president of the Chase Manhattan Bank, watched *United States* head out to sea from his office window, followed by *Queen Mary*—the old British liner was leaving New York for the last time. McCloy then composed one of the last notes Vera Gibbs received: "With your father who was the top of his profession and with your husband who was the leader of his, you have had a rich association with men of such outstanding quality. . . . We all know how much that ship was a part of his life. It brought back many memories as she went by."[21]

William Francis Gibbs was buried in Princeton, New Jersey, next to his mother, Frances. The body of his father, William Warren Gibbs, lay in faraway Hackettstown. Soon after the burial, the contents of Gibbs's desk in the Glass Menagerie—honorary degrees, sketches, newspaper clippings, his passport, a childhood book of psalms, keys, even a stray pack of chewing gum—were boxed up and shipped to the vaults of the Mariners' Museum in Newport News, Virginia.

The same month William Francis Gibbs died, *Queen Mary* sailed her last transatlantic voyage. After thirty-one years of service, the great Cunarder would return to Southampton and then to her final port of call: Long Beach, California. The city had purchased the ship for use as a permanently moored convention center. *Queen Elizabeth* would follow suit within a few months, destined, first, for Port Everglades, Florida, and then to Hong Kong to be refurbished as a floating university. But in 1972, just as the refit approached completion, the renamed *Seawise*

University burst into flames and capsized in Hong Kong harbor. The wreck was scrapped the following year.

On November 7, 1969, *United States* docked in New York after completing Voyage Number 400. Awaiting her was an ominous piece of news: the upcoming fifty-five-day Pacific cruise—during which the ship would pass through the Panama Canal for the first time—had been canceled.

Still, Commodore Alexanderson sailed the Big Ship down to Newport News for her annual overhaul. When she arrived at the shipyard, the usual army of workmen swarmed aboard, as they had every year for the past seventeen years. They checked the engines and winches, vacuumed staterooms, replaced wiring and lightbulbs, polished railings and doorknobs. Two teams high up in chairs hanging on ropes began repainting the funnels.

But then a foreman from the shipyard ordered the men to stop everything and leave the ship. Hundreds of perplexed men dropped their paintbrushes and toolboxes, and shuffled down the gangways. Slowly, the sound of voices died out in the public rooms, staterooms, corridors, and engine rooms, and the ship went silent.

They would be back tomorrow, the shipyard workers thought.

The next day, they found the gangway doors sealed shut.

Commodore Alexanderson had received a call from the superintendent of the yard. "He said, 'I want all the crew laid off,' " Alexanderson recalled. "Some of the crew never believed the ship was going to be stayed laid up, so left their things aboard the ship . . . figuring they'd be called back. They never were."[22]

SECRETS TOLD

The United States Lines was forced to retire *United States* when Congress did not renew operating subsidies for fiscal year 1970. "Our gross revenues today on the *United States* are $18-million, about the same as in 1953 when the vessel first came out," said one United States Lines executive. "But in the meantime our expenses have skyrocketed, wages and fringe benefits have skyrocketed, maintaining the ship has skyrocketed. We just can't make a go of it because of these high costs and the airline competition."[1]

Because of the ship's original and still potential connection to national defense, the Navy hermetically sealed her, installing a dehumidification system to preserve her interiors and machinery. The United States Maritime Administration took full title from the ailing company, which had stopped making payments to the government on the outstanding loan. After a short tow across the James River to Norfolk, Virginia, *United States* would sit undisturbed and forgotten for the next decade. In 1976, Norwegian Cruise Line, hoping to add a large ship to its hugely profitable Caribbean cruise fleet, made an offer to buy *United States* and refit her as a modern cruise ship. The Maritime Administration refused, citing the ship's still top-secret design features. Norwegian purchased and renovated the idled transatlantic liner *France* instead.

Renamed *Norway,* she became one of the most successful cruise ships of all time.

It was not until 1977 that the Navy decided to declassify the ship's design features. The man who made them public was John R. Kane, a retired vice president of Newport News Shipbuilding, who had almost lost his job once for refusing to divulge the ship's top speed to Admiral Hyman Rickover. Now, freed to talk, he presented a paper on November 3, 1977, to the Hampton Roads section of the Society of Naval Architects and Marine Engineers.

The first revelations were visual. For the first time, published drawings showed the ship's hull shape and propeller configuration. Kane illustrated his paper with previously top-secret photographs of the ship in dry dock, as well as diagrams of the model tests. The unique, below-the-waterline hull contours that Gibbs had guarded so zealously, the anticavitation propeller design he kept from public view, and his key engineering specifications were now available outside the "need to know" community.

Kane went on to give his thoughts about the ship in the light of the newly released information. Taken as a whole, he noted, the ship had the lines of a "rather conventional liner of that era, except that they are exceptionally lean and fine." But the "delicate balance" of the ship's form and the distribution of her displacement weight had enabled extraordinary performance—not only in terms of speed, but also fuel efficiency and stability in foul weather. This could only be the achievement of a master designer, one of "exceptional naval architectural art and skill."[2]

Kane summed up William Francis Gibbs's simple design philosophy by saying: "Combine the maximum driving power you can achieve, with the lightest displacement compatible with the work the ship must do, and with the longest, finest and cleanest lines that will serve to make a good wholesome sea-keeping ship."

But the naval architects, engineers, and reporters in his audience had only one thing on their minds: what was the classified top speed of *United States*?

The answer was an astonishing 38.32 knots, achieved during the

Big Ship's June 1952 trials—almost 45 land miles per hour. But experts knew that *United States* was capable of even higher speeds. The 38.32-knot performance was produced by her four Westinghouse turbines, which developed 241,785 shaft horsepower. The maximum engine output was more still, an estimated 247,785 horsepower. The newly released data on the ship's power also explained something else. It was already well-known that on her maiden voyage, *United States* averaged 35.59 knots eastbound, capturing the Blue Riband. During that voyage, Kane now pointed out, Commodore Manning did not push *United States* beyond two-thirds power. The ship had beaten all speed records by a substantial margin, "at less horsepower than either the *Queen Mary* or *Queen Elizabeth*." The two larger British ships, each of 80,000 gross tons, had maximum power ratings of about 158,000 shaft horsepower. At a relatively nimble 53,329 gross tons, *United States* proved herself to have the greatest power-to-weight ratio of any major commercial vessel in history.[3] In essence, Gibbs had done what no naval architect had done before: combine the size and luxury of an ocean liner with the lightness and speed of a destroyer. There was also a good economic reason why she wasn't operated at full engine capacity while in commercial service: to achieve that additional 2-knots difference between her maiden voyage speed (35.59 knots) and her highest trial speed (38.32 knots), the engines had to generate nearly 50 percent more horsepower.[4]

The declassification of *United States'* military secrets also announced to the world that the most advanced liner ever built had become irrelevant to American military needs. Ships would continue to dominate international cargo transport, both military and civilian. But it would be aircraft that would speed American troops to war zones from Vietnam to Iraq and Afghanistan, and the commercial airline industry that would take the American people anywhere they wanted to go.

By 1980, with her secrets now declassified, the military felt it had no possible future use for *United States.* The U.S. Maritime Administration put the now-rusting giant up for sale. Thirty years before, William Francis Gibbs claimed that "Joe Stalin would love to know what this ship will do." The asking price was $5 million, or about 7 percent of the

$78 million that it had cost to build her back in 1952.[5] Thus began a long saga of sale, stripping, stagnation, and sale again.

First, Richard Hadley, a Seattle-based real estate developer, bought the ship, and planned to turn her into floating condominiums. But the scheme was unworkable, and to pay off creditors, Hadley ordered all the ship's furniture, fittings, and artwork yanked out. Guernsey's of New York hosted the biggest auction in history—the murals from the Duck Suite, the Ostuni paintings from the Navajo Lounge, the kidney-shaped bar from the ballroom, the butterfly-and-tree-patterned bedspreads, United States Lines monogrammed china, the pots and pans from the galley, the desk from the commodore's cabin, even the ship's wheel—all were carted away and put on the block.

Hadley stubbornly held on to the ship itself—now a ransacked, musty shell of her former self—even as the elements began to seep through hatches and broken portholes into interior spaces.[6] In 1991, Earl Swift from the *Virginia Pilot-Ledger Star* boarded the vessel to look around, and was shocked. "Rust has invaded the mammoth black hull," he wrote, and "boilers are ruined; pumps, motors, and turbines have atrophied. Cabins are stripped or strewn with trash, soggy bedding, [and] paperwork. Bathrooms are littered with smashed porcelain. Hydraulic fluid stands inches deep in some machinery spaces. Pigeon droppings are inches deep on the weather decks. Weeds grow from the superstructure." A shipping executive was quoted indignantly saying that the owner "has taken what was rightfully a national monument and the vessel has been raped."[7]

By the early 1990s Hadley was bankrupt, and *United States* was on a fast track to the scrap yard. But in 1992, the ship was auctioned to Marmara Marine. Owner Fred Mayer, who had immigrated to America from Turkey on *United States* in 1963, bought the vessel on the Newport News courthouse steps. The price had dropped to $2.6 million. The new owner hoped to operate *United States* as a running mate to Cunard's *Queen Elizabeth 2,* the only ship left in regular transatlantic service. He had the great ship towed to Turkey, and then the Ukraine, where her asbestos-laden interiors were stripped down to bare metal.

The toxic material was everywhere: in marinite wallboard, floor tile, and pipe insulation.

In 1996, the rust-scarred and now-gutted *United States* was towed to Philadelphia, where she was tied up on the Delaware River. Mayer's deal with Cunard collapsed, his syndicate went bankrupt, and the ship was left to languish again. New Jersey developer Edward Cantor then purchased the ship, made promises, and did nothing.

Many people in Philadelphia were unhappy to see the ship docked on their waterfront. "Meet the *SS United States,* mother of all abandoned vehicles," the *Philadelphia Daily News* complained.[8]

When Cantor died in 2002, his heirs put *United States* up for sale, hoping to send her to the scrap yard once and for all. At the last minute, Norwegian Cruise Line, which had tried to buy the ship twenty-five years earlier, reappeared and barely outbid an Indian scrapper. "When we discovered this American icon was in jeopardy, we saw a unique opportunity and acted immediately," said Norwegian Cruise Line CEO Colin Veitch in 2003. "The ship is a classic, she was built in America and is eligible to operate in domestic service under existing law and regulation."[9]

NCL America's stated goal was to run a modernized *United States* on cruises between the West Coast and Honolulu. The Merchant Marine Act—passed by Congress in 1920 and still in effect today—stipulates that only American-flagged ships can carry passengers and cargo directly between two U.S. ports. *United States* is one of the few ships afloat that meet those requirements. But refitting *United States* with the features cruise passengers expect today—stateroom balconies, large outdoor swimming pools, onboard shopping malls—could cost up to $500 million, comparable to the cost of a new cruise ship. NCL let the ship languish. In early 2009, the Great Recession brought on by the financial crisis hit the cruise industry hard. In March of the following year, the company announced that they were putting *United States* up for sale, and that they were accepting offers from scrappers.

If purchased by a scrapper after forty years of waiting for a reprieve, *United States* will be towed away from the city of her designer's birth and dragged by her anchor chains onto a beach on the Gulf of Mexico. Here the greatest American ship ever built will be ripped to pieces and melted down to make razor blades and bedsprings.

She would be demolished in the reverse of how she was built. The wreckers will start their cutting away at the aluminum superstructure, sending the radar mast and mighty smokestacks tumbling down. The bow and stern will be next; with a shower of sparks, large sections of the graceful steel hull will fall away from the ship and land with a thud onto the beach. As more of the ship is demolished, she will be hauled higher and higher on land. Finally, the mighty engines and bronze propellers will be harvested from the skeletal remains of the ship, and then hauled away by truck to be melted down. After a year or so, the keel plates, laid down on the Newport News slipway #10 on a blustery February morning back in 1950, will be pulled from the mud and carted away.

During the spring of 2010, *United States* awaited a decision about her future at Philadelphia's Pier 82, surrounded by a maze of security cameras and barbed wire. The once-busy Delaware River waterfront scene, where Willy Gibbs watched the *St. Louis* christened in 1894, is long gone. In place of piers and shipyards are strip malls, big-box stores, parking lots, and highways. Each day thousands speed by *United States* on Columbus Boulevard, Interstate 95, and the Walt Whitman Bridge, scarcely noticing the faded red, white, and blue funnels that loom over South Philadelphia.

But close up, the great ship's immense size is still overwhelming. Stripped of fittings, she rides higher in the water than intended. Her once-lustrous exterior paint is cracked and falling in sheets; red antifouling paint above the waterline has faded to a ghoulish gray. The black hull is scarred with rust. Red, white, and blue chips from the two funnels lie scattered like leaves on the pier's loading dock. The wind pushes the liner back and forth from the pier, causing groans from the ship and shrieks from the lines tying her down.

On board, a visitor walks on broken glass shards scattered on cracked decks. All four propellers lie on the fantail; a nearby section of stern railing is mangled from a mishap during the hoist of the propellers up and over. A few shuffleboard courts survive on the first-class sports deck, their painted white lines barely visible, their surfaces pockmarked. All of the ship's aluminum lifeboats, which once buffered the sports decks from the wind and spray of the North Atlantic, have long been removed and sold for scrap.

Inside, Dorothy Marckwald's crisp, classic interiors have been stripped to bare metal. Windows that crewmen once polished to a shine are streaked with grime. There are a few bits and pieces of what once was: a crew bulletin board, a few swivel chairs in a crew lounge, a Ping-Pong table sitting in the chartroom, a stained fragment of a first-class dinner menu wedged in a storage locker. On the bridge, all of the original instrumentation, including the ship's wheel, navigation equipment, and engine room telegraphs, are gone, leaving raised stubs and severed wiring poking up from the deck. The first-class ballroom—once the epitome of 1950s postwar chic—now sits vacant and silent. Pools of water sit on the stage once graced by the likes of Duke Ellington, Ethel Merman, and Meyer Davis. The circular dance floor remains, its black surface peeling up at the edges. Scrawled notes on metal bulkheads are ghostly reminders of the workers who built the Big Ship more than half a century ago. The floor plan of the famed Duck Suite can still be made out by the partition slots in the decks, as well as holes where plumbing fixtures used to be. A bit of midnight-blue paint still clings to a column in the darkened, musty-smelling cabin-class dining room.

The movie theater, now stripped of its blue plush seats and oyster-white walls, still has the stage where, shortly after the maiden voyage, William Francis Gibbs told the crew: "I have a great affection for this ship. It's a large object, but I think my affection is coextensive with its size, and likewise I have an affection for every man who has made possible and makes possible day by day, the extraordinary success of this ship so far."[10] Two sets of aluminum double doors, whose handles are still labeled "push" in stylized 1950s lettering, lead from the first-class grand entrance to the enclosed promenade decks. These two boulevards at sea were once lined with deck chairs and uniformed stewards serving tea and hot bouillon. They are now strewn with piles of feathers and bones. A number of pigeons found their way in through the broken windows. Unable to find their way out, they starved to death.

The days of the transatlantic liner as the only way to cross the Atlantic, when "getting there is half the fun," are long gone. Almost all the great liners are either lying on the bottom of the ocean or have been melted down for scrap metal.

Yet the passenger ship itself is far from dead. Today's cruise ships, which are so much bigger and more luxurious than Gibbs's masterpiece, do not have that same sense of speed or purpose. They are designed not for cold weather crossings, but for balmy vacation cruises. Cunard's *Queen Mary 2* and *Queen Victoria* share the oceans with ships named *Carnival Miracle, Fantasia,* and *Oasis of the Seas.* Ships intended for port-hopping holiday excursions don't need speed; they are built to maximize entertainment features and ocean-view balconies. Most look like condominium blocks, bloated and top-heavy. In rough weather, stabilizer fins jut out to eliminate rolling almost completely. Gone is the steady thump of the engines pulsating through the ship; propulsion units are now placed outside of the hull in pods, and insulation almost completely muffles vibration. In addition, standardization and prefabrication in design and construction has resulted in most of these ships looking alike. The visual distinction that made the profiles of transatlantic liners of the past instantly recognizable to seafarers and schoolchildren alike is gone.

United States would seem to have little in common with her successors, and yet the great ship left its lasting mark even here. Her construction techniques and standards continue to be the benchmark for ship design. She was the first big liner built and floated out of a dry dock, which is now the norm for new ship construction, not sent sliding down the ways. Her use of prefabricated components, manufactured off-site—a technique that Gibbs pioneered with the Liberty ship and used on

United States—is now universal. And *United States*' innovative use of aluminum construction is the method that makes possible the high, balconied superstructures of modern cruise ships.

Unfortunately, cruise ship passenger safety standards fall short of those used on *United States*. The cruise ship is not required to follow stringent Navy fire codes. A number of recent fires on board proved William Francis Gibbs's wisdom in using nonflammable construction materials and furnishings in vulnerable passenger areas. In the other important area of ship safety—protection from sinking in the event of a collision—the new 2010 Safety of Life at Sea (SOLAS) standards are tighter than ever, but still short of those set by Gibbs on *United States*. SOLAS requires every new ship built today to be able to remain afloat with as many as three compartments flooded; *United States* was designed to remain afloat with as many as five breached. In fact, the ship's seventeen-year career was both accident- and fire-free.

The Big Ship incorporated all of Gibbs's design breakthroughs: the troop-carrying capacity of *Leviathan*; the extra compartmentalization that kept *Malolo* from sinking in 1927; the high-pressure, high-temperature steam turbines of the *Mahan*-class destroyers of the 1930s; the sweeping lines and fire prevention measures of *America*; the prefab parts assembly of the Liberty ship. William Francis Gibbs saw that commercial innovation can have military application—and deserves equal support and secrecy.

Yet the Big Ship that Gibbs willed into existence was more than a sum of the lessons he learned in a forty-year career, and definitely more than a mere machine. She was his masterpiece, a supreme expression of his artistic vision, a symbol of excellence for the world to see. The fastest, most beautiful ocean liner ever built had a soul—the soul of her creator and the nation whose name she bore. No other ocean liner—perhaps no other ship in history—was so bound up in the life of her designer.

In life and business, Gibbs defied convention and thrived, deftly working conventional corridors of power to finance his dream. But for him, the deepest satisfactions in life lay in work and creation. And to live that life he first had to re-create himself, from a shy young man who isolated himself in his college dorm room to a charismatic and self-assured leader of men—confident enough to win the long fight to build the Big Ship, a creation that reflected the United States in its full post-war glory, flush with prosperity, military might, and industrial supremacy. "This ship is the product of explosive power—American industry," Gibbs proudly said, adding that the workers who built her were "trustees for the people of the United States."[11]

Gibbs saw his creative journey as something larger than himself. "Humility and belief in divine guidance; belief in your fellow man," he once asserted late in life, "and what they have done for you, and the little part you can play yourself, are characteristics that all of us can embrace and employ."[12]

But in a private moment as *United States* neared completion, William Francis Gibbs took a breather from his relentless inspection tours. He found a secluded area on a deck and lay down. Clouds drifted by and seagulls wheeled above him. The clatter and din of the shipyard filled his ears, and the spring sun bathed the length of his thin frame.

He turned to a friend, an enormous grin on his face.

"Boy!" he shouted. "Don't we have fun!"[13]

Epilogue

The future of the great ship *United States* remains uncertain, but she continues to fascinate thousands of people. Those who helped create her, worked on her, traveled on her, or knew someone who did make the story of *United States* not one of a ship, but of the people whose lives she touched, transformed, and inspired.

Frederic Gibbs devoted his last years to preserving his brother's legacy. In 1978, after the ship's design features were declassified, he donated two immense volumes to the Mariners' Museum in Newport News, Virginia. Hundreds of memoranda and diagrams, which catalogue every aspect of the ship's design and construction, were titled Design 12201, "S.S. United States—Design Particulars and Information," and marked "confidential." He finished his work just in time. Shortly after Frederic's death, new Gibbs & Cox management destroyed most of William Francis Gibbs's personal archive, including his *Leviathan* scrapbook and correspondence not donated to the Mariners' Museum. Frederic also gave the Straitsmouth Island portion of the Rockport, Massachusetts, property to the Audubon Society as a wild bird sanctuary. It was here that his brother and best friend used to sit and watch the ocean for hours. Frederic Gibbs died in 1980.

Vera Cravath Gibbs, eleven years younger than her husband, remained a fixture in New York social and artistic life after his death. She remained devoted to the future of the opera music her soprano mother loved, as a supporter of the Metropolitan Opera. She moved to Rockport permanently in the 1970s, and died of a heart attack in 1985 at the age of eighty-nine.[1] After his parents died, son Francis Gibbs returned to the Northeast and moved to Maine, where in his off hours he would cruise the Atlantic in a Grand Banks motor yacht.[2] Granddaughter Susan Gibbs, born in 1962 and raised in near-total ignorance of the Big Ship and William Francis Gibbs's accomplishments, would become a passionate proponent of the ship's preservation. Susan is now executive director of the SS United States Conservancy, a nonprofit organization dedicated to "the protection and restoration of the SS *United States*."[3]

In 1960, General John M. Franklin retired from the United States Lines, and regretted it. According to his daughter Laura Franklin Dunn, the dynamic former shipping leader quickly became bored with life at his Hayfields estate, with bridge being his only diversion. "He was very sad when *United States* was no longer in service," his daughter Laura recalled. "It was a different world coming along."[4] Franklin died on June 2, 1975, at age seventy-nine.

After his forced retirement in August 1952, Commodore Harry Manning settled in Saddle River, New Jersey. He was unhappy that the United States Lines never built a sister ship for *United States*. "She should have had a sister ship," he said after his appointment as an admiral in the U.S. Maritime Service in 1967. "This business of putting it all in airplanes is nonsense."[5] He died in 1974.

In 1973, after thirteen terms as president, Big Joe Curran retired from the National Maritime Union under heavy criticism for his lavish lifestyle, including one of the highest salaries of any union leader, and allegations of corruption. Today very few ships sail under the American flag, and the unionized American sailor—there were once one hundred thousand of them—is almost extinct. Just before Curran's death in 1981 in Boca Raton, Florida, a court ruled that there had been "no evidence of improprieties" involving NMU funds.[6]

After teaching high school math on Long Island for many years, Elaine Kaplan returned to Gibbs & Cox as a marine engineer in the 1980s. She died in 1996, proud to the last of her "first baby." A model of the ship is displayed in her daughter Susan Caccavale's home in Smithtown, New York.

One of the ship's spare five-bladed propellers has been placed at the entrance of the Mariners' Museum in Newport News, Virginia—the repository of the William Francis Gibbs Collection—where it is set atop a granite fountain. After sunset, the floodlit eighteen-foot manganese bronze propeller casts a shimmering glow across the fountain basin.

The lay-up of *United States* did not save the United States Lines. The company floundered during the 1970s, burdened by outdated tonnage and high operating costs. The once-mighty shipping operation—with roots dating back to 1871 and stewarded by Clement Griscom, J. P. Morgan, the Franklins, and Vincent Astor—went bankrupt and sold off its remaining ships in 1986. In its last year, the company had an annual loss of $67 million and was burdened with $1.4 billion in debt.[7]

Gibbs & Cox, although smaller than it was in William Francis Gibbs's time, still exists. Its clients include not only the United States Navy, but also the navies of Spain, Norway, and Taiwan. The firm no longer designs passenger ships.

The Newport News Shipbuilding and Dry Dock Company continues to prosper as a division of the defense contractor Northrop Grumman. The shipyard that built *United States* continues to construct and service naval vessels, including nuclear-powered aircraft carriers, but like Gibbs & Cox it no longer builds passenger ships.

Few ships designed by William Francis Gibbs continue to sail today. After fourteen years of successful Australian service, the former *America* (renamed *Australis*) was laid up in 1980 in the Greek harbor of Piraeus. After rusting for over a decade, the ship was purchased in 1993 for conversion into a floating hotel off the coast of Thailand. During the tow, a violent gale lashed the ship off the Canary Islands. The towline snapped, and she drifted ashore onto a beach. After pounding by heavy surf, the old *America*—launched by Eleanor Roosevelt on the eve of World War II—split in half and slowly dropped into the Atlantic Ocean. Today, only a few fragments of twisted metal poke above the surface.

Santa Rosa, built for the Grace Line in 1958 and the last ocean liner designed by William Francis Gibbs and decorated by Dorothy Marckwald, still exists as *The Emerald.* Often described as a miniature *United*

States, she has been extensively altered in recent years. Laid up in 2009 after decades of successful cruising, she is currently awaiting her fate.

The magnificent yacht *Savarona,* designed by Gibbs & Cox in 1929 for the Cadwalader family, is owned by the Turkish government. Restored and modernized, she continues to sail on luxurious charter cruises.[8]

The tall ship *Sea Cloud,* built in 1931 as Marjorie Merriweather Post's private yacht, continues to sail the world as a high-end cruise ship.

Out of 2,700 Liberty ships built during World War II, only two survive: *John W. Brown,* berthed in Baltimore, and *Jeremiah O'Brien,* berthed in San Francisco. Both ships are fully operational, floating memorials to the men and women who served in World War II. The rest have been scrapped or sunk as artificial reefs.

On September 11, 2001, one of William Francis's ships again made history. On that day, the New York Fire Department's *Fire Fighter,* built in 1938 and still one of the most powerful fireboats in the world, was dispatched to the ruins of the World Trade Center to help extinguish the flames after the twin towers collapsed.

As for *United States,* the greatest ocean liner ever built, good fortune intervened one more time.

H. F. "Gerry" Lenfest is a billionaire Philadelphia philanthropist who made a fortune in the cable television business. Lenfest, now in his early eighties, had a soft spot for ships. His father was a naval architect who also owned a machine shop that built the watertight doors and bridge equipment for *United States* back in 1952. Growing up in Scarsdale, New York, young Gerry spent hours in his father's study—known as the "ship room"—where he would leaf through engineering periodicals and play with his father's model collection. After college, he served in the Navy as the captain of a destroyer.

When asked for help, Lenfest said that he didn't think saving the ship made financial sense, but he did offer to put up $300,000 in matching grant money to help purchase the ship. But no matching money was forthcoming. So in the winter of 2010, Norwegian Cruise Line put *United States* up for bid to scrappers. The estimated scrap value ranged from $3 million to $6 million. It looked like the ship had reached the end of the line.

The day before all bids were due, March 10, 2010, former crewman and conservancy board member Joe Rota got a phone call at his house in upstate New York.

"My name is Gerry Lenfest," the voice announced to Rota. "How much is it to buy the ship?"

A few months later, on July 1, 2010, a crowd of nearly one thousand people gathered in South Philadelphia where *United States* is docked, just downriver from the site of the old Cramp Shipyard on the Delaware River. The twilight sun shone on the National Flagship Celebration, as reporters and photographers roamed the ship's decks. The American flag, flying from the ship's radar mast, whipped in the breeze. Fifty-eight years earlier almost to the day, *United States* left New York on her maiden voyage. In the gathering were former crew members—Bill Krudener, Jim Green, and Joe Rota.

That day, Norwegian Cruise Line signed an exclusive purchase option with the SS United States Conservancy. Thanks to Lenfest's $5.8 million gift, the conservancy now has enough money to purchase the ship outright and maintain her for two years.

The plan is to raise enough money from private donors and for-profit developers to restore *United States* as a stationary hotel and convention center, to be moored in one of the two cities that William Francis Gibbs called home. If the plan is successful, the exterior will be restored to its original appearance, her two stacks repainted to a gleaming red, white, and blue. According to the preliminary plan, the interior will boast convention areas, dining, retail, and a boutique hotel, as well as a museum dedicated to the history of the ship and the transatlantic liner. Some spaces, such as the first-class ballroom and the bridge, will be restored to their original condition, giving people a taste of what it was like to travel on the fastest ship in the world.

If she is moved to New York, she will probably be docked at her old berth at Pier 86, next to the aircraft carrier USS *Intrepid.* If she remains in Philadelphia, the ship will be the centerpiece of a redeveloped South Philadelphia waterfront, a tourist draw and visual anchor on the river where her designer first fell in love with ships.

"We have a great opportunity here," Lenfest told the flag-waving crowd on July 1. "We are sitting across from the greatest ocean liner ever built. She's worth keeping, she's worth saving. If you look at her, at least from a distance, she is still the most majestic and most beautiful ship afloat."[9]

A recording of the ship's whistles then roared across the Delaware River, and powerful floodlights bathed her funnels, radar mast, and bridge in a golden glow.

For a brief moment, one could imagine William Francis Gibbs's beloved Big Ship—her decks lined with cheering people and streamers fluttering from her railings—slowly pulling away from her pier to sail the North Atlantic once more and forever.

Notes

1. SIZE, LUXURY, AND SPEED

1. "Technological Revolutionist," *Time,* September 28, 1942, http://www.time.com/time/magazine/article/0,9171,773637,00.html, accessed November 17, 2008.
2. Gail E. Farr and Brett F. Bostwick with the assistance of Merville Wess, *Shipbuilding at Cramp and Sons: A History and Guide to the Collections of the William Cramp and Sons Ship and Engine Building Company (1830–1927) and the Cramp Shipbuilding Company (1941–1946)* (Philadelphia: Philadelphia Maritime Museum, 1991), p. 7.
3. "Launch of the St. Louis: Mrs. Cleveland Christens the New Liner," *New York Times,* November 13, 1894.
4. Ibid.
5. William Francis Gibbs, "S.S. United States," *Journal of the Franklin Institute,* Philadelphia, December 1953, p. 547.
6. *New York Times,* August 12, 1883, as quoted by Stephen Fox, *Transatlantic: Samuel Cunard, Isambard Brunel, and the Great Atlantic Steamships* (New York: HarperCollins, 2003), pp. 124–28.
7. Charles MacIver, as quoted in Howard Johnson, *The Cunard Story* (London: Whittet Books, 1987), p. 55.
8. Johnson, *The Cunard Story,* p. 21.
9. Ibid.
10. Fox, *Transatlantic,* pp. 124–28.
11. Henry Adams, *The Education of Henry Adams*, as quoted in Fox, *Transatlantic,* p. 338.
12. Lamar Cecil, *Albert Ballin: Business and Politics in Imperial Germany 1888–1918* (Princeton, NJ: Princeton University Press, 1967), pp. 24–25, 28.

13. Mark D. Warren, ed., *The Cunard Turbine-Driven Quadruple-Screw Atlantic Liner "Mauretania"* (Wellingborough, UK: Patrick Stephens, 1987), p. C.
14. Fox, *Transatlantic,* p. 267.
15. "American Line Ends Long Mail Subsidy," *New York Times,* October 26, 1920.
16. "The Blue Riband of the Atlantic, Westbound," http://www.greatships.net/riband.html, accessed November 21, 2008.

2. ESCAPING THE RICH BOYS

1. Alva Johnson, "The Mysterious Mr. Gibbs—II," *Saturday Evening Post,* January 27, 1945, p. 20. Reprinted from *The Saturday Evening Post* magazine, © 1945 Saturday Evening Post Society. Reprinted with permission.
2. Ibid.
3. Winthrop Sergeant, "Profiles: The Best I Know How," *New Yorker,* June 6, 1964, p. 73.
4. E. Digby Balzell, *Philadelphia Gentlemen: The Making of a National Upper Class* (Glencoe, IL: Free Press, 1958), p. 125.
5. "The Man in the Street," *New York Times,* September 21, 1902.
6. "Cellulose in the Navy: Objections to Its Use to Keep Water Out of Ships," *New York Times,* January 23, 1893.
7. "Some Happenings in Good Society," *New York Times,* January 7, 1900.
8. "Spring Lake," *New York Times,* September 10, 1899.
9. "Elegant Wedding at St. James: Miss Augusta M. Gibbs Becomes the Wife of Mr. W.H.T. Huhn," *Philadelphia Inquirer,* April 9, 1899.
10. John J. McCloy, interview by Kai Bird, June 23, 1983, from Kai Bird, *The Chairman: John J. McCloy and the Making of the American Establishment* (New York: Simon & Schuster, 1991), p. 57.
11. Stuart Wells, "The Residence at 1733 Walnut Street," HSTVP 600 Documentation and Archival Research, Dr. Roger Moss, December 12, 1986, Philadelphia Athenaeum, HR 86.4., p. 8.
12. Sergeant, "Profiles: The Best I Know How," p. 73.
13. Ibid., p. 74.
14. Frank Braynard and Robert Hudson Westover, *S.S. United States: Fastest Ship in the World* (Paducah, KY: Turner, 2002), p. 13.
15. *De Lancey School, The Blight School Merged, Philadelphia 1914,* collection of the Episcopal Academy, Merion, PA, p. 16.
16. Ibid., p. 53.

17. Application of William Francis Gibbs, Harvard University Archives, UAIII 15.75.12.
18. As quoted in Ronald Steel, *Walter Lippmann and the American Century* (New York: Vintage Books, 1980), p. 13.
19. Johnson, "The Mysterious Mr. Gibbs—II," p. 96.
20. Ibid.
21. Frank O. Braynard, *By Their Works Ye Shall Know Them: The Life and Ships of William Francis Gibbs 1886–1967* (New York: Gibbs & Cox, 1968), pp. 10–11.
22. Four photographs, William Francis Gibbs Collection, MS 179, Mariners' Museum, Newport News, VA.
23. John Reed, "Almost Thirty," N/R 4/29/36, cited in Samuel Eliot Morison, *Three Centuries of Harvard* (Cambridge, MA: Harvard University Press, 1936), pp. 434–35, as quoted in Steel, *Walter Lippmann and the American Century,* p. 13.
24. Morison, *Three Centuries of Harvard,* p. 442.
25. Johnston, "The Mysterious Mr. Gibbs—II," p. 96.
26. "Football Ushers Appointed for Year," *Harvard Crimson,* October 9, 1909.
27. Transcript of William Francis Gibbs, Harvard University Archives, UAIII 15.75.12.
28. Alva Johnson, "The Mysterious Mr. Gibbs—III," *Saturday Evening Post,* February 3, 1945, p. 32. Reprinted from *The Saturday Evening Post* magazine, © 1945 Saturday Evening Post Society. Reprinted with permission.
29. William Francis Gibbs, as quoted by Winthrop Sergeant, "Profiles: The Best I Know How," *New Yorker,* June 6, 1964, p. 73.
30. George W. Cram to W. W. Gibbs, July 30, 1907, Harvard University Archives, UAIII 15.75.12.
31. Mitchell Charles Harrison, *Prominent and Progressive Americans* (New York: New York Tribune, 1901), p. 130.
32. "Skylarking Over; Now for Business," *Philadelphia Inquirer,* December 9, 1901.
33. "Suit Against W. W. Gibbs," *New York Times,* April 20, 1902.
34. "Seek Dissolution of American Alkali Co, Four Stockholders of the Corporation Charge Fraud," *New York Times,* March 12, 1902.
35. "Receivers Named for American Alkali," *Philadelphia Inquirer,* September 12, 1902.
36. "A Buoyant Market at the Close," *Philadelphia Inquirer,* April 13, 1905.
37. "Gibbs Mansion May Go Under the Hammer," February 19, 1911, publication unknown, Perkins Collection, Walnut Street, Walnut Street Volume III, from West of Broad, Historical Society of Pennsylvania, Philadelphia.

38. "The Merger of the Steamship Lines: Clement A. Griscom Gives a Description of the Combination," *New York Times,* April 21, 1902.
39. Lamar Cecil, Albert Bullin: *Business and Politics in Imperial Germany, 1888–1918* (Princeton, New Jersey: Princeton University Press, 1967), 57.
40. Mark D. Warren, ed., *Lusitania* (Wellingborough, UK: Patrick Stephens, 1986), p. 12, reprinted from *Engineering,* 1907.
41. Cunard Agreement with British Admiralty, as quoted by Mark D. Warren, ed., *The Cunard Turbine-Driven Quadruple-Screw Atlantic Liner "Mauretania"* (Wellingborough, UK: Patrick Stephens, 1987), p. 14.
42. Ibid., p. 13.
43. Warren, ed., *The Cunard Turbine-Driven Quadruple-Screw Atlantic Liner "Mauretania,"* pp. 2–3.

3. *MAURETANIA*

1. Mark D. Warren, ed., "The Trials of the Mauretania," in *The Cunard Turbine-Driven Quadruple-Screw Atlantic Liner "Mauretania"* (Wellingborough, UK: Patrick Stephens, 1987), p. vii.
2. John Maxtone-Graham, *The Only Way to Cross* (New York: Macmillan, 1978), p. 39.
3. Warren, ed., *The Cunard Turbine-Driven Quadruple-Screw Atlantic Liner "Mauretania,"* p. iii.
4. "Giant New Liner Gets Here in Fog," *New York Times,* November 23, 1907.
5. Alva Johnson, "The Mysterious Mr. Gibbs, Part I," *Saturday Evening Post,* January 20, 1945, p. 10.
6. Deutschland (1900), http://www.passagierdampfer.de/Schiffe/Liner/Deutschland_1900_/deutschland_1900_.htmla, accessed August 14, 2008.
7. Warren, ed., *The Cunard Turbine-Driven Quadruple-Screw Atlantic Liner "Mauretania,"* p. vii.
8. Mark D. Warren, ed., *The Cunard Turbine-Driven Quadruple-Screw Atlantic Liner "Lusitania"* (Wellingborough, UK: Patrick Stephens, 1987), p. 15.
9. Warren, ed., *The Cunard Turbine-Driven Quadruple-Screw Atlantic Liner "Mauretania,"* pp. ii, 31.
10. Ibid., pp. 28–29.
11. Charles Dickens, *American Notes* (New York: Fawcett, 1961), pp. 15–47, as quoted in Stephen Fox, *Transatlantic* (New York: HarperCollins, 2003), p. 96.
12. "Giant New Liner Gets Here in Fog," *New York Times,* November 23, 1907.
13. Ibid.
14. Ibid.

15. Winthrop Sergeant, "Profiles: The Best I Know How," *New Yorker,* June 6, 1964, p. 58.
16. "Giant New Liner Gets Here in Fog," *New York Times,* November 23, 1907.
17. Ibid.
18. Franklin Delano Roosevelt, "The Queen with a Fighting Heart," 1936, published in *Sea Breezes,* 1950, as quoted in Warren, ed., "The Trials of the Mauretania," in *The Cunard Turbine-Driven Quadruple-Screw Atlantic Liner "Mauretania,"* p. v.
19. "Giant New Liner Gets Here in Fog," *New York Times,* November 23, 1907.
20. Transcript of William Francis Gibbs, Harvard University Archives, UAIII 15.75.12.
21. "18th Street Entrance to Gibbs Mansion Where Boy Was Killed," Walnut Street, Volume III, Perkins Collection, compiled by Helen C. Perkins, Pennsylvania Historical Society, Philadelphia, from Stuart Wells, "The Residence at 1733 Walnut Street," HSTVP 600 Documentation and Archival Research, Dr. Roger Moss, December 12, 1986, Collection of the Philadelphia Athenaeum, HR 86.4, p. 11.
22. "Gibbs Mansion May Go Under the Hammer," February 19, 1911, publication unknown, Perkins Collection, Walnut Street, Walnut Street Volume III from West of Broad, Historical Society of Pennsylvania, Philadelphia, from Wells, "The Residence at 1733 Walnut Street."
23. Sergeant, "Profiles: The Best I Know How," p. 73.
24. Philadelphia Deed Book ELT 324, p. 28, Historical Society of Pennsylvania, Philadelphia, from Wells, "The Residence at 1733 Walnut Street," p. 12.
25. "Mr. Gibbs' Fountain: The Financier Startled by a Bill from the Water Bureau," *Philadelphia Inquirer,* April 3, 1900.
26. Sergeant, "Profiles: The Best I Know How," p. 73.
27. Frank O. Braynard, *By Their Works Ye Shall Know Them* (New York: Gibbs & Cox, 1968), p. 197.

4. J. P. MORGAN'S *TITANIC*

1. "Vincent Astor Dies In His Home at 67," *New York Times,* February 4, 1959.
2. Ibid.
3. Jack Alexander, "Profiles: The Golden Spoon II," *New Yorker,* March 12, 1938, p. 23.
4. Ibid., p. 26.
5. "Hamburg-American in Shipping Rate War," *New York Times,* August 20, 1904.

6. As quoted in Wyn Craig Wade, *Titanic: Death of a Dream* (New York: Rawson, Wade, 1979), p. 32.
7. Ibid., pp. 103–4.
8. Ibid., p. 112.
9. John P. Eaton and Charles A. Haas, *Titanic: Triumph and Tragedy* (New York: Norton, 1995), p. 206.
10. Wade, *Titanic,* p. 37.
11. Speech of Senator William Alden Smith, May 28, 1912, *United States Inquiry into the Titanic Disaster,* http://www.titanicinquiry.org/USInq/USReport/AmInqRepSmith01.php, accessed May 14, 2008.
12. "Titanic's Engines Were At Full Speed," *New York Times,* May 26, 1912.
13. Speech of Senator William Alden Smith, May 28, 1912.
14. Wade, *Titanic,* p. 293.
15. Ibid., pp. 304–5.
16. *Saturday Review,* as quoted in Wade, *Titanic,* p. 294.
17. Murken, *Linienreederei-Verbande,* p. 240, as quoted in Lamar Cecil, *Albert Ballin: Business and Politics in Imperial Germany 1888–1918* (Princeton, NJ: Princeton University Press, 1967), p. 58.
18. "Colonel John Jacob Astor IV," *Encyclopedia Titanica,* http://www.encyclopedia-titanica.org/titanic-biography/john-jacob-astor.html, accessed July 1, 2008.
19. Jack Alexander, "Profiles: The Golden Spoon II," *New Yorker,* March 12, 1938, p. 26.

5. PIPE DREAMERS AT WORK

1. Frank O. Braynard, *By Their Works Ye Shall Know Them: The Life and Ships of William Francis Gibbs 1886–1967* (New York: Gibbs & Cox, 1968), pp. 10–11.
2. "W. W. Gibbs Seriously Ill but Improving," *Philadelphia Inquirer,* July 25, 1911.
3. "Home of W. W. Gibbs Damaged by Fire," *Philadelphia Inquirer,* September 15, 1913.
4. Lamar Cecil, *Albert Ballin: Business and Politics in Imperial Germany 1888–1918* (Princeton, NJ: Princeton University Press, 1967), p. 51.
5. Interview with Vice Admiral Harry Manning (USMS), November 30, 1969, as quoted in Frank O. Braynard, *The World's Greatest Ship: The Story of the Leviathan,* vol. 1 (New York: South Street Seaport Museum, 1972), p. 184.
6. Ibid., pp. 85–86.

7. Sir Edward Grenfell to J. P. Morgan Jr., September 16, 1914, International Mercantile Marine Folder, J. Pierpont Morgan Library, New York.
8. *Report of the International Mercantile Marine Company for the Fiscal Year Ended December 31, 1916,* Proof No. 4 of July 16, 1917, 9104A-EB-1, International Mercantile Marine Folder, the J. Pierpont Morgan Library, New York.
9. Testimony of Edward Wilding, examined by Mr. Rowatt, Titanic Inquiry Day 19, pp. 20223–24, 20227, http://www.titanicinquiry.com/BOTInq/BOTInq19Wilding01.php, accessed February 28, 2008.
10. Andrew Gibson and Arthur Donovan, *The Abandoned Ocean: A History of United States Maritime Policy* (Columbia: University of South Carolina Press, 2000), pp. 109–10.
11. Richard Austin Smith, "The Love Affair of William Francis Gibbs," *Fortune,* August 1957, p. 158.
12. William L. R. Emmet, "Propelling Machinery for Collier Jupiter," *International Marine Engineering* 17, no. 8 (1913), p. 324.
13. Braynard, *By Their Works Ye Shall Know Them,* p. 11.
14. Frank O. Braynard and Robert Hudson Westover, *S.S. United States: Fastest Ship in the World* (Paducah, KY: Turner, 2002), p. 17.
15. Ibid., p. 17.
16. William Hovgaard, "Biographical Memoir of David Watson Taylor, 1864–1940," *National Academy of Sciences of the United States of America Biographical Memoirs,* vol. 22, 7th memoir, presented to the academy at the annual meeting, 1941, p. 135.
17. Waldemar Kaempffert, "The Race for Ocean Supremacy: What Is the Limit?" *New York Times,* August 18, 1929.
18. Braynard and Westover, *S.S. United States,* p. 16.
19. Hovgaard, "Biographical Memoir of David Watson Taylor, 1864–1940," p. 139.
20. Model and Notes, "1000-Foot Superliner, 'Project S-171,' Proposed American Passenger Ship," 1916 (?), accession 1971.0074.000001, collection of the Mariners' Museum, Newport News, Virginia.
21. Ibid.
22. Alva Johnson, "The Mysterious Mr. Gibbs—II," *Saturday Evening Post,* January 27, 1945, p. 97.
23. "Intruder Has Dynamite," *New York Times,* July 4, 1915.
24. J. P. Morgan Jr. to Owen Wister, August 18, 1915, Letter Press Book 17, p. 94, International Mercantile Marine Folder, J. Pierpont Morgan Library, New York.
25. Braynard, *By Their Works Ye Shall Know Them,* p. 15.
26. Ibid.

6. PRIZES OF WAR

1. Frank O. Braynard, *Leviathan,* vol. 1 (New York: South Street Seaport Museum, 1972), p. 113.
2. Edward Hungerford, *Saturday Evening Post,* November 22, 1919, as quoted in Frank O. Braynard, *The World's Greatest Ship: The Story of the Leviathan,* vol. 2 (New York: Fort Schuyler Press, 1976), p. 15.
3. Commander E. P. Jessop, "Repairing German Vandalism on Interned Vessels by Electric Welding," *Journal of the American Society of Naval Engineers,* March 18, 1918, p. 124, as quoted in Frank O. Braynard, *The World's Greatest Ship: The Story of the Leviathan,* vol. 1 (New York: South Street Seaport Museum, 1972), p. 120.
4. *The North American* (Philadelphia newspaper), July 5, 1923, as quoted in Braynard, *The World's Greatest Ship,* vol. 1, p. 121.
5. Herbert Hartley, *Home Is the Sailor* (Birmingham, AL: Vulcan Press, 1955), pp. 78–70, as quoted in Braynard, *The World's Greatest Ship,* vol. 1, p. 125.
6. General John M. Franklin, *Recollections of My Life* (Baltimore: Reese Press, 1973), p. 5, courtesy of Laura Franklin Dunn.
7. Ibid., p. 10.
8. Ibid., p. 22.
9. Ibid., pp. 11–19.
10. Ibid., p. 29.
11. Braynard, *The World's Greatest Ship,* vol. 1, p. 212.
12. Ballin to Ernst Francke, April 12, 1907, as quoted in Lamar Cecil, *Albert Ballin: Business and Politics in Imperial Germany 1888–1918* (Princeton, NJ: Princeton University Press, 1967), p. 32.
13. As quoted in Douglas R. Burgess Jr., *Seize the Trident: The Race for Superliner Supremacy and How It Altered the Great War* (New York: McGraw-Hill Professional, 2005), p. 265.

7. A GIANT LIVES AGAIN

1. Frank O. Braynard, *The World's Greatest Ship: The Story of the Leviathan,* vol. 2 (New York: Fort Schuyler Press, 1976), p. 18.
2. *New York Herald,* March 20, 1921, as quoted in Braynard, *The World's Greatest Ship,* vol. 2, p. 19.
3. *Shipping Board Operations, Hearings Before Select Committee on U.S. Shipping Board Operations, House of Representatives, Sixty-Sixth Congress, Second Session, Reconditioning of U.S.S. Leviathan, Part 4* (Washington, DC:

Government Printing Office, 1920), pp. 1346–47, National Archives, Washington, DC.

4. Ibid., pp. 1349–50.
5. Braynard, *The World's Greatest Ship,* vol. 2, p. 11.
6. November 13, 1919, memo from J. J. Flaherty, Secretary, U.S. Shipping Board, to Major Cushing, as quoted in Braynard, *The World's Greatest Ship,* vol. 2, p. 11.
7. U.S. Shipping Board, *Fourth Annual Report,* p. 129, in Braynard, *The World's Greatest Ship,* vol. 2, p. 18.
8. *New York American,* February 14, 1920, as quoted in Braynard, *The World's Greatest Ship,* vol. 2, p. 24.
9. October 4, 1921 statement by P. A. S. Franklin, in *Marine Journal,* November 12, 1912, as quoted in Braynard, *The World's Greatest Ship,* vol. 2, p. 25.
10. U.S. Senate, Document Number 231, 66th Congress, 2nd Session, as quoted in Braynard, *The World's Greatest Ship,* vol. 2, p. 28.
11. *New York Evening Sun,* September 10, 1920, as quoted in Braynard, *The World's Greatest Ship,* vol. 2, p. 61.
12. W. F. Gibbs to R. D. Gatewood, USSB, September 17, 1920, as quoted in Braynard, *The World's Greatest Ship,* vol. 2, p. 62.
13. John A. Morello, *Selling the President, 1920: Albert D. Lasker, Advertising, and the Election of Warren G. Harding* (Westport, CT: Praeger, 2001), pp. 1–2.
14. Ibid.
15. "Lasker Says Loss on Wartime Fleet Was Four Billions," *New York Times,* July 17, 1921.
16. Ibid.
17. Braynard, *The World's Greatest Ship,* vol. 2, p. 90.
18. William Francis Gibbs, plans and construction specifications for the reconditioning of the S.S. Leviathan, as quoted in Braynard, *The World's Greatest Ship,* vol. 2, p. 46.
19. Ibid.
20. Braynard, *The World's Greatest Ship,* vol. 2, p. 106.
21. Ibid., p. 112.
22. P. A. S. Franklin, as quoted in the *New York Journal of Commerce,* February 16, 1922, as quoted in Braynard, *The World's Greatest Ship,* vol. 2, p. 113.
23. Unidentified newspaper from Gibbs file, April 8, 1922, as quoted in Braynard, *The World's Greatest Ship,* vol. 2, p. 118.
24. Braynard, *The World's Greatest Ship,* vol. 2, p. 18.
25. "The Leviathan's Coming Trip," *New York Times,* March 26, 1922.

26. *New York Journal of Commerce,* April 11, 1922, as quoted in Braynard, *The World's Greatest Ship,* vol. 2, p. 121.
27. Braynard, *The World's Greatest Ship,* vol. 2, p. 128.
28. *Leviathan* construction specifications prepared by William Francis Gibbs, December 1921, collection of Frank O. Braynard, as quoted in Braynard, *The World's Greatest Ship,* vol. 2, p. 50.
29. *New York Journal of Commerce,* April 13, 1923, as quoted in Frank O. Braynard, *The World's Greatest Ship: The Story of the Leviathan,* vol. 4 (Newport News, VA: Mariners' Museum, 1978), p. 179.
30. "Southampton Keen to Greet Leviathan," *New York Times,* July 10, 1923.
31. W. F. Gibbs, as quoted in *New York Journal of Commerce,* February 16, 1922, as quoted in Braynard, *The World's Greatest Ship,* vol. 2, p. 111.
32. *Shipping Board Operations, Hearings Before Select Committee on U.S. Shipping Board Operations, House of Representatives, Sixty-Sixth Congress, Second Session, Reconditioning of U.S.S. Leviathan, Part 4,* p. 1366.
33. *Marine News,* May 1923, as quoted in Braynard, *The World's Greatest Ship,* vol. 2, pp. 216–17.
34. Frank O. Braynard and Robert Hudson Westover, *S.S. United States: Fastest Ship in the World* (Paducah, KY: Turner, 2002), p. 25.
35. *New York Journal of Commerce,* September 7, 1922, as quoted in Braynard, *The World's Greatest Ship,* vol. 2, p. 132.
36. P. A. S. Franklin to William Francis Gibbs, September 19, 1922, as quoted in Braynard, *The World's Greatest Ship,* vol. 2, p. 133.
37. William Francis Gibbs to P. A. S. Franklin, September 22, 1922, as quoted in Braynard, *The World's Greatest Ship,* vol. 2, p. 133.
38. William Francis Gibbs, "America's Ability To Do," *Management Engineering* 4, no. 5 (May 1, 1923), as quoted in Braynard, *The World's Greatest Ship,* vol. 2, pp. 212–13.
39. Winthrop Sergeant, "The Best I Know How," *New Yorker,* June 6, 1964, pp. 52–53.

8. THE PARVENU

1. Herbert Hartley, *Home Is the Sailor* (Birmingham, AL: Vulcan Press, 1955), p. 83.
2. Frank O. Braynard, *The World's Greatest Ship: The Story of the Leviathan,* vol. 2 (New York: Fort Schuyler Press, 1976), p. 271.
3. *New York Sun and Globe,* June 25, 1923, as quoted in Braynard, *The World's Greatest Ship,* vol. 2, p. 274.

4. "Leviathan Breaks World Speed Record in 25-Hour Spurt," *New York Times,* June 24, 1923.
5. "The S.S. *Leviathan*—Views and Information on the Ritz-Carlton Room, Library and Winter Garden," Gjenvick-Gjønvik Archives, http://www.gjenvick.com/UnitedStatesLines/1923-TheSteamshipLeviathan-Section-2.html, accessed October 19, 2009.
6. *New York Times* and *New York Tribune,* May 6, 1923, as quoted in Braynard, *The World's Greatest Ship,* vol. 2, p. 215.
7. "Leviathan Sails with 1,700 Aboard as 10,000 Cheer," *New York Times,* July 5, 1923.
8. Oran McCormack, *The World as Oran McCormack Saw It* (Boston, MA: printed by author, 1939), pp. 77–78, as quoted in Braynard, *The World's Greatest Ship,* vol. 2, pp. 310–11.
9. "Cheer Leviathan at Southampton," *New York Times,* July 11, 1923.
10. President Warren G. Harding, as quoted in "Leviathan Twice Passes the France," *New York Times,* July 6, 1923.
11. "Cheer Leviathan at Southampton," *New York Times,* July 11, 1923.
12. "Leviathan Guests Marvel at Change," *New York Times,* July 13, 1923.
13. William Francis Gibbs, as quoted in *New York Herald Tribune,* n.d., probably July 12, 1923, as quoted in Frank O. Braynard, *By Their Works Ye Shall Know Them: The Life and Ships of William Francis Gibbs 1886–1967* (New York: Gibbs & Cox, 1968), pp. 189–90.
14. "U.S. at War: Technological Revolutionist," *Time,* September 28, 1942, http://www.time.com/time/magazine/article/0,9171,773637-5,00.html, accessed August 4, 2010.
15. William Francis Gibbs, as quoted in *New York Herald Tribune,* n.d., probably July 12, 1923, as quoted in Braynard, *By Their Works Ye Shall Know Them,* pp. 189–90.
16. *Proposed Sale of Certain Ships by the United States Shipping Board, Hearings Before a Subcommittee of the Committee of Commerce United States Senate, Sixty-Ninth Congress, Second Session Pursuant to S. Res.294 Requesting the Shipping Board to Postpone the Consummation of the Sale or Charter of the "Leviathan" and Certain Other Vessels Operated by the Board, Part* 8 (Washington, DC: Government Printing Office, 1927), p. 251.
17. White Star Advertisement, ca. 1925, as quoted in John Maxtone-Graham, *The Only Way to Cross* (New York: Collier Books, 1978), p. 169.
18. "Leviathan Awakens American Interest in Shipping," *Marine Engineering and Shipping Age* 28, no. 8 (August 1923), p. 455.
19. Frank O. Braynard, *The World's Greatest Ship: The Story of the Leviathan,* vol. 4 (Newport News, VA: Mariners' Museum, 1978), p. 364.

20. Edward Ellsberg, "Structural Damage Sustained by S.S. Majestic," *International Marine Engineering* 29, no. 8 (August 1925), p. 430.
21. Interview of Norman Zippler by Frank O. Braynard, November 26, 1969, as quoted in Braynard, *The World's Greatest Ship,* vol. 4, p. 364.
22. George Horne, "Skipper of the Blue Ribbon Liner," *New York Times,* July 13, 1952.
23. Florence Manning to Steven Ujifusa, November 14, 2010.
24. "Invasion, 1952," *Time,* June 23, 1952, http://www.time.com/time/magazine/article/0,9171,859829-6,00.html, accessed December 12, 2008.
25. Ibid.
26. Ibid.
27. Ibid.
28. Ibid.
29. Interview between Frank O. Braynard and Captain J. Linder, December 26, 1969, as quoted in Frank O. Braynard, *The World's Greatest Ship: The Story of the Leviathan,* vol. 3 (New York: Fort Schuyler Press, 1976), p. 161.
30. Interview of Harry Manning by Frank O. Braynard, November 22, 1969, as quoted in Braynard, *The World's Greatest Ship,* vol. 4, p. 134.
31. William Francis Gibbs, untitled article, *Nautical Gazette,* early 1927, as quoted in Braynard, *By Their Works Ye Shall Know Them,* p. 47.
32. "Invents Aerial Torpedo with Uncanny Precision," *Philadelphia Inquirer,* January 22, 1917.
33. *Social Register, Philadelphia 1924,* vol. 38, no. 3 (New York: Social Register Association, 1923), p. 102; *Social Register, Philadelphia 1926,* vol. 40, no. 3 (New York: Social Register Association, 1925), p. 106.
34. Letters of Administration for the Estate of W. W. Gibbs, Register of Wills, City of Philadelphia, Number 465, Book 27, p. 113.
35. Winthrop Sergeant, "Profiles: The Best I Know How," *New Yorker,* June 6, 1964, p. 56.

9. MRS. WILLIAM FRANCIS GIBBS

1. "Dinner Guests Hear Mrs. Larkin is Wed," *New York Times,* January 7, 1927.
2. Olga Samaroff Stokowski, *An American Musician's Story* (New York: Read Books, 1939), p. 259.
3. "Profiles: Public Man," *New Yorker,* January 2, 1932, p. 21.
4. Richard Austin Smith, "The Love Affair of William Francis Gibbs," *Fortune,* August 1957, p. 154.

5. "Dinner Guests Hear Mrs. Larkin is Wed," *New York Times,* January 7, 1927.
6. Ibid.
7. "Notes of Social Activities in New York and Elsewhere," *New York Times,* August 14, 1929.
8. "U.S. at War: Technological Revolutionist," *Time,* September 28, 1942, http://www.time.com/time/magazine/article/0,9171,773637-5,00.html, accessed August 4, 2010.
9. Winthrop Sergeant, "Profiles: The Best I Know How," *New Yorker,* June 6, 1964, p. 62.
10. Smith, "The Love Affair of William Francis Gibbs," p. 158.
11. Vera Cravath Gibbs to Carl Van Vechten, November 14, 1939, Carl Van Vechten Correspondence, JWJ Carl Van Vechten, Beinecke Library, Yale University, New Haven, CT.
12. Sergeant, "Profiles: The Best I Know How," p. 64.
13. Christmas card from Francis C. Gibbs to William Francis Gibbs (n.d.), William Francis Gibbs Collection, Mariners' Museum, MS 179 Box 16, Newport News, VA.
14. Paula Gibbs, "Dogs I Have Known and the People They Have Loved," *Wiscasset Newspaper,* January 15, 2004.
15. Interview with Lawrence W. Ward, September 21, 2010.
16. Robert T. Swaine, *The Cravath Firm and Its Predecessors, 1819–1948,* vol. 2 (privately printed, 1948), p. 4, as quoted in Kai Bird, *The Chairman: John J. McCloy, The Making of the American Establishment* (New York: Simon & Schuster, 1992), p. 62.
17. Sergeant, "Profiles: The Best I Know How," p. 64.
18. Interview by Durward Primrose, as quoted in Frank O. Braynard, *By Their Works Ye Shall Know Them: The Life and Ships of William Francis Gibbs, 1886–1967* (New York: Gibbs & Cox, 1968), p. 36.

10. *MALOLO*

1. Pacific Marine Review, as quoted by Theodore E. Ferris, in *Hearings Before the Commerce Committee, United States Senate, Sixty-Fifth Congress, Second Session, on S. Res.170, Directing the Committee on Commerce to Investigate All Matters Connected with the Building of Merchant Vessels Under the Direction of the United States Shipping Board Emergency Fleet Corporation and Report Its Findings to the Senate, Together with Its Recommendations Thereon* (Washington, DC: Government Printing Office, 1918), p. 2311.
2. William Hovgaard, "Biographical Memoir of David Watson Taylor, 1864–1940," *National Academy of Sciences of the United States of America*

Biographical Memoirs, vol. 22, 7th memoir, presented to the academy at the annual meeting, 1941, p. 143.

3. Richard Austin Smith, "The Love Affair of William Francis Gibbs," *Fortune,* August 1957, p. 139.
4. "Malolo Docks Here with Gash in Hull," *New York Times,* May 29, 1927.
5. Ibid.
6. Frederic Gibbs, as quoted in Frank O. Braynard, *By Their Works Ye Shall Know Them: The Life and Ships of William Francis Gibbs, 1886–1967* (New York: Gibbs & Cox, 1968), p. 39.
7. "Malolo Docks Here with Gash in Hull," *New York Times,* May 29, 1927.
8. William Francis Gibbs, "Malolo Collision Vindicates Safety Measures," *Marine Engineering and Shipping Age,* July 1927, p. 400.
9. Ibid., pp. 499–500.

11. A GERMAN SEA MONSTER

1. John Malcolm Brinnin, *The Sway of the Grand Saloon: A Social History of the North Atlantic* (New York: Delacorte Press, 1971), pp. 441, 471.
2. William H. Miller Jr., *The Great Luxury Liners, 1927–1954: A Photographic Record* (Mineola, NY: Dover, 1981), p. 10.
3. Brinnin, *The Sway of the Grand Saloon,* p. 446.
4. Commodore Herbert Hartley, *Home Is the Sailor* (Birmingham, AL: Vulcan Press, 1955), pp. 147–48.
5. Jean dal Piaz, as quoted in John Maxtone-Graham, *Normandie* (New York: Norton, 2007), p. 34.
6. *Proposed Sale of Certain Ships by the United States Shipping Board, Hearings Before a Subcommittee of the Committee of Commerce United States Senate, Sixty-Ninth Congress, Second Session Pursuant to S. Res.294 Requesting the Shipping Board to Postpone the Consummation of the Sale or Charter of the "Leviathan" and Certain Other Vessels Operated by the Board, Part 8,* January 4, 1927 (Washington, DC: Government Printing Office, 1927), pp. 248–50.
7. "Full Board Votes Ships to Chapman," *New York Times,* February 15, 1929.
8. "Proposed Ship Sale Hit by Rival Bidder," *New York Times,* February 14, 1929.
9. Ibid.
10. *New York Times,* February 15, 1929, as quoted in Frank O. Braynard, *The World's Greatest Ship: The Story of the Leviathan,* vol. 4 (Newport News, VA: Mariners' Museum, 1978), p. 256.
11. Pacific Marine Review, as quoted by Theodore E. Ferris, in *Hearings Before the Commerce Committee, United States Senate, Sixty-Fifth Congress, Second*

Session, on S. Res.170, Directing the Committee on Commerce to Investigate All Matters Connected with the Building of Merchant Vessels Under the Direction of the United States Shipping Board Emergency Fleet Corporation and Report Its Findings to the Senate, Together with Its Recommendations Thereon (Washington, DC: Government Printing Office, 1918), p. 2311.

12. "Morro Castle: Turbo-Electric Ward Liner," *Marine Engineering and Shipping Age,* September 1930, p. 474.
13. Ibid., p. 486; *Investigation of the Burning of the Steamer Morro Castle: Copy of Testimony, Report, and Recommendations of Dickerson N. Hoover, Assistant Director, Made to the Secretary of Commerce,* Department of Commerce, Bureau of Navigation and Steamboat Inspection, New York, September 10–28, 1934, p. 11.
14. *Investigation of the Burning of the Steamer Morro Castle,* p. 10.
15. Wythe Williams, "Giant Bremen Sails on Maiden Voyage," *New York Times,* July 17, 1929.
16. Waldemar Kaempffert, "The Race for Ocean Supremacy: What Is the Limit?" *New York Times,* August 18, 1929.
17. Dr. Ernst Foerster, "Speed and Power of Ships: The Effect of Recent Progress in Shipbuilding upon Speed, Power, and Economy in Propulsion," *Marine Engineering and Shipping World,* May 1930, pp. 257–58.
18. "Talk of the Town: Toot! Toot! Toot!" *New Yorker,* January 19, 1935, p. 11.
19. Brinnin, *The Sway of the Grand Saloon,* p. 454.
20. Ibid.
21. Ibid.
22. "The Blue Riband of the North Atlantic, Westbound," http://www.greatships.net/riband.html, accessed November 28, 2008.
23. Kaempffert, "The Race for Ocean Supremacy."
24. Archibald Horka, as quoted in Frank O. Braynard, *The World's Greatest Ship: The Story of the Leviathan,* vol. 5 (Newport News, VA: The Mariners Museum, 1981), p. 233.
25. Cunard advertising brochure, ca. 1930, as quoted in Brinnin, *The Sway of the Grand Saloon,* p. 446.
26. Letter from Captain S. Reed to Frank O. Braynard, October 16, 1969, as quoted in Braynard, *The World's Greatest Ship,* vol. 5, pp. 363–64.
27. Braynard, *The World's Greatest Ship,* vol. 5, pp. 8–9.
28. Interview of Norman Zippler by Frank O. Braynard, November 26, 1969, as quoted in Frank O. Braynard, *The World's Greatest Ship: The Story of the Leviathan, vol. 4* (Newport News, VA: Mariners' Museum, 1978), p. 364.
29. Kaempffert, "The Race for Ocean Supremacy."

30. Foerster, "Speed and Power of Ships," p. 259.
31. Ibid., p. 260.
32. Richard Austin Smith, "The Love Affair of William Francis Gibbs," *Fortune,* August 1957, p. 153.
33. Braynard, *By Their Works Ye Shall Know Them,* pp. 191, 195.
34. Theodore E. Ferris, "Design of American Super Liners," *Transactions of the Society of Naval Architects and Marine Engineers* 39 (1931), p. 327.
35. John Emery, "Theodore Ferris: Portrait of a Naval Architect," *PowerShips,* no. 275 (Fall 2010), p. 20.
36. Ferris, "Design of American Super Liners," p. 341.
37. Ibid., p. 348.
38. Ibid., p. 349.
39. John M. Franklin. *Recollections of My Life* (Baltimore: Reese Press, 1973), p. 38.
40. "Huge Ship Merger May Unite 12 Lines," *New York Times,* October 9, 1931.
41. John J. Miller, Acting Director of Finances, United States Maritime Commission, to Haskins & Sells, February 11, 1937, RG 178, Records of the U.S. Maritime Commission, File 130-29, Box 926, HM 2005, National Archives, College Park, MD.

12. DEATH BY FIRE

1. *Tragic Disaster Which Befell the American Liner "Morro Castle,"* British Pathé Limited newsreel, September 17, 1934, www.britishpathe.com, accessed September 19, 2008.
2. Ibid.
3. Brian Hicks, *When the Dancing Stopped: The Real Story of the Morro Castle and Its Deadly Wake* (New York: Free Press, 2006), p. 163.
4. *Morro Castle Aftermath,* British Pathé Limited newsreel, September 20, 1934, www.britishpathe.com, accessed September 19, 2008.
5. Hicks, *When the Dancing Stopped,* p. 89.
6. Harold B. Burton, *The Morro Castle: Tragedy at Sea* (New York: Viking Press, 1973), p. 151.
7. Hicks, *When the Dancing Stopped,* p. 103.
8. "Lack of Discipline in Emergency Contributed to Morro Castle Disaster," *Marine Engineering and Shipping Age,* October 1934, p. 386.
9. Franklin D. Mooney, as quoted in ibid., p. 387.
10. Ibid.

11. Hicks, *When the Dancing Stopped,* pp. 218, 283.
12. Ibid., p. 212.
13. "Saturnalia," Time, July 1, 1935, http://www.time.com/time/magazine/article/0,9171,770043,00.html, accessed September 1, 2008.
14. Hicks, *When the Dancing Stopped,* p. 209.
15. Theodore Ferris, as interviewed by Durward Primrose, editor of *Marine Journal,* as quoted in Burton, *The Morro Castle,* p. 180.

13. FDR'S NEW NAVY

1. "Good, Very Good!" *Time,* August 28, 1934, http://www.time.com/time/magazine/article/0,9171,930095,00.html, accessed October 7, 2008.
2. Ibid.
3. Gregory Norris, "The Race for the Blue Riband," presentation given at the annual meeting of the SS United States Conservancy, Long Beach, CA, May 3, 2008.
4. Frank O. Braynard, *The World's Greatest Ship: The Story of the Leviathan,* vol. 6 (King's Point, NY: American Merchant Marine Academy, n.d.), p. 109.
5. Harvey Ardman, "The Ship That Died of Carelessness," *American Heritage,* December 1983, http://www.americanheritage.com/articles/magazine/ah/1983/1/1983_1_60.shtml, accessed October 15, 2008.
6. Interview of Norman Zippler by Frank O. Braynard, November 26, 1969, cited in Braynard, *The World's Greatest Ship: The Story of the Leviathan,* vol. 3 (New York: Fort Schuyler Press, 1976), p. 106.
7. John Maxtone-Graham, *Normandie* (New York: Norton, 2006), p. 65.
8. *Investigation of the Progress of the War Effort: Hearings Before the Committee on Naval Affairs, House of Representatives, Seventy-Eighth Congress, Second Session, Pursuant to H.Res.30 A Resolution Authorizing and Directing an Investigation of the Progress of the War Effort (Gibbs & Cox, Incorporated), Volume 5, May 8 and 9th, 1944* (Washington, DC: U.S. Government Printing Office, 1944), p. 3976.
9. Alva Johnson, "The Mysterious Mr. Gibbs—II," *Saturday Evening Post,* January 27, 1945, p. 98.
10. Dorothy Marckwald, as quoted in Gordon R. Ghareeb, "A Woman's Touch: The Seagoing Interiors of Dorothy Marckwald," undated article, *The Ocean Press,* Steamship Historical Association of America—South California Chapter, http://home.pacbell.net/steamer/sshsa_socal.html, accessed August 11, 2009.
11. Frank O. Braynard, *By Their Works Ye Shall Know Them: The Life and Ships of William Francis Gibbs 1886–1967* (New York: Gibbs & Cox, 1968), p. 190.

12. Vice Admiral Harold G. Bowen, *Ship Machinery and Mossbacks: The Autobiography of a Naval Engineer* (Princeton, NJ: Princeton University Press, 1954), p. 60.
13. Ibid., pp. 32–33.
14. Ibid., p. 56.
15. Ibid., p. 58.
16. Ibid., p. 60.
17. *Investigation of the Progress of the War Effort,* p. 3973.
18. William Hovgaard, "Biographical Memoir of David Watson Taylor, 1864–1940," *National Academy of Sciences of the United States of America Biographical Memoirs,* vol. 22, 7th *memoir,* presented to the academy at the annual meeting, 1941, p. 137.
19. Ibid., p. 143.
20. Interview with William Garzke Jr., March 9, 2008.
21. *Investigation of the Progress of the War Effort,* pp. 3973–74.
22. Ibid., pp. 3977–77.
23. Frances Perkins, *The Roosevelt I Knew* (New York: Harper Colophon, 1946), p. 33.
24. Bowen, *Ship Machinery and Mossbacks,* p. 60.
25. Alva Johnson, "The Mysterious Mr. Gibbs—III," *Saturday Evening Post,* February 3, 1945, p. 84.
26. Bowen, *Ship Machinery and Mossbacks,* p. 69.
27. Braynard, *By Their Works Ye Shall Know Them,* p. 191.
28. George Home, "Designer Keeps New Superliner a Structural Mystery to the World," *New York Times,* May 13, 1949.
29. James H. Davidson to William Francis Gibbs, June 1936, as quoted in Braynard, *By Their Works Ye Shall Know Them,* p. 63.
30. Bowen, *Ship Machinery and Mossbacks,* p. 85.
31. Ibid., p. 102.
32. Frederic C. Lane, *Ships for Victory: A History of Shipbuilding under the U.S. Maritime Commission* (Baltimore: Johns Hopkins University Press, 2001), p. 99.
33. "Technological Revolutionist," *Time,* September 28, 1942, http://www.time.com/time/magazine/article/0,9171,773637,00.html, accessed December 21, 2008.
34. Ibid.
35. Ellen Hanlon to Vera Cravath Gibbs, September 21, 1967, William Francis Gibbs Collection, Mariners' Museum, Newport News, VA.
36. *Investigation of the Progress of the War Effort,* p. 3790.

37. Horne, "Designer Keeps New Superliner a Structural Mystery to the World."
38. Alva Johnson, "The Mysterious Mr. Gibbs—I," *Saturday Evening Post,* January 20, 1945, p. 10.
39. Richard Austin Smith, "The Love Affair of William Francis Gibbs," *Fortune,* August 1957, p. 148.
40. "Giant New Liner Gets Here in Fog," *New York Times,* November 23, 1907.

14. THE *QUEEN* AND THE *AMERICA*

1. "*The Queen Mary*: The Creative Years (1926–1936)," http://www.queenmary.com/1929-1936.aspx, accessed March 15, 2011.
2. *Architect Builder and News,* as quoted in John Malcolm Brinnin, *The Sway of the Grand Saloon: A Social History of the North Atlantic* (New York: Delacorte Press, 1971), p. 484.
3. "RMS Queen Mary," http://www.ocean-liners.com/ships/qm.asp, accessed October 23, 2008.
4. Howard Johnson, *The Cunard Story* (London: Whittet Books, 1987), p. 107.
5. Ibid.
6. John Maxtone-Graham, *Normandie* (New York: Norton, 2007), pp. 157, 163.
7. Robert Winter, as quoted in Johnson, *The Cunard Story,* p. 116.
8. Johnson, *The Cunard Story,* p. 116.
9. "Berlin Is Angered by Ship Riot Here," *New York Times,* July 28, 1935.
10. Frank O. Braynard, *The World's Greatest Ship: The Story of the Leviathan,* vol. 6 (King's Point, NY: American Merchant Marine Academy, n.d.) p. 94.
11. Franklin Delano Roosevelt, "The Queen with a Fighting Heart," 1936, published in *Sea Breezes,* 1950, as quoted in Mark D. Warren, ed., "The Trials of the Mauretania," in *The Cunard Turbine-Driven Quadruple-Screw Atlantic Liner "Mauretania"* (Wellingborough, UK: Patrick Stephens, 1987), p. v.
12. Drew Pearson, "The Washington Merry-Go-Round," *Dover (Ohio) Reporter,* June 24, 1935, as quoted in Braynard, *The World's Greatest Ship,* vol. 6, p. 112.
13. "Saturnalia," Time, July 1, 1935, http://www.time.com/time/magazine/article/0,9171,770043,00.html, accessed September 1, 2008.
14. United States Maritime Commission, *Economic Survey of the American Merchant Marine,* November 10, 1937, pp. 22–23.
15. *St. Louis Star & Times,* March 22, 1935, as quoted in Braynard, *The World's Greatest Ship,* vol. 6, p. 105.
16. Harold F. Norton and John F. Nicholas, "The United States Liner 'America,'" *Transactions, Volume 48, 1940, The Society of Naval Architects and Marine*

Engineers (New York: Society of Naval Architects and Engineers, 1941), p. 45, collection of William duBarry Thomas.

17. Secretary of the Navy Claude Swanson to Secretary of Commerce Daniel C. Roper, November 8, 1935, RG 178, Records of the U.S. Maritime Commission, File 502-2, Box 1770, HM 2005, National Archives, College Park, MD.
18. Frank Braynard and Robert Hudson Westover, *S.S. United States: Fastest Ship in the World* (Paducah, KY: Turner, 2002), p. 28.
19. Norton and Nicholas, "The United States Liner 'America,' " p. 27.
20. Ibid., p. 25.
21. *Proposed Plan for the Future Development of the United States Lines Company,* 1937, p. 5, RG 178, Records of the U.S. Maritime Commission, File 130–29, Box 927, HM 2005, National Archives, College Park, MD.
22. Interview of Clifford D. Mallory Jr., March 12, 1973, by Frank O. Braynard, and letter dated November 2, 1981, as quoted in Braynard, *The World's Greatest Ship,* vol. 6, p. 291.
23. Ibid.
24. John Merryman Franklin, *Recollections of My Life* (Baltimore: Reese Press, 1973), p. 44.
25. Elgen M. Long and Marie K. Long, *Amelia Earhart: The Mystery Solved* (New York: Simon & Schuster, 1999), p. 102.
26. "Technological Revolutionist," *Time,* September 28, 1942, http://www.time.com/time/magazine/article/0,9171,773637,00.html, accessed December 21, 2008.
27. "Politics and Pork Chops," *Time,* June 17, 1946, http://www.time.com/time/magazine/article/0,9171,793041-4,00.html, accessed September 24, 2008.
28. Ibid.
29. Ibid.
30. "Ethical Question," *Time,* November 20, 1939, http://www.time.com/time/magazine/article/0,9171,762795-1,00.html, accessed March 5, 2009.
31. Ibid.
32. Interview with Laura Franklin Dunn, June 16, 2008.
33. Richard Austin Smith, "The Love Affair of William Francis Gibbs," *Fortune,* August 1957, p. 153.

15. A FULL MEASURE OF TOIL

1. William H. Miller Jr., *The Great Luxury Liners, 1927–1954: A Photographic Record* (Mineola, NY: Dover, 1980), p. 103.
2. Peter Huchthausen, "Shadow Voyage—Escape of the German Liner S.S. *Bremen,*" Maritime Network, http://www.freewebs.com/tmnarticles/bremen.htm, accessed December 17, 2008.

3. Frank O. Braynard, *By Their Works Ye Shall Know Them: The Life and the Ships of William Francis Gibbs 1886–1967* (New York: Gibbs & Cox, 1968), p. 95.
4. Miller, *The Great Luxury Liners, 1927–1954*, p. 108.
5. "Invasion, 1952," *Time*, June 23, 1952, p. 86, http://www.time.com/time/magazine/article/0,9171,859829-4,00.html.
6. Harry Manning, "Ten Minutes," *U.S. Naval Institute Proceedings* 66, no. 453, (November 1940), p. 1591.
7. Ibid., p. 1592.
8. John Merryman Franklin, *Recollections of My Life* (Baltimore: Reese Press, 1973), p. 48.
9. Braynard, *By Their Works Ye Shall Know Them*, pp. 81–83.
10. Frederic C. Lane, *Ships for Victory: A History of Shipbuilding under the U.S. Maritime Commission* (Baltimore: Johns Hopkins University Press, 2001), p. 82.
11. Ibid., p. 93.
12. Braynard, *By Their Works Ye Shall Know Them*, pp. 86–87.
13. Lane, *Ships for Victory*, p. 79.
14. Ibid., pp. 98–99.
15. *House of Representatives, Independent Offices Appropriation Bill for 1943, Hearings, before the Subcommittee of the Committee of Appropriations, 77th Congress, 2nd Session, on the Independent Office Appropriations Bill for 1943*, December 9, 1941, p. 277, as quoted in Lane, *Ships for Victory*, p. 99.
16. William Francis Gibbs to Gibbs & Cox staff, December 8, 1941, as quoted in Braynard, *By Their Works Ye Shall Know Them*, p. 95.
17. John Maxtone-Graham, *Normandie* (New York: Norton, 2006), p. 217.
18. Ibid., pp. 221–23.
19. Braynard, *By Their Works Ye Shall Know Them*, p. 69.
20. Frank O. Braynard, *Picture History of the Normandie* (New York: Dover, 1987), p. 93.
21. Ibid.
22. Ibid., p. 97.
23. Ibid., p. 119.
24. Ibid., p. 123.
25. "Technological Revolutionist," *Time*, September 28, 1942, http://www.time.com/time/magazine/article/0,9171,773637,00.html, accessed December 21, 2008.
26. George Horne, "Interview of the Week: Meet William F. Gibbs, Designer of the Superliner 'United States,' to Whom One Lifetime Is Not Enough," *Senior Scholastic*, April 30, 1952.

27. Richard Austin Smith, "The Love Affair of William Francis Gibbs," *Fortune,* August 1957, p. 139.
28. Braynard, *By Their Works Ye Shall Know Them,* p. 194.
29. Smith, "The Love Affair of William Francis Gibbs," p. 138.
30. Ibid., p. 139.
31. Ibid.
32. Ibid., p. 158.
33. Braynard, *By Their Works Ye Shall Know Them,* p. 133.
34. Governor Charles Edison to President Franklin Roosevelt, December 31, 1941, 18–Misc. Naval Building File, Franklin D. Roosevelt Library, Hyde Park, NY.
35. Ibid.
36. President Franklin Roosevelt to Governor Charles Edison, January 7, 1942, 18-Misc. Naval Building File, Franklin D. Roosevelt Library, Hyde Park, NY.
37. Lane, *Ships for Victory,* p. 609.
38. Ibid.
39. George Horne, "Designer Keeps New Superliner a Structural Mystery to the World," *New York Times,* May 13, 1949.
40. Miller, *The Great Luxury Liners, 1927–1954,* p. 114.
41. "L.I. Properties in New Ownership," *New York Times,* March 11, 1950.
42. "Women, Children & Horses," *Time,* November 10, 1941, http://www.time.com/time/printout/0,8816,851440,00.html, accessed July 20, 2007.
43. "Technological Revolutionist," *Time,* September 28, 1942, http://www.time.com/time/magazine/article/0,9171,773637-1,00.html, accessed August 27, 2008.
44. "Cravath Bequests Exceed $2,000,000," *New York Times,* July 12, 1940.
45. John G. Sharp, *History of the Washington Navy Yard Civilian Workforce, 1799–1962* (Washington, DC: Naval District Washington, Washington Navy Yard, 2005), p. 72, http://www.history.navy.mil/books/sharp/WNY_History.pdf, accessed March 19, 2011.
46. "The Vinson Naval Plan," http://www.cvn70.navy.mil/vinson/vinson2.htm, accessed December 27, 2008.
47. "Baubles," *Time,* September 23, 1946, http://www.time.com/time/magazine/article/0,9171,777097,00.html, accessed December 27, 2008.
48. "Chronology, Harry S. Truman's Life and Presidency," Harry S. Truman Library and Museum, Independence, MO, http://www.trumanlibrary.org/truman-c.htm, December 27, 2008.
49. *Investigation of the Progress of the War Effort,* p. 3937.

50. Ibid., p. 3938.
51. Ibid., pp. 3940–41.
52. Ibid., p. 3955.
53. Ibid., p. 3971.
54. Ibid., p. 3974.
55. Ibid., p. 3978.
56. Ibid., p. 3978.
57. Ibid., p. 3995.
58. Ibid., p. 3997.
59. Ibid., pp. 3998–4000.
60. Ibid., p. 4027.
61. Ibid., p. 4036.
62. "New Suit Against Alkali," *Philadelphia Inquirer,* March 12, 1902.
63. Alva Johnson, "The Mysterious Mr. Gibbs—I," *Saturday Evening Post,* January 20, 1945, p. 39.
64. Address by William Francis Gibbs to Gibbs & Cox staff, June 1944, as quoted in Braynard, *By Their Works Ye Shall Know Them,* p. 118.

16. A VERY PLEASING APPEARANCE

1. Frank O. Braynard, *By Their Works Ye Shall Know Them: The Life and Ships of William Francis Gibbs* (New York: Gibbs & Cox, 1968), pp. 145–46.
2. Ibid., p. 107.
3. *Superliner Legislation: Hearings Before the Committee on Merchant Marine and Fisheries, House of Representatives, Eight-Fifth Congress, Second Session on H.R. 9342 to Authorize the Construction and Sale of a Superliner Passenger Vessel Equivalent to the Steamship United States, Identical and Similar Bills; H.R. 9432 to Authorize the Construction and Sale by the Federal Maritime Board of a Passenger Vessel for Operation in the Pacific Ocean, and Identical Bills* (Washington, DC: U.S. Government Printing Office, 1958), p. 158.
4. "Winged Victory Presentation Aboard the S.S. United States, July 1, 1952," *Compass Points,* published by Gibbs & Cox, n.d. (ca. July 10, 1952), pp. 7–8, collection of Susan Caccavale.
5. Johnson, "The Mysterious Mr. Gibbs—Part I," p. 10.
6. Winthrop Sergeant, "Profiles: The Best I Know How," *New Yorker,* June 6, 1964, p. 62.
7. Memorandum from William Francis Gibbs to Frederic H. Gibbs, December 11, 1944, 12201 SI-I (16-13400) Confidential, Subject: Design

12201—S.S. United States—Design Particulars and Information, January 24, 1950 Edition, Revised to June 27, 1950, March 19, 1951, and April 24, 1951, vol. 1 of 2, pp. 1–678-B, VM383 U5 G52, p. 4, Mariners' Museum Library, Newport News, VA.

8. William Francis Gibbs, "S.S. United States," *Journal of the Franklin Institute*, Philadelphia, December 1953, p. 549.
9. Interview of Thomas Buermann, Raymond Foster Options, Ltd., *The S.S. United States: From Dream to Reality,* produced by the Mariners' Museum, Newport News, VA, 1992.
10. Ibid.
11. John Merryman Franklin, *Recollections of My Life* (Baltimore: Reese Press, 1973), p. 49.
12. *Exhibit A: Agreement of Merger Between International Mercantile Marine Company, a New Jersey Corporation and United States Lines Company, a Nevada Corporation, Merging United States Lines Company into International Mercantile Marine Company, the New Jersey Corporation (To be known as the United States Lines Company, the Surviving Corporation), Proof No. 5, April 8, 1943* (New York: Bowne, 1943), p. 3, RG 178, Records of the United States Maritime Commission, 130-29, Exhibit 6 to 130-29-4 (PT-1), Box 928, HM 2005, National Archives, College Park, MD.
13. Basil Harris, "Preamble to Post War Shipping: Excerpt from an address by Basil Harris President of the United States Lines Company before the faculty of Rutgers University, New Brunswick, N.J.," December 13, 1944, unknown publication, File No. 130-29, Part 7, RG 178, Records of the United States Maritime Commission, 130-29 (PT-5) to 130-29, Exhibit 5, Box 927, HM 2005, National Archives, College Park, MD.
14. "President Urges Maritime Planning," *New York Times,* October 26, 1944.
15. Truman to Advisory Committee on the Merchant Marine, March 11, 1947, from *Report of the President's Advisory Committee of the Merchant Marine,* Washington, DC, November 1947, p. (d).
16. Basil Harris to Vice Admiral Emory S. Land, May 16, 1945, File No. 130-29, Part 7, RG 178, Records of the United States Maritime Commission, 130-29 (PT-5) to 130-29, Exhibit 5, Box 927, HM 2005, National Archives, College Park, MD.
17. Vice Admiral Emory S. Land to Basil Harris, May 30, 1945, File No. 130-29, Part 7, RG 178, Records of the United States Maritime Commission, 130-29 (PT-5) to 130-29, Exhibit 5, Box 927, HM 2005, National Archives, College Park, MD.
18. John R. Kane. "The Speed of the S.S. United States," *Marine Technology,* April 1978, p. 121.

19. Memorandum from William Francis Gibbs to Frederic H. Gibbs, June 26–27, 1951, 12201 SI-I (16-13400) Confidential, Subject: Design 12201—S.S. United States—Design Particulars and Information, January 24, 1950 Edition, Revised to June 27, 1950, March 19, 1951, and April 24, 1951, vol. 2 of 2, pp. 679–1214, VM383 U5 G52, p. 1232, Mariners' Museum Library, Newport News, VA.
20. Ibid.
21. Matthew G. Forrest, *Notes on the Design of the Transatlantic Liners, Comment on the S.S. United States,* Presented at the Meeting of SNA & ME, September 17, 1952, collection of Lawrence W. Ward.
22. John R. Kane. "The Speed of the SS United States," p. 122.
23. Henry Billings, *Superliner S.S. United States* (New York: Viking Press, 1952), p. 101.

17. PRIMACY ON THE SEAS

1. Memorandum from William Francis Gibbs to Frederic H. Gibbs, February 6, 1946, 12201 SI-I (16-13400) Confidential, Subject: Design 12201—S.S. United States—Design Particulars and Information, January 24, 1950 Edition, Revised to June 27, 1950, March 19, 1951, and April 24, 1951, vol. 1 of 2, pp. 1–678-B, VM383 U5 G52, p. 5, Mariners' Museum Library, Newport News, VA.
2. Ibid.
3. "Mr. Gibbs' Baby," *New Yorker,* November 16, 1957.
4. Memorandum from William Francis Gibbs to Frederic H. Gibbs, March 4, 1946, 12201 SI-I (16-13400) Confidential, Subject: Design 12201—S.S. United States—Design Particulars and Information, January 24, 1950 Edition, Revised to June 27, 1950, March 19, 1951, and April 24, 1951, vol. 1 of 2, pp. 1–678-B, VM383 U5 G52, p. 5, Mariners' Museum Library, Newport News, VA.
5. Memorandum from John M. Franklin to the U.S. Maritime Commission, March 26, 1946, 12201 SI-I (16-13400) Confidential, Subject: Design 12201—S.S. United States—Design Particulars and Information, January 24, 1950 Edition, Revised to June 27, 1950, March 19, 1951, and April 24, 1951, vol. 1 of 2, pp. 1–678-B, VM383 U5 G52, pp. 7–8, Mariners' Museum Library, Newport News, VA.
6. Memorandum from Henry C. E. Meyer to William Francis Gibbs, May 12, 1947, Subject: Design 12201—S.S. United States—Design Particulars and Information, January 24, 1950 Edition, Revised to June 27, 1950, March 19, 1951, and April 24, 1951, vol. 2 of 2, pp. 679–1214, VM383 U5 G52, p. 461, Mariners' Museum Library, Newport News, VA.

7. Memorandum from Henry C. E. Meyer to William Francis Gibbs, June 3, 1947, Subject: Design 12201—S.S. United States—Design Particulars and Information, January 24, 1950 Edition, Revised to June 27, 1950, March 19, 1951, and April 24, 1951, vol. 2 of 2, pp. 679–1214, VM383 U5 G52, p. 461, Mariners' Museum Library, Newport News, VA.
8. Memorandum from William Francis Gibbs to Frederic Gibbs, May 12, 1947, Subject: Design 12201—S.S. United States—Design Particulars and Information, January 24, 1950 Edition, Revised to June 27, 1950, March 19, 1951, and April 24, 1951, vol. 2 of 2, pp. 679–1214, VM383 U5 G52, p. 465, Mariners' Museum Library, Newport News, VA.
9. Memorandum from Admiral Frederick E. Haeberle to Admiral E. W. Mills, April 25, 1947, Subject: Design 12201—S.S. United States—Design Particulars and Information, January 24, 1950 Edition, Revised to June 27, 1950, March 19, 1951, and April 24, 1951, vol. 2 of 2, pp. 679–1214, VM383 U5 G52, p. 384, Mariners' Museum Library, Newport News, VA.
10. Truman to Advisory Committee on the Merchant Marine, March 11, 1947, from *Report of the President's Advisory Committee of the Merchant Marine,* Washington, DC, November 1947, p. (d).
11. "Tattered Ensign," *Time,* July 24, 1950, http://www.time.com/time/magazine/article/0,9171,812855,00.html#ixzz0mXIBNzDD, accessed May 10, 2010.
12. "American Queen," *New York Times,* November 16, 1946.
13. George Horne, "Skipper of the Blue-Ribbon Liner," *New York Times,* July 13, 1952.
14. Richard Austin Smith, "The Love Affair of William Francis Gibbs," *Fortune,* August 1957, p. 153.
15. William H. Miller Jr., *The Great Luxury Liners, 1927–1954: A Photographic Record* (Mineola, NY: Dover, 1981), p. 131.
16. "Ships Carried Major Portion of Trans-Atlantic Passengers in 1947," *Marine Age,* January 1948, as quoted in memorandum from William Francis Gibbs to Frederic H. Gibbs, 1948, 12201 SI-I (16-13400) Confidential, Subject: Design 12201—S.S. United States—Design Particulars and Information, January 24, 1950 Edition, Revised to June 27, 1950, March 19, 1951, and April 24, 1951, vol. 1 of 2, pp. 1–678-B, VM383 U5 G52, p. 25, Mariners' Museum Library, Newport News, VA.
17. Memorandum from William Francis Gibbs to Frederic H. Gibbs, ibid.
18. Memorandum from William Francis Gibbs to Frederic H. Gibbs, May 11, 1946, 12201 SI-I (16-13400) Confidential, Subject: Design 12201—S.S. United States—Design Particulars and Information, January 24, 1950 Edition, Revised to June 27, 1950, March 19, 1951, and April 24, 1951, vol. 1 of 2, pp. 1–678-B, VM383 U5 G52, p. 41, Mariners' Museum Library, Newport News, VA.

19. Winston Churchill, *The Sinews of Peace,* quoted in Mark A. Kishlansky, ed., *Sources of World History* (New York: HarperCollins, 1995), pp. 298–302, as quoted in *The History Guide: Lectures on Twentieth Century Europe,* http://www.historyguide.org/europe/churchill.html, accessed January 20, 2009.
20. George Kennan, "The Long Telegram," February 22, 1946, http://www.gwu.edu/~nsarchiv/coldwar/documents/episode-1/kennan.htm, accessed July 29, 2008.
21. *Report of the President's Advisory Committee of the Merchant Marine,* Washington, D.C., November 1947, p. 5.
22. *Report of the President's Advisory Committee of the Merchant Marine,* p. 12.
23. *Report of the President's Advisory Committee of the Merchant Marine,* p. 64.
24. Center for Defense Information, "U.S. Military Spending 1945–1996," http://www.cdi.org/issues/milspend.html, accessed November 15, 2009.
25. James V. Forrestal to William Francis Gibbs, approximately September 1, 1945, as quoted in Frank O. Braynard, *By Their Works Ye Shall Know Them,* p. 135.
26. "Methods of Financing US Lines' Ship Based on Newport News Bid, December 14, 1948," memorandum from William Francis Gibbs to Frederic H. Gibbs, 12201 SI-I (16-13400) Confidential, Subject: Design 12201—S.S. United States—Design Particulars and Information, January 24, 1950 Edition, Revised to June 27, 1950, March 19, 1951, and April 24, 1951, vol. 1 of 2, pp. 1–678-B, VM383 U5 G52, p. 104.
27. Frank O. Braynard, "Captions for the illustrations for United States book, #4," ca. 1980, Frank O. Braynard Collection, Sea Cliff, NY, courtesy of Noelle Hollander.
28. Memorandum from Attorney General James P. McGranery to President Harry Truman, December 4, 1952, p. 76, from *The Case of the United States Lines Company,* vol. 1, box 2, Robert L. Dennison Papers, Harry S. Truman Presidential Library, Independence, MO.
29. *Inquiry into the Operations of the Maritime Commission,* Fourth Intermediary Report of the Committee on Expenditures, 81st Congress, 1st Session, House Report No. 1423, October 13, 1949, p. 11, from *The Case of the United States Lines Company,* vol. 1, box 2, Robert L. Dennison Papers, Harry S. Truman Presidential Library, Independence, MO.
30. George Horne, "Liner United States Is Beautiful, Fast, Powerful, and Broke," *New York Times,* September 21, 1969.
31. "Ready to Ratify Superliner Pact," *New York Times,* April 4, 1949.
32. Frank O. Braynard and Robert Hudson Westover, *S.S. United States: Fastest Ship in the World* (Paducah, KY: Turner, 2002), p. 36.
33. William Francis Gibbs, "The SS *United States,*" delivered at the 1953 Medal Day Meeting, October 21, 1952, in acceptance of the Franklin Medal,

Journal of the Franklin Institute, ca. 1953, p. 549, Mariners' Museum, Newport News, VA.

34. William Francis Gibbs, "The Superliner," *Compass Points* no. 4 (April 1949), p. 8, collection of Susan Caccavale.
35. Smith, "The Love Affair of William Francis Gibbs," p. 160.
36. As quoted in Kevin D. McFarland and David L. Roll, *Louis Johnson and the Arming of America: The Roosevelt and Truman Years* (Bloomington: Indiana University Press, 2005), p. 178.
37. *Inquiry into the Operations of the Maritime Commission with Particular Reference to Allowances for National Defense Features and Construction-Differential Subsidies Under Title V of the Merchant Marine Act of 1936, as Amended, Based on the Special Report of the Comptroller, Fourth Intermediate Report of the Committee on Expenditures in the Executive Departments* (Washington, DC: U.S. Government Printing Office, 1949), p. 31, Robert L. Dennison Papers, vol. 1, Harry S. Truman Presidential Library, Independence, MO.
38. Ibid., p. 47.
39. Ibid., p. 11.
40. Ibid.
41. Ibid.
42. Memorandum from William Francis Gibbs to Frederic H. Gibbs, July 27, 1949, 12201 SI-I (16-13400) Confidential, Subject: Design 12201—S.S. United States—Design Particulars and Information, January 24, 1950 Edition, Revised to June 27, 1950, March 19, 1951, and April 24, 1951, vol. 1 of 2, p. 86, VM383 U5 G52, p. 78, Mariners' Museum Library, Newport News, VA.
43. Raymond Foster Options, Ltd., *The S.S. United States: From Dream to Reality* (Newport News, VA: Mariners' Museum, 1992).

18. A MIGHTY SWEET BABY RISES

1. Raymond Foster Options, Ltd., *The S.S. United States: From Dream to Reality* (Newport News, VA: Mariners' Museum, 1992).
2. C. B. Palmer, "The Building of the S.S. United States," *New York Times Magazine,* n.d. (ca. 1951), p. 15, Mariners' Museum, Newport News, VA.
3. "Design and Construction of the Superliner *United States,*" reprinted from the September 1952 issue of *Marine Engineering and Shipping* Review, p. 20, collection of Lawrence W. Ward.
4. Palmer, "The Building of the S.S. United States," p. 15.
5. Interview with Lawrence W. Ward, August 7, 2008.

6. Interview of Susan Caccavale by Steven Ujifusa, April 2, 2009.
7. Record of the Proceedings at Presentation of Medals by Mr. W. F. Gibbs to Members of the Staff at Gibbs & Cox, Inc., Whose Efforts Contributed to the Design and Construction of the S.S. United States, Making Possible a World Record-Breaking Maiden Voyage by the Ship, Ceremonies Held at Downtown Athletic Club, Monday, December 14, 1953, and in Gibbs & Cox 11th Floor Exhibition Room, 21 West Street, Monday, June 7, 1954, p. 14, collection of Susan Caccavale.
8. Interview of Susan Caccavale by Steven Ujifusa, April 2, 2009.
9. John R. Kane, "The Speed of the SS *United States,*" *Marine Technology* 15, no. 2 (April 1978), p. 123.
10. Memorandum from William Francis Gibbs to Frederic Gibbs, May 12, 1947, 12201 SI-I (16-13400) Confidential, Subject: Design 12201—S.S. United States—Design Particulars and Information, January 24, 1950 Edition, Revised to June 27, 1950, March 19, 1951, and April 24, 1951, vol. 1 of 2, p. 96, VM383 U5 G52, p. 410, Mariners' Museum Library, Newport News, VA.
11. Record of the Proceedings at Presentation of Medals, p. 14.
12. *Newport News Daily Press,* August 14, 1950, as quoted in Frank Braynard and Robert Hudson Westover, *S.S. United States: Fastest Ship in the World* (Paducah, KY: Turner, 2002), p. 40.
13. Braynard and Westover, *S.S. United States,* p. 43.
14. Walter Hamshar, "U.S. Will Pay Building Costs on Superliner," *New York Herald Tribune,* October 29, 1950.
15. Kevin D. McFarland and David L. Roll, *Louis Johnson and the Arming of America: The Roosevelt and Truman Years* (Bloomington: Indiana University Press, 2005), p. 343.
16. Braynard and Westover, *S.S. United States,* p. 43.
17. W. MacDonald, "Ships Lead Marine," publication unknown, ca. November 1950, Frank O. Braynard Collection, Sea Cliff, NY.
18. Interview of Tom Paris in Raymond Foster Options, Ltd. *The S.S. United States.*
19. "SS United States Scrapbook," *Richmond Times-Dispatch,* February 18, 1951, Mariners' Museum, Newport News, VA.
20. Interview of Tom Paris in Raymond Foster Options, Ltd. *The S.S. United States.*
21. Memorandum from William Francis Gibbs to Frederic H. Gibbs, December 14, 1950, 12201 SI-I (16-13400) Confidential, Subject: Design 12201—S.S. United States—Design Particulars and Information, January 24, 1950 Edition, Revised to June 27, 1950, March 19, 1951, and April 24, 1951, vol. 1 of 2, pp. 1–678-B, VM383 U5 G52, p. 1091, Mariners' Museum Library, Newport News, Virginia.

22. Bill Lee, "A Lesson for Mr. Gibbs: How the Famous Naval Architect got Up-Staged on his Uptakes," Newport News Apprentice School, Northrup Grumman Shipbuilding, Newport News, VA, November 2003, http://www.nnapprentice.com/alumni/letter/Apprentice_Trivia_120903.pdf, accessed March 26, 2009.
23. "World's Largest Stacks," *Shipyard Bulletin, S.S. United States Issue,* Newport News Shipbuilding and Dry Dock Company, vol. 14, no. 4 (May–June 1951), p. 12, collection of Susan Caccavale.
24. Interview of Tom Paris in Raymond Foster Options, Ltd. *The S.S. United States.*
25. Palmer, "The Building of the S.S. United States," p. 15.
26. *Richmond Times-Dispatch,* Sunday, February 18, 1951, and "News Release: Newport News Ship Building and Dry Dock Company, Virginia," June 21, 1952, S.S. United States Collection, Mariners' Museum, Newport News, VA.
27. Ibid.

19. AMERICAN MODERNE

1. Memorandum from William Francis Gibbs to Frederic H. Gibbs, April 30, 1951, 12201 SI-I (16-13400) Confidential, Subject: Design 12201—S.S. United States—Design Particulars and Information, January 24, 1950 Edition, Revised to June 27, 1950, March 19, 1951, and April 24, 1951, vol. 1 of 2, pp. 1–678-B, VM383 U5 G52, p. 8A-6, Mariners' Museum Library, Newport News, VA.
2. Diary of Vera Cravath Gibbs, as read by Susan Gibbs in *S.S. United States: Lady in Waiting,* directed by Robert Radler and produced by Mark B. Perry, Big Ship Films LLC, 2008.
3. Richard Austin Smith, "The Love Affair of William Francis Gibbs," *Fortune,* August 1957, p. 154.
4. Interview of Susan Caccavale by Steven Ujifusa, April 2, 2009.
5. *S.S. United States: Christening Ceremonies, Newport News, Virginia, June 23, 1951,* publication unknown, collection of Susan Caccavale.
6. "Remarks by Major General John M. Franklin, President, United States Lines Company, at the Launching of the S.S. United States, Newport News, Virginia, June 23, 1951," William Francis Gibbs Collection, Mariners' Museum, Newport News, VA.
7. "Connally, Thomas Terry (1877–1963)," *The Handbook of Texas Online,* http://www.tshaonline.org/handbook/online/articles/CC/fco36.html, accessed March 24, 2009.
8. Stanley Schill, "SS United States Christened Fleet Queen," *Newport News Daily Press,* June 24, 1951.

9. "Gibbs & Cox, Inc. Dinner Party, Fort Room—Chamberlin Hotel, Old Point Comfort, Virginia, Celebrating the Launching of the S.S. United States, Saturday, June 23, 1951, 6:00 P.M. E.S.T.," collection of Susan Caccavale.
10. Diary of Vera Cravath Gibbs, as read by Susan Gibbs in *S.S. United States: Lady in Waiting.*
11. Frank Braynard and Robert Hudson Westover, *S.S. United States: Fastest Ship in the World* (Paducah, KY: Turner, 2002), p. 34.
12. *Superliner Legislation: Hearings Before the Committee on Merchant Marine and Fisheries, House of Representatives, Eight-Fifth Congress, Second Session on H.R. 9342 to Authorize the Construction and Sale of a Superliner Passenger Vessel Equivalent to the Steamship United States, Identical and Similar Bills. H.R. 9432 To Authorize the Construction and Sale by the Federal Maritime Board of a Passenger Vessel for Operation in the Pacific Ocean, and Identical Bills* (Washington, DC: U.S. Government Printing Office, 1958), p. 153.
13. Dorothy Marckwald, as quoted in Gordon R. Ghareeb, "A Woman's Touch: The Seagoing Interiors of Dorothy Marckwald," undated article, *Ocean Press,* Steamship Historical Association of America—South California Chapter, http://home.pacbell.net/steamer/sshsa_socal.html, accessed August 11, 2009.
14. Ibid.
15. Promotional material for Walter Jones, March 8, 1951, Frank O. Brayard Collection, Sea Cliff, NY, courtesy of Noelle Hollander.
16. Memorandum from William Francis Gibbs to Frederic H. Gibbs, January 17, 1950. *12201 SI-I (16-13400) Confidential, Subject: Design 12201—S.S. United States—Design Particulars and Information, January 24, 1950 Edition, Revised to June 27, 1950, March 19, 1951, and April 24, 1951,* vol. 2 of 2, pp. 679–1214, VM383 U5 G52, p. 1028, Mariners' Museum Library, Newport News, VA.
17. Charles Anderson, as interviewed in *S.S. United States: Lady in Waiting.*
18. Interview of Thomas Buermann, in Raymond Foster Options, Ltd., *The S.S. United States: From Dream to Reality,* produced by the Mariners' Museum, Newport News, VA, 1992.
19. Mark Stebbins, *Flowering Trees of Florida* (Sarasota, FL: Pineapple Press, 1999), p. 80.
20. "Fireproofing All Furnishings for Liner Test of Ingenuity," *Newport News Daily Press,* February 13, 1952.
21. "S.S. United States Lounge Fittings Made Here, Ellison Workmen Proud of Job," *Jamestown Post-Journal,* April 9, 1952.
22. William H. Miller Jr. *S.S. United States: The Story of America's Greatest Ocean Liner* (New York: Norton, 1991), p. 86.

23. Sanka Knox, "Rigid Safety Rules Guided Liner's Art," *New York Times,* n.d., Mariner's Museum, Newport News, VA.
24. Passenger List and Shipboard Information: List of First Class Passengers from New York, Thursday, March 21, 1968 to Havre, Southampton and Bremerhaven, MS15.19515, United States Lines, Steamship Ephemera, Mariners' Museum, Newport News, VA.
25. "Fireproofing All Furnishings for Liner Test of Ingenuity," *Newport News Daily Press,* February 13, 1952.
26. "Theater Equipment Installed on New Liner United States," *Newport News Daily Press,* September 30, 1951, Mariners' Museum, Newport News, VA.
27. Granville Parkinson, "Report on the SS. United States," *Compass Points* 10, no. 6 (June 1952), p. 3.
28. "Fireproofing All Furnishings for Liner Test of Ingenuity," *Newport News Daily Press,* February 13, 1952.
29. Braynard and Westover, *S.S. United States,* p. 69.
30. Passenger Fares, No. 1 March 1952, MSIS 19738-MSIS 19738, United States Lines, Steamship Ephemera Collection, United States Lines, Mariner's Museum, Newport News, VA.
31. Knox, "Rigid Safety Rules Guided Liner's Art."
32. Ibid.
33. "A Woman's Touch: The Seagoing Interiors of Dorothy Marckwald," http://home.pacbell.net/steamer/marckwald.html, accessed August 30, 2007.
34. Parkinson, "Report on the SS. United States," pp. 2–3.
35. Memorandum from William Francis Gibbs to Frederic H. Gibbs, April 15, 1952, 12201 SI-I (16-13400) Confidential, Subject: Design 12201—S.S. United States—Design Particulars and Information, January 24, 1950 Edition, Revised to June 27, 1950, March 19, 1951, and April 24, 1951, vol. 1 of 2, pp. 1–678-B, VM383 U5 G52, pp. 81-C–81-C2, Mariners' Museum Library, Newport News, VA.

20. TURN HER UP

1. George Horne, "United States, 990 Foot Long Superliner, Will Start 3 Day Sea Trials Today," *New York Times,* May 14, 1952, Mariners' Museum, Newport News, VA.
2. Ibid.
3. Frank O. Braynard, "Captions for illustrations for *United States* book, #27," ca. 1980, Frank O. Braynard Collection, Sea Cliff, NY, courtesy of Noelle Hollander.

4. Walter Hamshar, "Aboard the Liner United States," *New York Herald Tribune,* May 14, 1952.
5. John McCarten, "Talk of the Town: Trial by Raydist," *New Yorker,* May 31, 1952, p. 18.
6. Photo-request No. 52-274/24, First Trial Run, from the Frank O. Braynard Collection, Sea Cliff, NY.
7. Hamshar, "Aboard the Liner United States."
8. McCarten, "Talk of the Town: Trial by Raydist," p. 18.
9. John R. Kane, "The Speed of the S.S. United States," *Marine Technology* 15, no. 2 (April 1978), pp. 119–43.
10. McCarten, "Talk of the Town: Trial by Raydist," p. 18.
11. Ibid., p. 19.
12. "A Woman's Touch: The Seagoing Interiors of Dorothy Marckwald," http://home.pacbell.net/steamer/marckwald.html, accessed August 30, 2007.
13. George Horne, "Superliner Ends Successful Test," *New York Times,* May 17, 1952.
14. Ibid.
15. George Horne, "Superliner Mishap Delays Speed Run," *New York Times,* May 16, 1952.
16. "A Woman's Touch: The Seagoing Interiors of Dorothy Marckwald."
17. Frank O. Braynard, *By Their Works Ye Shall Know Them: The Life and Ships of William Francis Gibbs 1886–1967* (New York: Gibbs & Cox, 1968), p. 192.
18. Geoffrey T. Hellman, "Talk of the Town: And the Big Ship," *New Yorker,* June 14, 1952, p. 24.
19. "Many Travel to Area to View Great Liner," *Newport News Times-Herald,* June 20, 1952.
20. Diary of Vera Cravath Larkin Gibbs, as quoted by Susan L. Gibbs in foreword to Frank Braynard and Robert Hudson Westover, *S.S. United States: Fastest Ship in the World* (Paducah, KY: Turner, 2002), p. 4.
21. Ibid.
22. Kane, "The Speed of the SS United States," p. 137.
23. "Design and Construction of the Superliner United States," reprinted from the September 1952 issue of *Marine Engineering and Shipping Review,* p. 75, collection of Lawrence W. Ward.
24. Diary of Vera Cravath Larkin Gibbs, as quoted by Gibbs in foreword to in Braynard and Westover, *S.S. United States,* p. 4.
25. Ibid. p. 5.
26. Ibid.

21. CREWING UP THE BIG SHIP

1. Meyer Davis, *Saturday Evening Post,* April 20, 1963, http://www.parabrisas.com/d_davism.php, accessed August 30, 2007.
2. "Liner's Stewards Near Shipshape," *New York Times,* June 6, 1952.
3. Joe Curran to General John Franklin, January 5, 1949, as quoted in Frank Braynard and Robert Hudson Westover, *S.S. United States: Fastest Ship in the World* (Paducah, KY: Turner, 2002), p. 35.
4. Interview of William Krudener by Steven Ujifusa, March 4, 2008.
5. "Liner's Stewards Near Shipshape," *New York Times,* June 6, 1952.
6. Charles W. Puffenbarger, "Neptune's Queen Rushing to Keep her May 14 Date," *Norfolk Virginian-Pilot,* April 27, 1952.
7. Interview of William Krudener by Steven Ujifusa, March 4, 2008.
8. Ibid.
9. Braynard and Westover, *S.S. United States,* pp. 46–47.
10. Interview of William Krudener by Steven Ujifusa, March 4, 2008.

22. TRUMAN ON THE ATTACK

1. Lindsay Warren to Charles Sawyer, May 27, 1952, p. 40, Case of the United States Lines, vol. 2, Robert L. Denison Papers, Harry S. Truman Presidential Library, Independence, MO. Correspondence cited in this chapter is from the same volume of the Denison Papers at the Truman Library.
2. Admiral Robert L. Dennison to President Harry Truman, "Memorandum for the President: Maritime Board Construction Subsidy Redeterminations," May 23, 1952.
3. Charles Sawyer to Lindsay Warren, March 31, 1952.
4. Lindsay Warren to Charles Sawyer, April 21, 1952.
5. Charles Sawyer to Lindsay Warren, May 6, 1952.
6. Lindsay Warren to Charles Sawyer, May 27, 1952, p. 7.
7. Charles Sawyer to President Harry Truman, June 7, 1952, p. 2.
8. General John Franklin to Admiral Edmund Cochrane, July 10, 1952.
9. Charles Sawyer to President Harry Truman, June 13, 1952.
10. President Harry Truman to Charles Sawyer, June 13, 1952.
11. Charles Sawyer to John F. Shelley, June 20, 1952.
12. Ibid.
13. Ibid.
14. President Harry Truman to James P. McGranery, June 20, 1952.

15. Luther Huston, "U.S. Delivers Superliner; Truman Challenges Costs," *New York Times,* June 21, 1952.
16. Lindsay Warren to President Harry Truman, June 23, 1952.
17. "Big Liner Delivered to U.S. Lines, Probe of Cost is Ordered," *Newport News Times-Herald,* June 20, 1952.
18. James P. McGranery to President Harry Truman, June 27, 1952.
19. Luther Huston, "U.S. Delivers Superliner; Truman Challenges Costs," *New York Times,* June 21, 1952.

23. THE NEW SEA QUEEN ARRIVES

1. Frank O. Braynard and Robert Hudson Westover, *S.S. United States: Fastest Ship in the World* (Paducah, KY: Turner, 2002), p. 66.
2. "Thousands Line Up to Visit Sea Queen," *New York Times,* June 29, 1952.
3. "Christoffersen Appointed Field Marshall," *Compass Points,* July 1952, p. 11, collection of Lawrence W. Ward.
4. "Thousands Line Up to Visit Sea Queen," *New York Times,* June 29, 1952.
5. "Christoffersen Appointed Field Marshal," *Compass Points,* July 1952.
6. "Thousands Line Up to Visit Sea Queen," *New York Times,* June 29, 1952.
7. Interview of William Krudener by Steven Ujifusa, March 4, 2008.
8. "Thousands Line Up to Visit Sea Queen," *New York Times,* June 29, 1952.
9. Confidential Gibbs & Cox memo to Elaine Kaplan, dated June 24, 1952, collection of Susan Caccavale.
10. "Winged Victory Presentation Aboard the S.S. United States, July 1, 1952," *Compass Points,* July 1952, pp. 7–8, collection of Susan Caccavale.
11. "Thousands Line Up to Visit Sea Queen," *New York Times,* June 29, 1952.
12. Memorandum from William Francis Gibbs to Frederic H. Gibbs, September 18, 1951, 12201 SI-I (16-13400) Confidential, Subject: Design 12201—S.S. United States—Design Particulars and Information, January 24, 1950 Edition, Revised to June 27, 1950, March 19, 1951, and April 24, 1951, vol. 1 of 2, p. 40(6), VM383 U5 G52, p. 34, Mariners' Museum Library, Newport News, VA.
13. Records of Proceedings at the Presentation of Medals by W. F. Gibbs to Crew Members of the S.S. United States Who Served Aboard the Vessel During Its Record Breaking Maiden Voyage, July 13–15, 1952, Ceremony Held in First Class Theatre, S.S. United States, 11:00 AM, Wednesday, July 22, 1953, MS 179, Gibbs Awards, William Francis Gibbs Collection, Mariners' Museum, Newport News, VA.

24. A VERY FAST LADY

1. Henry Billings, *Superliner S.S. United States* (New York: Viking Press, 1952), p. 65.
2. George Horne, "Superliner Begins Her First Crossing," *New York Times*, July 4, 1952.
3. Laura Franklin Dunn to Steven Ujifusa, February 18, 2009.
4. Ibid.
5. Ibid.
6. George Horne, "Superliner Begins Her First Crossing," *New York Times*, July 4, 1952.
7. Interview of Walter Scott, Raymond Foster Options, Ltd., *The S.S. United States: From Dream to Reality* (Newport News, VA: Mariners' Museum, 1992).
8. Don Iddon, "Don Iddon's Blue Riband Diary," *London Daily Mail*, July 8, 1952, Mariners' Museum, Newport News, VA.
9. "Report from Rainbow Land," *Time*, May 28, 1951, http://www.time.com/time/magazine/article/0,9171,890088,00.html?promoid=googlep, accessed March 17, 2009.
10. Iddon, "Don Iddon's Blue Riband Diary."
11. Sidney Malmquist, "A Visit to the Machinery Spaces of the S.S. United States as the Ship Leaves New York," *Compass Points*, n.d. (ca. July 10, 1952), p. 1, collection of Susan Caccavale.
12. Ibid.
13. Malmquist, "A Visit to the Machinery Spaces."
14. Interview of William Krudener and Joseph Rota by Steven Ujifusa, March 4, 2008.
15. First Class Menus, July 4, 5, and 6, 1952, collection of Susan L. Gibbs, Washington, DC.
16. Granville Parkinson, "Report on the SS. United States," *Compass Points*, June 1952, p. 3, collection of Lawrence W. Ward.
17. Iddon, "Don Iddon's Blue Riband Diary."
18. "Invasion, 1952," *Time*, June 23, 1952, http://www.time.com/time/magazine/article/0,9171,859829-6,00.html, accessed December 12, 2008.
19. Iddon, "Don Iddon's Blue Riband Diary."
20. Joe Rota, Crew Recollections, S.S. United States Conservancy Annual Meeting, Independence Seaport Museum, Philadelphia, June 9, 2007.
21. Interview of Walter Scott, Raymond Foster Options, Ltd., *The S.S. United States: From Dream to Reality* (Newport News, VA: Mariners' Museum, 1992).

22. Iddon, "Don Iddon's Blue Riband Diary."
23. *The Ocean Press: World Wide News of United Press, Transmitted by Radiomarine Corporation of America,* Friday, July 4, 1952, collection of Susan L. Gibbs.
24. Iddon, "Don Iddon's Blue Riband Diary."
25. Ibid.
26. *The Ocean Press: World Wide News of United Press, Transmitted by Radiomarine Corporation of America,* Saturday, July 5, 1952, collection of Susan L. Gibbs.
27. Interview of Bill Krudener and Joseph Rota by Steven Ujifusa, March 4, 2008.
28. Iddon, "Don Iddon's Blue Riband Diary."
29. Frank O. Braynard and Robert Hudson Westover, *S.S. United States: Fastest Ship in the World* (Paducah, KY: Turner, 2002), p. 70.
30. Horne, "Superliner Begins Her Crossing."
31. George Horne, "Liner United States Breaks Speed Record First Day Out," *New York Times,* July 5, 1952.
32. George Horne, "35-Knot Mark Hit by American Liner," *New York Times,* July 6, 1952.
33. Captain Harry Grattidge, as quoted in Braynard and Westover, *S.S. United States,* p. 70.
34. Larry Driscoll, "The Race for the Blue Riband," http://united-states-lines.org/luxury%20liner%20row.htm, compiled from *New York Times* articles and magazine article by W. Kaiser, "Power for a Winner," and edited by Fay Richards, accessed June 9, 2010.
35. *S.S. United States: Lady in Waiting,* directed by Robert Radler and produced by Mark B. Perry, Big Ship Films LLC, 2008.
36. Horne, "Liner United States Breaks Speed Record First Day Out."
37. North American Newspaper Alliance, "The Queen Mary Seas Her," July 6, 1952.
38. Larry Driscoll, "Captain of the Line: An Interview with Captain John S. Tucker," *Steamboat Bill* 268 (Summer 2008), p. 8.
39. Walter Bachman to Elaine Kaplan, July 7–8, 1952, collection of Susan Caccavale.
40. Braynard and Westover, *S.S. United States,* p. 71.
41. Walter Bachman to Elaine Kaplan, July 7–8, 1952, collection of Susan Caccavale.
42. *The Ocean Press: World Wide News of United Press, Transmitted by Radiomarine Corporation of America,* Sunday, July 6, 1952, collection of Susan L. Gibbs.

43. Interview with Laura Franklin Dunn by Steven Ujifusa, June 29, 2009.
44. Interview of Jim and Frieda Green by Steven Ujifusa, May 4, 2008.
45. S.S. United States Foundation, http://www.ssunitedstates.org/theship.htm.
46. Wilfred F. Slocum, "Top Secret," *Compass Points,* July 1952, p. 9, collection of Lawrence W. Ward.
47. Laura Franklin Dunn, as quoted in Braynard and Westover; *S.S. United States,* p. 154.
48. John Merryman Franklin, *Recollections of My Life* (Baltimore: Reese Press, 1973), p. 78.
49. "Arrival Soton," *Compass Points,* n.d. (ca. July 16, 1952), p. 7, collection of Lawrence W. Ward.
50. "Britain Outdoes U.S. In Hailing Sea Queen," *New York Times,* July 6, 1952.
51. "Queen of the Seas," *Time,* July 14, 1952, http://www.time.com/time/magazine/article/0,9171,822294,00.html, accessed June 6, 2009.
52. "Arrival Soton," p. 11.
53. "Selected originals (offcuts, selected scenes, out-takes, rushes) for story 'Blue Riband First Try,' " newsreel, British Pathé Limited, 1952, http://www.britishpathe.com/record.php?id=29991, accessed July 28, 2010.

25. MASTERS OF VICTORY

1. Diary of Vera Cravath Gibbs (ca. July 1952), as quoted by Susan L. Gibbs in foreword to Frank O. Braynard and Robert Hudson Westover, *S.S. United States: Fastest Ship in the World* (Paducah, KY: Turner, 2002), p. 95.
2. Milton Lewis, "SS United States Docks to Cry of Welcome Home," *New York Herald-Tribune,* July 16, 1952, S.S. United States Scrapbook, Mariners' Museum, Newport News, VA.
3. Larry Driscoll, "Captain of the Line: An Interview with Captain John S. Tucker," *Steamboat Bill* 266 (Summer 2008), p. 8.
4. "New Speed Dashes for Liner Barred," *New York Times,* July 16, 1952, S.S. United States Scrapbook, Mariners' Museum, Newport News, VA.
5. Meyer Berger, "Manning and Crew Get Rainy Greeting," *New York Times,* July 19, 1952.
6. "Manning Quits Superliner's Bridge; Master of America Takes Command," *New York Times,* August 5, 1952.
7. William Francis Gibbs to Vera Cravath Gibbs (ca. 1952), as quoted by Susan L. Gibbs in foreword to Braynard and Westover, *S.S. United States,* p. 95.
8. Records of Proceedings at the Presentation of Medals by W. F. Gibbs to Crew Members of the S.S. United States Who Served Aboard the Vessel During its

Record Breaking Maiden Voyage, July 13–15, 1952, Ceremony Held in First Class Theatre, S.S. United States, 11:00 AM, Wednesday, July 22, 1953, William Francis Gibbs Collection, Mariners' Museum, Newport News, VA.

9. Ibid.
10. Ibid.
11. Ibid.
12. "Manning Quits Superliner's Bridge; Master of America Takes Command," *New York Times,* August 5, 1952.
13. Braynard and Westover, *S.S. United States,* p. 95; George Horne, "Liner United States Is Beautiful, Fast, Powerful, and Broke," *New York Times,* September 21, 1969.
14. Wilfred F. Slocum, "Top Secret," *Compass Points,* July 1952, p. 9, collection of Lawrence W. Ward.
15. Winthrop Sergeant, "Profiles: The Best I Know How," *New Yorker,* June 6, 1964, p. 50.

26. THE HALCYON YEARS

1. Untitled *New York Times* clipping, dated March 12, 1952, William Francis Gibbs Collection, Mariners' Museum, Newport News, VA.
2. Walter C. Bachman, "William Francis Gibbs, 1886–1967," *Memorial Tributes: National Academy of Engineering* (Washington, DC: National Academies Press, 1979), p. 83.
3. Winthrop Sergeant, "Profiles: The Best I Know How," *New Yorker,* June 6, 1964, p. 56.
4. Interview of Susan Caccavale by Steven Ujifusa, April 2, 2009.
5. Richard Goldrick, "Ventilation and Hull Machinery Section," *Compass Points,* May 1952, p. 7.
6. Interview with Lawrence W. Ward by Steven Ujifusa, August 7, 2008.
7. Frank O. Braynard, *By Their Works Ye Shall Know Them: The Life and Ships of William Francis Gibbs 1886–1967* (New York: Gibbs & Cox, 1968), pp. 195–96.
8. Geoffrey T. Hellman, "Talk of the Town: And the Big Ship," *New Yorker,* June 14, 1952, p. 23.
9. Frank O. Braynard and Robert Hudson Westover, *S.S. United States: Fastest Ship in the World* (Paducah, KY: Turner, 2002), p. 122.
10. George Horne, "His Fair Lady: After four years, the superliner United States is still a consuming passion with the Pygmalion who created her," *New York Times,* June 22, 1956.

11. As quoted in Braynard, *By Their Works Ye Shall Know Them,* p. 194.
12. Larry Driscoll, "Captain of the Line: An Interview with Captain John S. Tucker," *Steamboat Bill* 266 (Summer 2008), p. 8.
13. Interview of Jim and Frieda Green by Steven Ujifusa, May 4, 2008.
14. William Krudener, as interviewed in S.S. *United States: Lady in Waiting,* directed by Robert Radler and produced by Mark B. Perry, Big Ship Films LLC, 2008.
15. William H. Miller Jr., S.S. *United States: The Story of America's Greatest Ocean Liner* (New York: Norton, 1991), p. 116.
16. William Krudener, as interviewed in S.S. *United States: Lady in Waiting.*
17. Excerpt from journal of Vera Cravath Gibbs, September 1955, as quoted in 2006 S.S. United States Conservancy Calendar (Raleigh, NC: SS United States Conservancy, 2006).
18. Ibid.
19. Eugenia McCrary, as quoted in 2006 S.S. United States Conservancy Calendar.
20. Interview of Cissy Levy by Steven Ujifusa, September 22, 2008.
21. Miller, S.S. *United States,* p. 142.
22. Lauren E. Landers-Kirk, "The Windsors Are Aboard," http://uncommonjourneys.com/pages/windsors.htm, accessed September 1, 2007.
23. John S. Tucker, as quoted in Driscoll, "Captain of the Line," p. 8.
24. As quoted in Braynard, *By Their Works Ye Shall Know Them,* p. 175.
25. Interview of William Krudener and Joseph Rota by Steven Ujifusa, March 4, 2008.
26. Ibid.
27. Ibid.
28. Jim and Frieda Green, as interviewed in S.S. *United States: Lady in Waiting.*
29. Recollection by the author's grandmother, Judith Follmann, ca. 2005.
30. Interview of Frank X. Nolan III by Steven Ujifusa, August 24, 2010.
31. Interview with Kurt Wich by Steven Ujifusa, October 3, 2010.
32. Interview of Joseph Rota by Steven Ujifusa, January 28, 2008.
33. Richard Austin Smith, "The Love Affair of William Francis Gibbs," *Fortune,* August 1957, p. 148.
34. "Andrea Doria," *Lost Liners,* PBS, http://www.pbs.org/lostliners/andrea.html, accessed June 14, 2009.
35. Smith, "The Love Affair of William Francis Gibbs," p. 148.

27. TROUBLE ASHORE

1. Richard Austin Smith, "The Love Affair of William Francis Gibbs," *Fortune,* August 1957, p. 160.
2. Ibid.
3. Ibid., p. 158.
4. Ibid.
5. Ibid.
6. Ibid.
7. Frank O. Braynard, *By Their Works Ye Shall Know Them: The Life and Ships of William Francis Gibbs 1886–1967* (New York: Gibbs & Cox, 1968), p. 197.
8. Ibid.
9. "A Decade and Miss Cornell," *New York Times,* March 1, 1936.
10. Interview of Katharine Cornell by Frank O. Braynard, in Braynard, *By Their Works Ye Shall Know Them,* pp. 189–90.
11. Braynard, *By Their Works Ye Shall Know Them,* p. 1977.
12. William Francis Gibbs, transcription of speech given at the Thomas Alva Edison Foundation Mass Media Awards Dinner, Waldorf-Astoria Hotel, January 27, 1960, William Francis Gibbs Collection, Mariners' Museum, Newport News, VA.
13. Smith, "The Love Affair of William Francis Gibbs," pp. 138–39.
14. Excerpt from journal of Vera Cravath Gibbs, September 1955, as quoted by 2006 S.S. United States Conservancy Calendar (Raleigh, NC: SS United States Conservancy, 2006).
15. Winthrop Sergeant, "Profiles: The Best I Know How," *New Yorker,* June 6, 1964, p. 62.
16. *New Yorker,* November 16, 1957, as quoted in Braynard, *By Their Works Ye Shall Know Them,* p. 177.
17. Sergeant, "Profiles: The Best I Know How," *New Yorker,* June 6, 1964, p. 66.
18. Edna Herman, "Vacation Ho!" *Compass Points,* July–August 1952, p. 9, collection of Lawrence W. Ward.
19. Sergeant, "Profiles: The Best I Know How," *New Yorker,* June 6, 1964, p. 68.
20. Interview with Lindsay Falck by Steven Ujifusa, August 18 and 24, 2009.
21. Smith, "The Love Affair of William Francis Gibbs," p. 154.
22. Sergeant, "Profiles: The Best I Know How," p. 66.
23. Smith, "The Love Affair of William Francis Gibbs," p. 154.
24. Sergeant, "Profiles: The Best I Know How," p. 64.

25. Susan Gibbs, as interviewed in *S.S. United States: Lady in Waiting.*
26. Paula Gibbs, "Dogs I Have Known and the People They Have Loved," *Wiscasset Newspaper,* January 15, 2004.
27. Christoper Gibbs to Vera Cravath Gibbs, September 11, 1967, William Francis Gibbs Collection, Mariners' Museum, Newport News, VA.
28. Sergeant, "Profiles: The Best I Know How," p. 58.
29. Ibid.
30. St. Clement's Episcopal Church, Parish Profile, August 2004, p. 1, www.stclementsnyc.org/profile04.doc, accessed June 23, 2009.
31. Smith, "The Love Affair of William Francis Gibbs," p. 160.

28. NO TIME TO SAY GOOD-BYE

1. Interview of Joseph Rota and William Krudener by Steven Ujifusa, March 4, 2008.
2. Interview of Joseph Rota by Steven Ujifusa, January 27, 2008.
3. Dwight D. Eisenhower, "Statement by the President Upon Signing Bill Authorizing the Construction of Two Superliners," July 15, 1958, in John T. Woolley and Gerhard Peters, eds., *The American Presidency Project,* http://www.presidency.ucsb.edu/ws/index.php?pid=11130, accessed April 15, 2009.
4. Jean Marie, as quoted in Frank O. Braynard and Robert Hudson Westover, *S.S. United States: Fastest Ship in the World* (Paducah, KY: Turner, 2002), p. 115.
5. John Malcolm Brinnin, *The Sway of the Grand Saloon: A Social History of the North Atlantic* (New York: Delacorte Press, 1971), p. 539.
6. Interview of Joseph Rota by Steven Ujifusa, January 27, 2008.
7. Werner Bamberger, "Walkout Delays S.S. United States," *New York Times,* June 13, 1958.
8. Edward A. Morrow, "Deadlock Stands in Shipping Strike," *New York Times,* October 3, 1958.
9. William H. Miller Jr., *S.S. United States: The Story of America's Greatest Ocean Liner* (New York: Norton, 1991), p. 171.
10. Braynard and Westover, *S.S. United States,* p. 139.
11. McCandlish Phillips, "Superliner Gets Landlubber Help," *New York Times,* January 31, 1965.
12. Ibid.
13. Braynard and Westover, *S.S. United States,* p. 142.
14. Mark B. Perry, "Liner Memorabilia—Highlights from My Collection," http://www.shipgeek.com/linercollection.htm, accessed August 30, 2007.

15. George Horne, "Liner United States Is Beautiful, Fast, Powerful, and Broke," *New York Times,* September 21, 1969.
16. Reverend Frederick M. Morris to Vera Cravath Gibbs, September 9, 1969, September 18, 1967, William Francis Gibbs Collection, Mariners' Museum, Newport News, VA.
17. Lewis Lapham to Vera Cravath Gibbs, September 11, 1967, William Francis Gibbs Collection, Mariners' Museum, Newport News, VA.
18. Christopher Gibbs to Vera Cravath Gibbs, September 11, 1969, William Francis Gibbs Collection, Mariners' Museum, Newport News, VA.
19. Lewis Lapham to Vera Cravath Gibbs, September 11, 1967, William Francis Gibbs Collection, Mariners' Museum, Newport News, VA.
20. George Horne, "A Sea Dog Making His Last Voyage," *New York Times,* May 19, 1968, Mariners' Museum, Newport News, VA.
21. John J. McCloy to Vera Cravath Gibbs, September 22, 1967, William Francis Gibbs Collection, Mariners' Museum, Newport News, VA.
22. Interview of Leroy R. Alexanderson, WGAL-TV, Lancaster, PA, 1997, from *S.S. United States: Lady in Waiting,* directed by Robert Radler and produced by Mark B. Perry, Big Ship Films LLC, 2008.

29. SECRETS TOLD

1. Bernard Weinraub, "Liner United States Laid Up; Competition from Jets a Factor," *New York Times,* November 15, 1969.
2. John R. Kane, "The Speed of the SS United States," *Marine Technology* 15, no. 2 (April 1978), p. 123.
3. Ibid., p. 121.
4. Ibid., pp. 119–43.
5. Richard Austin Smith, "The Love Affair of William Francis Gibbs," *Fortune,* August 1957, p. 160.
6. "SS United States Conservancy—History," http://www.ssunitedstates conservancy.org/History.html, accessed September 1, 2007.
7. Ibid.
8. "Owner of the SS United States is a real skipper: Rusting hulk? Hey, ship happens," *Philadelphia Daily News,* February 9, 2000, http://www.ssunited statesconservancy.org/History.html.
9. Norwegian Cruise Line Acquires U.S. Flagship S/S United States," http://philadelphia.about.com/cs/travel/a/norwegiancruise.htm, accessed September 1, 2007.
10. Records of Proceedings at the Presentation of Medals by W. F. Gibbs to Crew Members of the S.S. United States Who Served Aboard the Vessel During its

Record Breaking Maiden Voyage, July 13–15, 1952. Ceremony Held in First Class Theatre, S.S. United States, 11:00 AM, Wednesday, July 22, 1953, William Francis Gibbs Collection, Mariners' Museum, Newport News, VA.

11. William Francis Gibbs, "The SS United States," delivered at the 1953 Medal Day Meeting, October 21, 1952, in acceptance of the Franklin Medal, *Journal of the Franklin Institute,* ca. 1953, p. 549, collection of the Mariners' Museum, Newport News, VA; Don Iddon, "Life Aboard the Aluminum Anne," publication unknown, date approximately July 5, 1952, from the collection of the Mariners' Museum, Newport News, VA.
12. William Francis Gibbs, "S.S. United States," *Journal of the Franklin Institute,* Philadelphia, December 1953, p. 549.
13. Smith, "The Love Affair of William Francis Gibbs," p. 160.

EPILOGUE

1. "Vera Cravath Gibbs, 89, Dies; Was Active in Opera Groups," *New York Times,* July 30, 1985.
2. Paula Gibbs, "Dogs I Have Known and the People They Have Loved," *Wiscasset Newspaper,* January 15, 2004.
3. "About the Conservancy," SS United States Conservancy, http://www.ssunitedstatesconservancy.org/SSUS/About.html, accessed June 29, 2009.
4. Laura Franklin Dunn, as interviewed in *S.S. United States: Lady in Waiting,* directed by Robert Radler and produced by Mark B. Perry, Big Ship Films LLC, 2008.
5. Harry Manning, as quoted in Frank O. Braynard and Robert Hudson Westover, *S.S. United States: Fastest Ship in the World* (Paducah, KY: Turner, 2002), p. 138.
6. Josh Barnabel, "Joseph Curran, 75, Founder of National Maritime Union," *New York Times,* August 15, 1981.
7. William H. Miller Jr., *Picture History of the SS United States* (Mineola, NY: Dover, 2003), p. 30.
8. M/Y *Savarona,* http://www.mysavarona.com/history.php, accessed June 30, 2009.
9. Steven B. Ujifusa, "Lenfest Keeps Big U Alive," *PlanPhilly.com,* July 2, 2010, http://planphilly.com/lenfest-keeps-big-u-alive, accessed August 15, 2010.

Acknowledgments

As a child I was fascinated by history, ships, and the sea, but Jonathan Karp, my editor at Simon & Schuster, has enabled me to write a book about William Francis Gibbs and the SS *United States.* He had faith in the project.

From Jonathan, I have learned how one must feel and think to write popular history. Also thanks to Karen Thompson and Colin Shepherd for their invaluable editorial guidance. Vivien Ravdin, in addition to her remarkable ear for the English language, provided invaluable insights into the flow of the narrative, as well as rich stories about Rittenhouse Square, where I have lived for the past several years . . . a stone's throw away from the site of Gibbs's childhood home.

My agents David Kuhn and Billy Kingsland at Kuhn Projects made sure the book proposal was well polished before it was sent to publishers. Thanks for being such tireless advocates for the past five years.

My book required a great deal of archival research, and I want to thank the staff at the Mariners' Museum in Newport News, Virginia; the Harry S. Truman Museum and Library in Independence, Missouri; SUNY Maritime Academy at Fort Schuyler in Bronx, New York; the Philadelphia Athenaeum; and the U.S. Naval Archives in Washington, D.C. Special thanks to Claudia Jew at the Mariners' Museum for helping me search through the enormous collection of SS *United States* material housed there.

Several very special people opened their personal collections, giving me access to rare materials that would otherwise be unavailable: Larry and Grace Ward, Laura Franklin Dunn, Bill and Tillie Krudener, Jim and Frieda Green, Susan Caccavale, Joseph Rota and Bonnie Davis, Noelle Braynard Hollander, Dr. William T. Flayhart III, and William duBarry Thomas.

I owe a debt of gratitude to the late maritime historian Frank O. Braynard, who dedicated his life to writing about great ships and preserving the legacy of William Francis Gibbs. Without him, much would have been lost forever.

I am grateful to Matthew Golas, editor of PlanPhilly.com, who gave me a start as a freelance writer on his website in April 2007, with the debut piece "Hidden Treasure on Delaware," which was about the SS *United States*. Out of that grew my book proposal, as well as a wonderful friendship with him and his wife Jane. Thanks also to Robert and Rachel Cheetham at *Philly*History.org, and Chris Satullo at WHYY's Newsworks.

Special thanks to the fellow members of the Orpheus Club of Philadelphia for the support and encouragement you have provided me since 2006. *Ecce quam bonum!*

Thanks also to the Green Line Cafe in West Philadelphia, where so much of this book was written and edited.

A number of friends have taken the time to read and critique portions of the manuscript in its various iterations: thank you Travis Logan, Andrew Kelly, Reverend Sean Mullen, Gregory Nickerson, Bryan Fields, and Andrew Fink for your proofreading and editorial suggestions.

Thanks also to my professors at Harvard College and the Penn School of Design, who taught me how to think and write about the history of the built environment: the late William Gienapp, Stephan Thernstrom, Brian Domitrovic, Randall Mason, John C. Keene, Frank Matero, and Donovan Rypkema.

The Board of the SS United States Conservancy has become my extended family. In July 2010, a $5.8 million grant from philanthropist H. F. "Gerry" Lenfest (USNR retired) saved the ship from certain destruction. Hats off to Dan McSweeney, Susan Gibbs (granddaughter of William Francis), Jeff Henry, Mark Perry, Joe Rota, Greg Norris, and the rest of the board for taking on the immense challenge of saving an irreplaceable American treasure. The determination of Judge Thomas Watkins and the vision of Gerry Lenfest have given the *United States* a new lease on life. The Conservancy is now faced with the biggest and most challenging historic preservation project in America today.

I must acknowledge the residents of a special seaside community on the South Shore of Massachusetts, my writing (and procrastination) haven for the past several summers. Thanks especially to Captain Peter B. Adams and his sister Ramelle Adams—direct descendants of President John Adams, founder of the American Navy. Peter and Ramelle's careful reading of the manuscript greatly improved its clarity and flow.

My father and mother, Grant and Amy Ujifusa, have always supported my interests in ships and the sea. Their love for me made the book possible. I regret that my paternal grandparents Mary and Tom Ujifusa, who for years lived and worked on a Wyoming farm, did not live to see the completion of my manuscript; I thank them for the work ethic that they instilled in me. Likewise, my late maternal grandfather Jerry Brooks and step-grandfather Joe Follmann had deep interests in history, music, and the arts that I absorbed as a member of the family. My journalist brother Andrew provided me with excellent advice about how to write sentences that conveyed precise meaning. My youngest brother, John, the builder in the family, continues the great American tradition of making great things. My uncle Jeffrey Brooks, professor of history at Johns Hopkins, has been a constant source of inspiration.

I want to dedicate my book to my grandmother Judith Follmann, world traveler and lady of culture who first introduced me to the story of the *Titanic* when I was six. It was with her that I first spotted the faded SS *United States* while we sped across the Walt Whitman Bridge in 1996. My grandmother is now ninety-five, and I love her very much. I am proud to have walked the same decks my grandmother did over half a century ago, when she was a passenger on the finest, fastest, most beautiful ship ever built.

Index

About the Author

Steven Ujifusa is a historian living in Philadelphia. He serves on the Advisory Council of the S.S. United States Conservancy. He received his master's degree in historic preservation and real estate from the University of Pennsylvania and his B.A. in history from Harvard University.